# Lecture Notes in Electrical Engineering

Volume 629

AF810796

The book series *Lecture Notes in Electrical Engineering* (LNEE) publishes the latest developments in Electrical Engineering—quickly, informally and in high quality. While original research reported in proceedings and monographs has traditionally formed the core of LNEE, we also encourage authors to submit books devoted to supporting student education and professional training in the various fields and applications areas of electrical engineering. The series cover classical and emerging topics concerning:

- Communication Engineering, Information Theory and Networks
- Electronics Engineering and Microelectronics
- Signal, Image and Speech Processing
- Wireless and Mobile Communication
- Circuits and Systems
- Energy Systems, Power Electronics and Electrical Machines
- Electro-optical Engineering
- Instrumentation Engineering
- Avionics Engineering
- Control Systems
- Internet-of-Things and Cybersecurity
- Biomedical Devices, MEMS and NEMS

For general information about this book series, comments or suggestions, please contact leontina.dicecco@springer.com.

To submit a proposal or request further information, please contact the Publishing Editor in your country:

**China**

Jasmine Dou, Associate Editor (jasmine.dou@springer.com)

**India, Japan, Rest of Asia**

Swati Meherishi, Executive Editor (Swati.Meherishi@springer.com)

**Southeast Asia, Australia, New Zealand**

Ramesh Nath Premnath, Editor (ramesh.premnath@springernature.com)

**USA, Canada:**

Michael Luby, Senior Editor (michael.luby@springer.com)

**All other Countries:**

Leontina Di Cecco, Senior Editor (leontina.dicecco@springer.com)

**** Indexing: The books of this series are submitted to ISI Proceedings, EI-Compendex, SCOPUS, MetaPress, Web of Science and Springerlink ****

More information about this series at http://www.springer.com/series/7818

G. Di Francia · C. Di Natale · B. Alfano ·
S. De Vito · E. Esposito · G. Fattoruso ·
F. Formisano · E. Massera · M. L. Miglietta ·
T. Polichetti
Editors

# Sensors and Microsystems

## Proceedings of the 20th AISEM 2019 National Conference

Springer

*Editors*
G. Di Francia
ENEA
Portici, Napoli, Italy

C. Di Natale
Department of Electronic Engineering
University of Rome Tor Vergata
Rome, Italy

B. Alfano
ENEA
Portici, Napoli, Italy

S. De Vito
ENEA
Portici, Napoli, Italy

E. Esposito
ENEA
Portici, Napoli, Italy

G. Fattoruso
ENEA
Portici, Napoli, Italy

F. Formisano
ENEA
Portici, Napoli, Italy

E. Massera
ENEA
Portici, Napoli, Italy

M. L. Miglietta
ENEA
Portici, Napoli, Italy

T. Polichetti
ENEA
Portici, Napoli, Italy

ISSN 1876-1100 ISSN 1876-1119 (electronic)
Lecture Notes in Electrical Engineering
ISBN 978-3-030-37560-7 ISBN 978-3-030-37558-4 (eBook)
https://doi.org/10.1007/978-3-030-37558-4



This Springer imprint is published by the registered company Springer Nature Switzerland AG
The registered company address is: Gewerbestrasse 11, 6330 Cham, Switzerland

# Preface

This book contains a selection of the papers presented at the 20th AISEM ("Associazione Italiana Sensori e Microsistemi") National Conference on Sensors and Microsystems, held in Napoli, on February 11–13, 2019, organized by ENEA, the Italian National Agency for New Technologies, Energy and Sustainable Economic Development, with the acknowledged support of the University of Napoli Federico II.

This edition of the conference was characterized by the presentation of 110 regular contributions, organized in 4 sessions: materials and processing technologies, sensor and microsystem devices, sensor systems and device applications that highlighted state-of-the-art results from both theoretical and applied researches in the field of sensors and related technologies.

During this conference edition, four start-up characterized by strongly innovative products, selected by the General Confederation of Italian Industry, Campania chapter and the AISEM Steering Committee, were awarded with the first edition of the "Confindustria-AISEM Innovation Prize."

Deep thanks are due both to Dr. Sara Zerbini, STMicroelectronics, and to Prof. Fabio Sciarrino, University of Roma "La Sapienza" for their very interesting special lectures.

The editors would finally acknowledge the numerous participants and the Local Organizing Committee for the organization of the event.

Portici, Italy

G. Di Francia
General Chairman AISEM 2019

# Contents

# Biometric Parameters Assessment for Foot Ulcers Prevention Through Wearable Devices

**Gabriele Rescio, Alessandro Leone, Chiara De Pascali, Luca Francioso and P. Siciliano**

**Abstract** A fast and reliable identification of foot pressure loads and temperature distributions changes on the plantar surface allows to prevent and reduce the consequences of ulceration of the diabetic foot. This work presents a smart insole in which both temperature and pressure data in 8 reading points are monitored in remote way for the assessment of the health foot conditions by a caregiver. Minimally invasive and low power temperature and force sensors have been chosen and integrated into two antibacterial polyurethane-based layers architecture. In this work the attention was focused on the heat transfer between the insole and the foot. Finite element simulations were performed to evaluate the effectiveness of the sensor array to detect, from thermal gradients measured on the plantar surface, inflammatory events that can be attributed to early signs of foot ulceration. The results demonstrated that small differences of temperature between the eight sensor nodes of the array can be discriminated and used to prevent the onset of ulcerative lesions, also giving a supplementary information about the position closer to a potential inflamed region of the foot.

**Keywords** Foot ulcer · Finite element analysis · Smart system · Diabetes

G. Rescio (✉) · A. Leone · C. De Pascali · L. Francioso · P. Siciliano
National Research Council of Italy, Institute for Microelectronics and Microsystems, Via Monteroni Presso Campus Universitario Palazzina A3, Lecce, Italy
e-mail: gabriele.rescio@cnr.it

A. Leone
e-mail: alessandro.leone@cnr.it

C. De Pascali
e-mail: chiara.depascali@le.imm.cnr.it

L. Francioso
e-mail: lucanunzio.francioso@cnr.it

P. Siciliano
e-mail: pietroaleardo.siciliano@cnr.it

 
G. Di Francia et al. (eds.), *Sensors and Microsystems*, Lecture Notes in Electrical Engineering 629, https://doi.org/10.1007/978-3-030-37558-4_1

# 1 Introduction

Diabetes represents one of the main problems in health system and a global public health threat that has increased dramatically over the past 2 decades [1]. One of the common and serious complication in diabetic patients is the foot ulcers that may affect up to 25% of patients in the course of their lifetime [2]. The main causes of the ulcers arising are due to the neuropathy, peripheral artery diseases and vascular alterations. When the ulcers appear, the treatment is challenging, expensive and a long term treatment is required. In this scenario the prevention of ulcers assumes a prominent rule. The use of technologies for the monitoring of relevant vital parameters on the plantar, as temperature and foot load distribution, may be very important to prevent foot ulcers. They may identify the prolonged and excessive pressure at a point of the foot or recognize anomalies in skin conditions and in bloody circulation. Another useful parameter for assessing the diabetic foot is the temperature. Progressive degeneration of sensory nerve pathways affect thermoreceptors and mechanoreceptors. High temperatures under the foot coupled with reduced or complete loss of sensation can predispose the patient to foot ulceration. So, the foot thermal monitoring may facilitate detection of diabetic foot problems [3]. The most part of the temperature foot measurements for the diabetic foot regarding with non-invasive, and accurate thermal images analysis or thermography inspections. In [4] infrared thermal imaging technique was exploited to study the plantar foot temperatures of 112 subjects with type 2 diabetes. The obtained results demonstrated that patients with diabetic neuropathy tend to have a higher foot temperature (32–35 °C) compared to patients without neuropathy (27–30 °C). The loss of sensation in the foot is another clinical manifestation of the diabetic neuropathy, which coupled with the higher mean temperature under the foot can even more predispose the patient to foot ulceration. A localized tissue inflammation is the most plausible explanation for the increase of temperature in at risk diabetic foot [5]. Literature does not give a precise indication about the geometry and the depth of inflammation associated with impending foot ulceration; the inflammation model proposed in [6] for the thermal analysis of deep injury of heel tissue considers a depth of the lesion of about 9 mm, underneath the epidermis up to inside the fat and muscle layers. As regards the at risk plantar regions, a study conducted on 87 patients with existing foot ulcers [7] demonstrated that the lesions are localized mostly under the metatarsal heads (56.3%), under the toes (32.1%) and at the heel (1.9%). Further, the majority of diabetic foot ulcers are developed on the foot plantar surface at sites of high pressure [8].

This work presents a smart insole in which both temperature and pressure data in 8 reading points are acquired and then transmitted through a wireless protocol to a gateway in order to monitor foot conditions and inform the caregiver about the health status. The attention was focused on the heat transfer between the insole and the foot, with the aim to evaluate the effectiveness of the sensor array to detect inflammatory events attributable to early signs of foot ulceration. The results of finite element simulations on a 3D model of insole-foot demonstrated that small differences of temperature measured between the eight sensor nodes of the array can be discriminated and used to prevent the onset of ulcerative lesions, also giving a supplementary information about the position closer to a potential inflamed region of the foot.

## 2 Smart Insole System

### *2.1 Design and Realization*

The architecture of the developed smart insole system consists of three subsystems: (1) sensors system for the temperature and pressure parameters acquisition; (2) wireless module transmission; (3) data aggregation and processing module. The first two subsystem are integrated in the insole, whereas the third subsystem is implemented on a Embedded PC able to send the data to a caregiver through a cloud service. For the evaluation of the pressure and temperature sensors placement, the distribution of load foot pressure was analyzed through the baropodometry P-Walk platform, produced by BTS Bioengineering [9]. For the sensor positioning and sensor dimensions have to be find the best trade-off through the covering area and accurately measurements at specific points of the plantar. According to the Ferber et al. [10] and the local load foot analysis, performed by using the BTS G-studio [9], the placing of the pressure sensors was designed as shown in Fig. 1. The temperature sensors were located close the pressure sensor in order to monitor the main pressure point of the foot.

For the temperature monitoring, the low power Maxim MAX30205 sensor was chosen since it presents a greater accuracy (0.1 °C) [11] than works belonging to the state-of-the-art and a low current consumption (600 μA in operative mode and 5 nA of leakage current), useful for the long-term monitoring. For the load pressure acquisition, the IEE CP151 FSR sensor (0.43 mm thick and 6% for the accuracy) has been chosen for the low invasive integration in the flexible thin-film layer (polyamide support with a dielectric and copper thickness of 360 μm and 18 μm respectively, as best trade-off in terms of invasiveness, ergonomics and robustness). The sensors placement was studied and optimized in order to obtain an accurate monitoring of foot in the 38–43 size range by using only the 39 and 42 foot sizes. To measure the

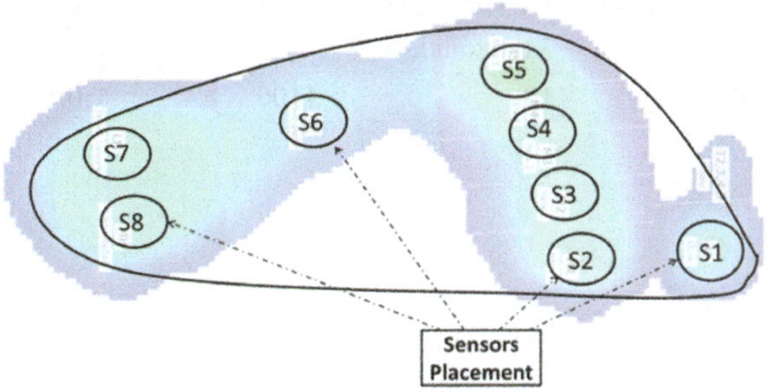


**Fig. 1** Temperature and FSR sensors positioning

foot temperature, a via hole has been realized on the board in correspondence of the sensing pad of each temperature sensor, assuring the thermal transfer between the bottom and the top of the insole, through the application of a silver-based conductive paste. The flexible circuit was incorporated between the first and the second layer of the two antibacterial polyurethane-based layers architecture, designed in accordance with the typical requirements of diabetic foot insoles.

## 2.2 *Heat Transfer Analysis of the Foot-Insole Model*

With the aim to evaluate the capability of the sensor array integrated into the engineered insole to detect early impending ulcerative conditions of the diabetic foot, a finite element thermal analysis (FEA) using Comsol Multiphysics was performed in this work. FEA simulations were used to evaluate the heat distribution within a 3D model of foot-insole, due to a simulated inflammatory lesion localized deep in the foot and associated to the diabetic neuropathy. The thermal insulation of footwear was considered in the model by including a 15 mm thick rubber layer under the insole.

Bioheat transfer within the biological tissue was solved by applying the Pennes' model into the standard thermal diffusion equation, for including the effects of blood perfusion and metabolism [12]. Convective heat transfer was assumed to occur at the interface between the footwear and the surrounding, with heat transfer coefficient equal to 5 $W/m^2$ K at environment temperature of 20 °C [13].

The thermo-physical properties of various tissue layers and inflammatory lesion used in the FEA simulations are reported in [6], those of the remaining materials constituting the insole are summarized in Table 1.

**Table 1** Thermo-physical properties of the insole materials

| | Thickness (mm) | Density ($kg/m^3$) | Thermal conductivity (W/m K) | Heat capacity at constant pressure (J/kg K) |
|---|---|---|---|---|
| Polyamide | 0.36 | 1300 | 0.15 | 1100 |
| Copper | 0.018 | 8960 | 401 | 384 |
| Polyurethane | 1 (upper layer) | 374 | 0.06 | 1337 |
| | 3 (bottom layer) | | | |
| Rubber | 15 | 1100 | 0.13 | 2010 |

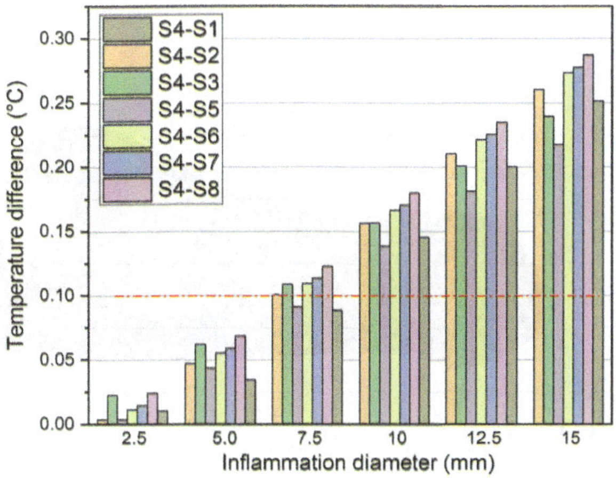


**Fig. 2** Thermal gradients measured between the sensor S4 (positioned on the inflammation centrally) and the other sensor nodes of the array integrated onto the insole

# 3 Results

## *3.1 FEA Simulation Results*

The inflamed tissue was modelled with circular shape and was localized in the metatarsal head area of the foot, underneath the epidermis layer, for a depth of 3.5 mm. The distribution of heat though the foot skin and the engineered insole was analyzed at varying of the diameter of the inflamed tissue, ranging from 2.5 to 15 mm. The study was finalized to investigate the thermal gradients developed between the various sensor nodes, at the varying of the inflammation extension, with the aim to evaluate the capability of the sensor array to detect early impending ulceration.

Figure 2 shows the differences of temperature obtained between the eight sensor nodes of the array integrated onto the modelled insole, for different sizes of the inflamed tissue. The sensor S4 is closer than the other ones to the inflammation site. As can be seen from the graph, inflammation with diameter smaller than 7.5 mm does not detected by the sensor array, being the thermal gradients smaller than the accuracy of the same temperature sensors (0.1 °C). Inflammations with diameter equal and higher than 7.5 mm can be detected by considering to be meaningful the temperature difference measured between the closest and the most distant sensors, with respect to the inflamed site (S4 (the closest) and S6, S7, S8 (the most distant) in the example reported in Fig. 2). Figure 3 shows the 2D temperature contour plot for the foot-insole model, in presence of an inflammation of diameter 10 mm localized in the metatarsal area of the foot. The results demonstrate that the designed temperature sensors array is effective to early detect small differences of temperature, which monitored in the time and also compared with those acquired on the opposite foot,

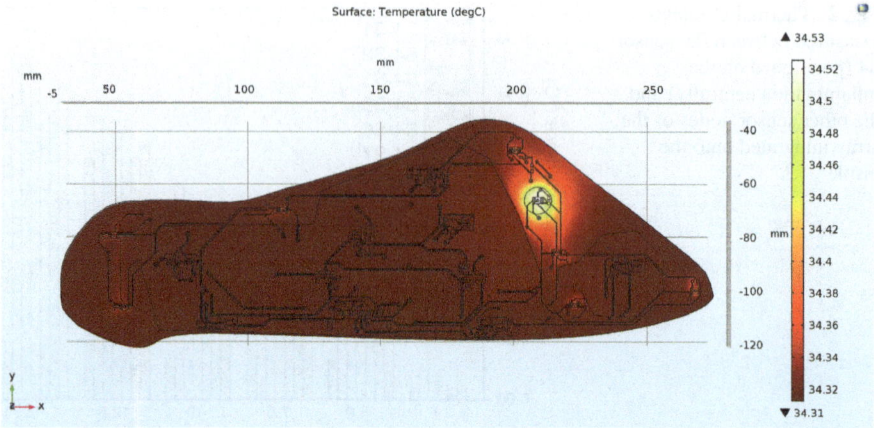


**Fig. 3** 2D temperature contour plot for the foot-insole model, in presence of an inflammation of diameter 10 mm localized in the metatarsal area of the foot

can be used to early alert about the onset of ulcerative lesions. Further, the evaluation of the difference of temperature between all the possible pairs of the sensors array can give a supplementary information about the position closer to a potential inflamed region of the foot.

## 4 Conclusion

A smart insole system for continuous temperature and load pressure foot monitoring in daily activities was presented for diabetes patients. The data were acquired from eight different points on the foot plantar both for temperature and load pressure parameters useful for the ulcer prevention. The attention was focused on the heat transfer between the insole and the foot. FEA results demonstrated that the proposed smart insole is able to detect inflammatory events that can be attributed to early signs of foot ulceration. Small differences of temperature between the eight sensor nodes of the array can be discriminated and used to prevent the onset of ulcerative lesions, also giving a supplementary information about the position closer to a potential inflamed region of the foot.

## References

1. Ogurtsova K, da Rocha Fernandes JD, Huang Y, Linnenkamp U, Guariguata L, Cho NH, Cavan D, Shaw JE, Makaroff LE (2017) IDF diabetes atlas: global estimates for the prevalence of diabetes for 2015 and 2040. Diabetes Res Clin Pract 128:40–50

2. Iraj B, Khorvash F, Ebneshahidi A, Askari G (2013) Prevention of diabetic foot ulcer. Int J Prev Med 4(3):373–376
3. Etehadtavakol M, Ng EYK, (2017) Assessment of foot complications in diabetic patients using thermography: a review. In: Application of infrared to biomedical sciences, pp 33–43
4. Subramnaiam B et al (2010) Correlation between plantar foot temperature and diabetic neuropathy: a case study by using an infrared thermal imaging technique. J Diabetes Sci Technol 4(6):1386–1392
5. Lavery LA et al (2004) Home monitoring of foot skin temperatures to prevent ulceration. Diabetes Care 27(11):2642–2647
6. Chanmugam A, Bhargava A, Herman C (2012) Heat transfer model and quantitative analysis of deep tissue injury. Int Mech Eng Congress Expo. 2012:717–723
7. Lavery L, Peters E, Armstrong D (2008) What are the most effective interventions in preventing diabetic foot ulcers? Internat Wound J 5(3):425–433
8. Katoulis E, Boulton A, Raptis S (1996) The role of diabetic neuropathy and high plantar pressures in the pathogenesis of foot ulceration. Horm Metab Res 28(4):159–164
9. https:// www.btsbioengineering.com
10. Ferber R, Webber T, Everett B, Groenland M (2013) Validation of plantar pressure measurements for a novel in-shoe plantar sensory replacement unit. J Diabetes Sci Technol 7(5):1167–1175
11. https://www.maximintegrated.com/en/products/sensors-and-sensor-interface/MAX30205.html
12. Pennes H (1948) Analysis of tissue and arterial blood temperatures in the resting human forearm. J Appl Physiol 1:91–122
13. Kuklane K, Afanasieva R, Burmistrova O, Bessonova N, Holmér I (1999) Determination of heat loss from the feet and insulation of the footwear. Int J Occup Saf Ergon 5(4):465–476

# Volume Phase Holographic Lenses for Efficient Planar Solar Track-Concentrators

**M. A. Ferrara, G. Bianco, V. Striano and G. Coppola**

**Abstract** Volume transmission phase holographic lenses were designed and recorded to obtain a simple, lightweight, compact and inexpensive planar solar concentrator. To avoid any mechanical movement, a passive solar tracking system is also proposed by using angular multiplexed holographic lenses. Furthermore, these solar concentrators have no overheating problems resulting from the absorption of infrared frequencies, because in this spectral region the proposed volume holographic lenses do not work. Finally, the realized samples were tested and a good efficiency was obtained.

**Keywords** Holographic lens · Solar concentration · Volume phase holographic optical elements

## 1 Introduction

Holographic PV concentrators were proposed for the first time in 80s [1–4], indeed holography as an optical technology presents many advantages respect to other concentrating optical systems (lenses or mirrors, for instance): it is much more versatile and cheaper than them. It can also eliminate the need for tracking, thus reducing the whole system complexity.

Among the different type of holograms, Volume Holographic Gratings (VHGs) show promise to enhanced performance in many applications respect to classical surface-relief grating technology [5–7]. A VHG is obtained inducing a periodical refractive index variation throughout the volume of an optically thick film [5, 6] and it can reach efficiency up to 100% for a selected wavelength. Furthermore, the hologram splits the solar spectrum so that the longer wavelength light can be separated to reduce unnecessary heating, while the shorter wavelengths spectra can

M. A. Ferrara (✉) · G. Bianco · G. Coppola
National Research Council, Institute for Microelectronics and Microsystems, Via Pietro Castellino 111, 80131 Naples, Italy
e-mail: antonella.ferrara@na.imm.cnr.it

V. Striano
OHB S.p.A, Via Tiengo snc, 82100 Benevento, Italy



G. Di Francia et al. (eds.), *Sensors and Microsystems*, Lecture Notes in Electrical Engineering 629, https://doi.org/10.1007/978-3-030-37558-4_2

be further divided into several bands to match the energy gaps of various solar cells [4, 8]. Thus, VHG allows obtaining high efficiency, controllable spectral response, and low scattering at the same time, and it is useful in many applications, such as Volume Holographic Optical Elements (V-HOEs) [9].

By implementing V-HOEs it is possible to develop holographic planar optics integrating the same functionalities of conventional concentrators (parabolic mirrors, Fresnel lens) reducing, in the meantime, cell volumes and costs. In addition, the versatility of holographic technology permits to cluster multiple holographic elements into the same panel, collecting solar rays with different incidence angles, although their efficiency could be lower if several holograms are multiplexed [10]. Such functionality can be exploited to develop the holographic passive solar tracker in order to increment the performance of single concentrators.

With this aim, we have been studied V-HOEs for controlling and directing the radiation of the sun with high potential for energy saving, taking into account all these important features.

## 2 Results and Discussions

V-HOEs have been recorded on a prototype of photopolymer sensitive to light at wavelength of 532 nm and deposited (thickness $\approx$ 30 $\mu$m) by the Doctor Blade method on a rigid support [11]. In order to obtain Volume Holographic Lens (VHL), a single lens was recorded with an in-line configuration by using a concave mirror with a focal length of 5 cm as object [12, 13]. The interference between a plane (reference) and a spherical (object) wavefronts allows to produce the desired refractive index modulation. The efficiencies of 40 VHLs were evaluated as the ratio between the power focused by the holographic lens and the power focused by a commercial Fresnel lens with the same focusing features. The mean value for the efficiency was 41.49% $\pm$ 1.35%. Additionally, each lens has been characterized at different angles of incidence, and an angular selectivity of about $\pm$8 was obtained.

Then, the passive solar tracking was realized by recording multiplexed holographic lenses. This structure allows directing the light in a given direction independently of the direction of the incoming light without any mechanical movement. In particular, a set of five lenses has been recorded on the same glass to focus the light on the same PV cell positioned close to the focus of the lenses system. The recording procedure was implemented for five angles: perpendicular illumination (0°), $\pm$15° and $\pm$30° illuminations. Holographic lenses at $\pm$15° and $\pm$30° were recorded by tilting of $\pm$15° and $\pm$30°, respectively, the photosensitive substrate. A scheme of the realized multiplexed holographic solar concentrator is reported in Fig. 1.

The five multiplexed holographic lenses system has been characterized in terms of degree of concentration, expressed in number of "suns". This parameter is evaluated as the ratio between the intensity of the light harvested by the PV cell with and without the solar concentrator. Preliminary results show a degree of concentration for the proposed system on average of 3 suns over the angular range of $\pm$30°.

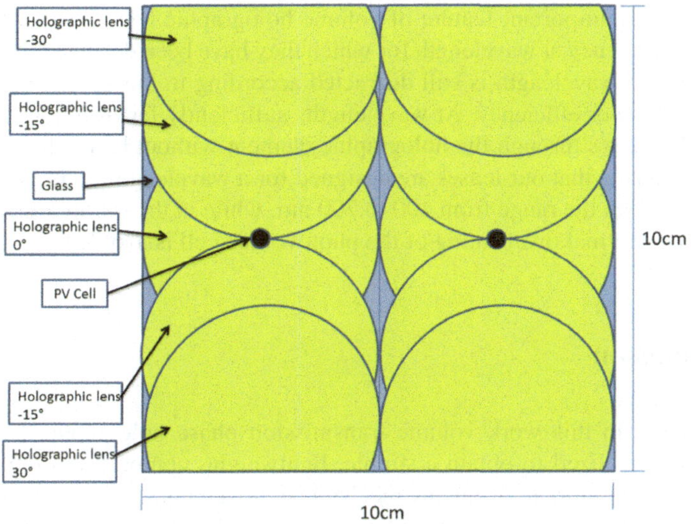


**Fig. 1** Angular multiplex lenses recorded at five angles: 0°, ±15°, and ±30°

Finally, the passive solar tracking collection was experimentally characterized in real operative conditions. The generated voltage of the photovoltaic system was continuously monitored during a clear day (24 h) and results are reported in Fig. 1. The measurement was performed with the system tilted 30° backwards in order to be able to measure system performance at solar heights between −90° and 90°. During the day monitored, no mechanical movement of the concentrator configuration with five lenses was performed, thus a passive solar tracking was obtained in the periods of the day in which the sun fulfils the recorded angle of each lens, as can clearly be seen in Fig. 2.

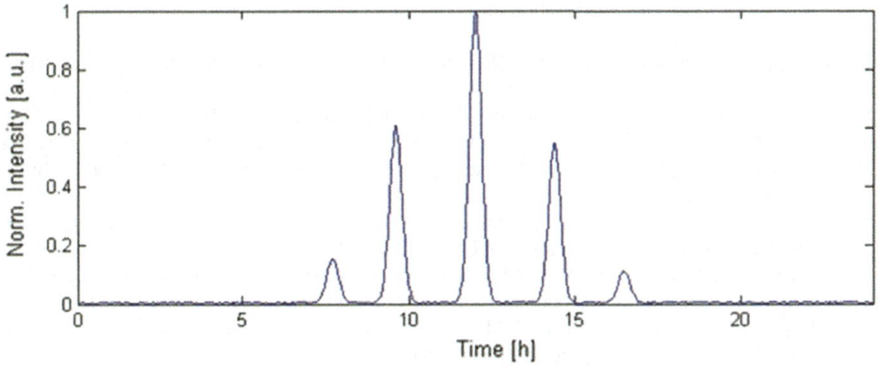


**Fig. 2** 24 h—measurement of the concentrated solar intensity in real operative condition

Moreover, an important feature of volume holographic lenses is that they show an optimal efficiency at wavelength for which they have been designed. Light nearly to the designed wavelength is still diffracted according to the grating equation but usually with lower efficiency. At wavelengths sufficiently far from the Bragg condition, light passes through the holographic element without being diffracted [14]. Thus, considering that our lenses are designed for a wavelength of 532 nm and that they well work in the range from 400 to 700 nm, while in the infrared region lenses do not work, thermal overheating of the photovoltaic cell is avoided.

## 3 Conclusion

In conclusion, in this work volume transmission phase holographic lenses were designed and realized to obtain a simple, lightweight, compact and low-cost planar solar concentrator. To avoid any mechanical movement, a passive solar tracking system was developed by using angular multiplexed holographic lenses. Moreover, our volume holographic lenses are not affected by thermal overheating due to the absorption of infrared frequencies, because in this spectral region they do not work. Our results show the possibility to realize solar concentrators for PV cells by using multiplexed VHL with a good efficiency and passive solar tracking.

## References

1. Shakher C, Ramamurthy V (1987) Thick transmission phase holograms for photovoltaic concentrator applications. Sol Energy Mater 16:215–221
2. Bainier C, Hernandez C, Courjon D (1988) Solar concentrating systems using holographic lenses. Sol Wind Technol 5:395–404
3. Ludman JE, Riccobono J, Semenova IV, Reinhand NO, Tai W, Li X, Syphers G, Rallis E, Sliker G, Martín J (1997) The optimization of a holographic system for solar power generation. Sol Energy 60:1–9
4. Zhang YW, Ih CS, Yan HF, Chang MJ (1988) Photovoltaic concentrator using a holographic optical element. Appl Opt 27:3556–3560
5. Kogelnik H (1969) Coupled-wave theory of thick hologram gratings. Bell Syst Tech J 48:2909
6. Barden SC, Arns JA, Colburn WS (1998) Volume-phase holographic gratings and their potential for astronomical applications. In: Proceedings of SPIE optical astronomical instrumentation, vol 3355, p 866
7. Blanche PA, Gailly P, Habraken S, Jamar LC (2004) Volume phase holographic gratings: large size and high diffraction efficiency. Opt Eng 43:2603–2612
8. Zhang D, Gordon M, Russo JM, Vorndran S, Kostuk RK (2013) Spectrum-splitting photovoltaic system using transmission holographic lenses. J Photonics Energy 3:034597-1–034597-12
9. Tedesco JM, Owen H, Pallister DM, Morris MD (1993) Principles and spectroscopic applications of volume holographic optics. Anal Chem 65:441 A–449 A
10. Chemisana D, Collados MV, Quintanilla M, Atencia J (2013) Holographic lenses for building integrated concentrating photovoltaics. Appl Energy 110:227–235

11. Ferrara MA, Borbone F, Striano V, Coppola G (2013) Characterization of photopolymers as optical recording materials by means of digital holography microscopy. Proc. SPIE 8792:87920Z
12. Bianco G, Ferrara MA, Borbone F, Roviello A, Striano V, Coppola G (2015) Photopolymer-based volume holographic optical elements: design and possible applications. J Eur Opt Soc Rapid 10:15057
13. Bianco G, Ferrara MA, Borbone F, Roviello A, Pagliarulo V, Grilli S, Ferraro P, Striano V, Coppola G (2015) Multiplexed holographic lenses: realization and optical characterization. In: IET conference publications, Issue CP667. Fotonica AEIT 2015, Turin (Italy)
14. Baldry IK, Bland-Hawthorn J, Robertson JG (2004) Volume phase holographic gratings: polarization properties and diffraction efficiency. Publ Astron Soc Pac 116:403–414. https://doi.org/10.1086/383622

# VCII-Based Sensor Interface for Silicon Photomultiplier

**G. Barile, A. Leoni, M. Muttillo and L. Pantoli**

**Abstract** We here propose a voltage conveyor-based sensor interface for silicon photomultipliers (SiPMs). The solution addresses a mixed-mode (voltage/current) design taking advantage from the current-mode approach to increase the response time and the drive capability. This solution, based on a voltage conveyor (VCII) shows very high transimpedance gain that is independent from the bandwidth and is able to work also with a very high input parasitic capacitive impedance.

**Keywords** WSN · Sensor interface · Environmental monitoring

## 1 Introduction

Silicon PhotoMultipliers (SiPMs) are solid-state single-photon-sensitive devices based on Single-Photon Avalanche Diode (SPADs) implemented on common silicon substrate. They are considered as an innovative technology that is rapidly becoming popular and intends to replace the traditional photomultiplier tubes in many scientific areas. In particular, it finds its use in a variety of fields, from chemical to astrophysical and medical applications and in general in portable applications. Inheriting the traditional design approach of photomultiplier tubes, SiPMs interfaces are usually designed with a voltage-mode approach due to the electrical characteristics of the SiPM [1–12]. A current-approach design is usually discouraged since even if it provides a faster response time, usually suffers of worsen noise performance, so preventing a practical use of these kinds of interface receiving low-power signals. In this paper we propose a voltage conveyor VCII-based sensor interface for SiPMs. The proposed solution is formed by a mixed-mode (voltage/current) design, based on a voltage conveyor (VCII), showing very high transimpedance gain that is independent from the bandwidth and is able to work also with very high input parasitic capacitive impedance. Simulation and theoretical results are presented in the following.

G. Barile (✉) · A. Leoni · M. Muttillo · L. Pantoli
Department of Industrial and Information Engineering and Economics,
University of L'Aquila, L'Aquila, Italy
e-mail: Gianluca.barile@univaq.it



G. Di Francia et al. (eds.), *Sensors and Microsystems*, Lecture Notes
in Electrical Engineering 629, https://doi.org/10.1007/978-3-030-37558-4_3

## 2 The Proposed Solution

The VCII is the dual block of CCII (Current Conveyor) [13–20]. It is formed by an input current buffer followed by a voltage buffer, as shown in the simplified schematic of Fig. 1. A preliminary investigation of this design approach has been already briefly illustrated in [13]. The proposed analog block has a current input at *Y* pin and a current output at *X* terminal, while the *Z* port is an output voltage terminal. This solution allows to achieve low input and output impedances, suitable for the current input signal to be converted into voltages. The VCII so designed has been connected to a SiPM has reported in Fig. 2, where the silicon photomultiplier is modelled with a current source, a series switch and a parallel parasitic impedance that is typically capacitive and very large. On the other side, the *X* pin is connected to an external resistor that is useful to fix the transimpedance gain. In fact, as demonstrated in [4], the output voltage of the circuit is:

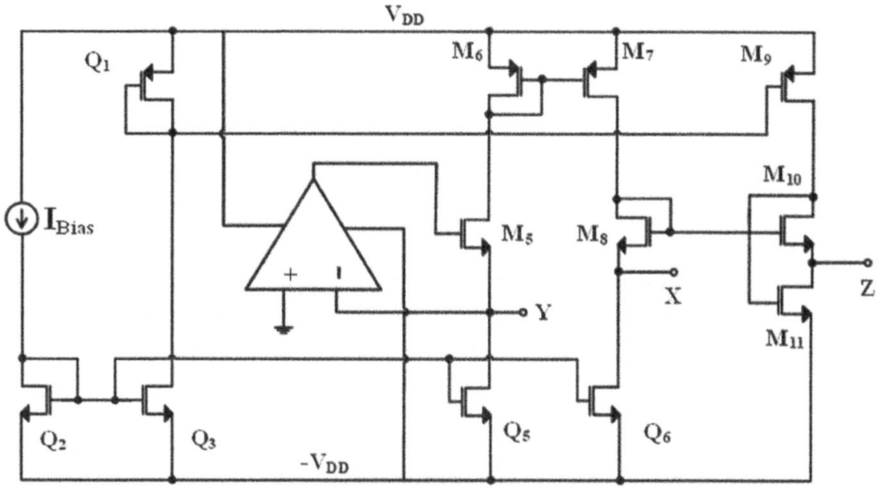


**Fig. 1** The proposed VCII base scheme

**Fig. 2** Simplified architecture of the VCII-based sensor interface

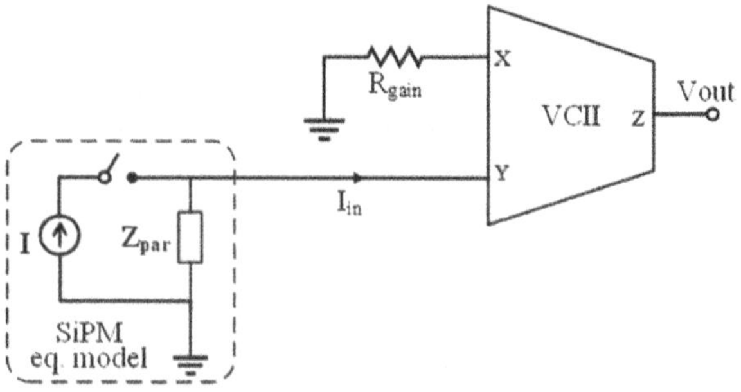

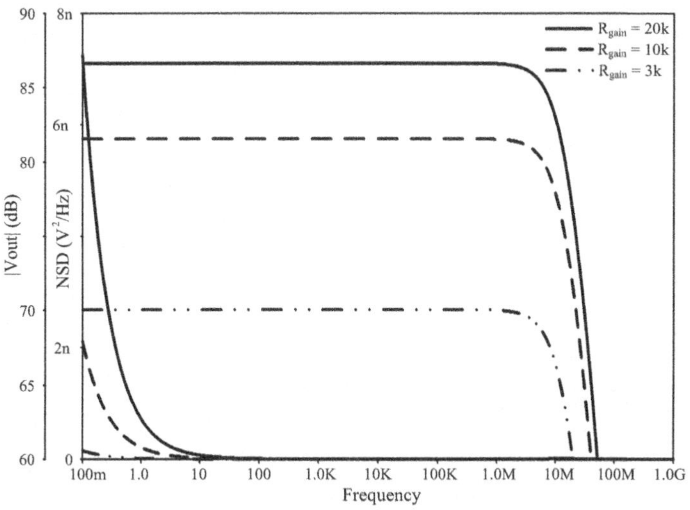


**Fig. 3** Transfer function and noise contribution for different gain levels of the proposed VCII-based interface

$$V_{\text{out}} = \alpha V_x \approx \pm \alpha \beta R_{\text{gain}} I_{\text{in}} \tag{1}$$

being $\alpha$ and $\beta$ voltage and current gain of the VCII, respectively, that are ideally equal to 1, is our case for a frequency range up to several MHz through a suitable microelectronic design.

The proposed interface has been simulated with the 0.35 μm CMOS technology process from AMS Foundry. As an example of application, in Figs. 3 and 4 some preliminary results are reported. Simulations shows a transimpedance gain up to 86 dB with a very large bandwidth (around 10 MHz) and a reduced power consumption (700 μW). The circuit is biased with ±1.65 V, the considered input current is about 30 μA, while a parasitic input capacitance up to 300 pF has been taken into account. The input and output port impedances are around 50 Ω and also noise performance are satisfactory (Fig. 3). Time domain response in Fig. 4 shows the feasibility of the proposed solution to be adopted in practical low voltage low power applications.

## 3 Conclusions

In this paper we have proposed a VCII-based sensor interface for silicon photomultiplier. Simulation results have shown that thanks to the voltage/current approach it is possible to achieve very high transimpedance gain that is independent from the bandwidth and is able to work also with very high input parasitic capacitive impedance.

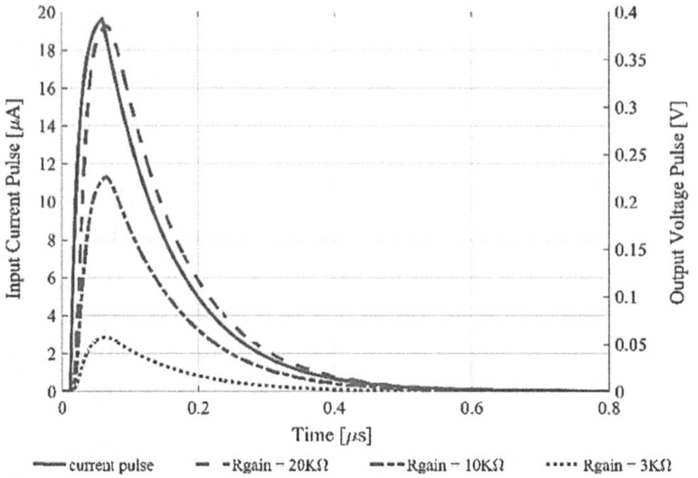


**Fig. 4** Input current pulse and voltage response at different gain levels (pulse responses refer to right axis)

# References


1. Kaplon J, Kulis S (2015) Review of input stages used in front end electronics for particle detectors (Online)
2. de Medeiros Silva M, Oliveira LB (2014) Regulated common-gate transimpedance amplifier designed to operate with a silicon photo-multiplier at the input. IEEE Trans Circ Syst I Reg Pap 61(3):725–735
3. Dorosz P, Baszczyk M, Kucewicz W, Mik Ł (2018) Low-power front-end ASIC for silicon photomultiplier. IEEE Trans Nucl Sci 65(4):1070–1078
4. Liu J, Sun Q, Fan Z, Jia Y (2018) TOF lidar development in autonomous vehicle. In: IEEE optoelectronics global conference, pp 185–190
5. Santillan J, Makinano-Santillan M, Cutamora L (2016) Integrating LiDAR and flood simulation models in determining exposure and vulnerability of buildings to extreme rainfall-induced flood hazards. In: International geoscience and remote sensing symposium, pp 7585–7588
6. Gargoum S, El-Basyouny K (2017) Automated extraction of road features using LiDAR data: a review of LiDAR applications in transportation. In: International conference on transportation information and safety, pp 563–574
7. Pantoli L, Barile G, Leoni A, Muttillo M, Stornelli V (2018) A novel electronic interface for micromachined Si-based photomultipliers. Micromachines 9(10):507
8. Barile G, Leoni A, Pantoli L, Safari L, Stornelli V (2018) A new VCII based low-power low-voltage front-end for silicon photomultipliers. In: International conference on smart and sustainable technologies (SpliTech), pp 1–4
9. Pantoli L, Barile G, Leoni A, Muttillo M, Stornelli V (2019) Electronic interface for lidar system and smart cities applications. J Commun Softw Syst 15(2):118–125
10. Zhang Y, Gu S, Yang J, Jose Alvarez M, Kong H (2018) Fusion of LiDAR and camera by scanning in LiDAR imagery and image-guided diffusion for urban road detection. In: Intelligent vehicles symposium, pp 579–584
11. Zhang Y, Xiong X, Zheng M, Huang X (2015) LiDAR strip adjustment using multifeatures matched with aerial images. IEEE Trans Geosci Remote Sens 53(2):976–987

12. Nam ES, Oh MS, Kim HY, Chong YJ (2008) Eye safe laser radar using a microchip laser, 2-dimensional InGaAs/InP photodiode arrays and the bi-axial optical lens system. In: 2008 Asia-Pacific microwave conference, pp 1–4
13. Safari L, Barile G, Stornelli V, Ferri G (2019) An overview on the second generation voltage conveyor: features, design and applications. IEEE Trans Circ Syst II Express Briefs 66(4):547–551
14. Stornelli V, Ferri G, Pantoli L, Barile G, Pennisi S (2018) A rail-to-rail constant-gm CCII for instrumentation amplifier applications. AEU Int J Electron Commun 91:103–109
15. Safari L, Barile G, Ferri G, Stornelli V (2018) High performance voltage output filter realizations using second generation voltage conveyor. Int J RF Microwave Comput Aided Eng 28(9):e21534
16. Stornelli V, Ferri G (2014) A single current conveyor-based low voltage low power bootstrap circuit for ElectroCardioGraphy and ElectroEncephaloGraphy acquisition systems. Analog Integr Circ Sig Process 79(1):171–175
17. Ferri G, Parente FR, Stornelli V (2017) Current conveyor-based differential capacitance analog interface for displacement sensing application. AEU Int J Electron Commun 81:83–91
18. Barile G et al (2017) Power-efficient dynamic-biased CCII. In: 2017 European conference on circuit theory and design, ECCTD
19. Stornelli V, Ferri G (2013) A 0.18 μm CMOS DDCCII for portable LV-LP filters. Radioengineering 22(2):434–439
20. Falconi C, Ferri G, Stornelli V, De Marcellis A, Mazzieri D, D'Amico A (2008) Current-mode high-accuracy high-precision CMOS amplifiers. IEEE Trans Circ Syst II Express Briefs 55(5):394–398

# New Chemometric Approach Using Data Obtained by a DMFC Device to Qualitatively and Quantitatively Determine Organic Molecules

**Mauro Tomassetti, Federico Marini, Riccardo Angeloni, Mauro Castrucci and Luigi Campanella**

**Abstract** In this communication first results are reported: seven organic molecules were considered: (3 alcohols) methanol, ethanol, propanol, (2 antibiotics) chloramphenicol, imipenem, (and 2 important biologic molecules) atropine and cortisone. For quantitative analysis seven different calibration curves were built (one for each molecule), showing different calibration sensitivity, which allows the quantitative determination of the seven organic molecules. For the qualitative analysis data from each of curves, obtained through the fuel cell, which represent the sensor's response to each of the molecules considered, where processed by chemometric methods, so that it was possible to directly identify and recognize each of the seven organic molecules.

**Keywords** Direct catalytic fuel cell · Coupled yeast cells · Glucose as fuel · Analytical and energetic purposes

## 1 Introduction

Organic molecules very different from one another, practically having in common only one –OH group, can be determined not only quantitatively, but also qualitatively, making use of a small commercial Direct Methanol Fuel Cell (DMFC), used as an analytical sensor [1]. In the first results reported in this communication, the following 7 organic molecules were considered: methanol, ethanol, propanol, chloramphenicol, imipenem, atropine and cortisone; that is, 3 alcohols, 2 antibiotics and 2 molecules very important from the bio-pharmaceutical point of view. From a quantitative point of view, the traditional approach was followed: seven different calibration curves were built (one for each molecule), which show very different calibration sensitivity. Each of them allows, therefore, the quantitative determination of the corresponding organic molecule, even if with very different sensitivities among them. For the qualitative analysis of single molecules, the approach has been much more innovative. In fact, by processing the data from each of the individual curves obtained through the

M. Tomassetti (✉) · F. Marini · R. Angeloni · M. Castrucci · L. Campanella
Department of Chemistry, University of Rome "La Sapienza", P.le a. Moro, 5, 00185 Rome, Italy
e-mail: mauro.tomassetti@uniroma1.it



G. Di Francia et al. (eds.), *Sensors and Microsystems*, Lecture Notes in Electrical Engineering 629, https://doi.org/10.1007/978-3-030-37558-4_4

fuel cell, which represent the sensor's response to each of the molecules considered and processing them using chemometric methods, it is possible to directly identify and recognize each of the 7 organic molecules.

## 2 Results

The recorded signal as a function of time (s) of the fuel cell for each one of seven studied organic molecules, are shown in Fig. 1.

### 2.1 Quantitative Analysis

From a quantitative point of view, the traditional approach was followed: 7 different calibration curves were constructed (one for each studied molecule), each of which allows the quantitative determination of one of the molecules. Seven calibration curves have been obtained and reported in Fig. 2. Also the different calibration curves equations have been reported in this figure, that evidences significant differences in the sensitivity of molecules to the seven analytes considered.

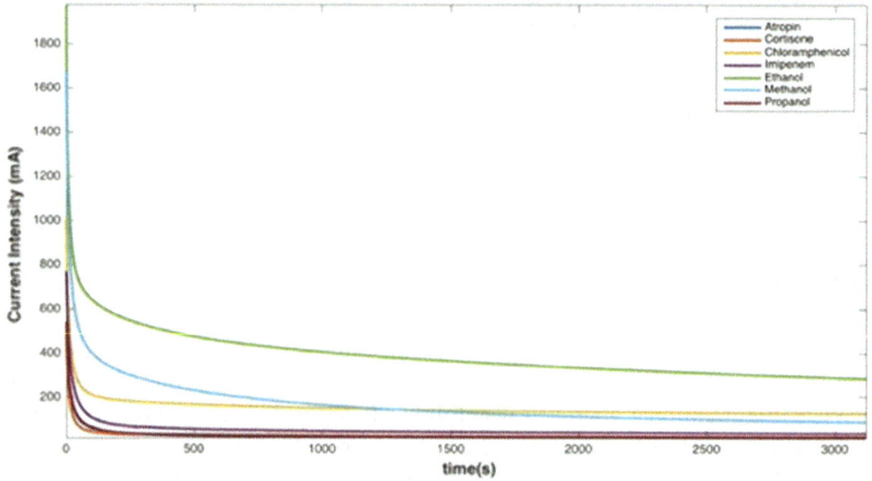


**Fig. 1** Current decays of the seven investigated solutions constituting the data set under investigation

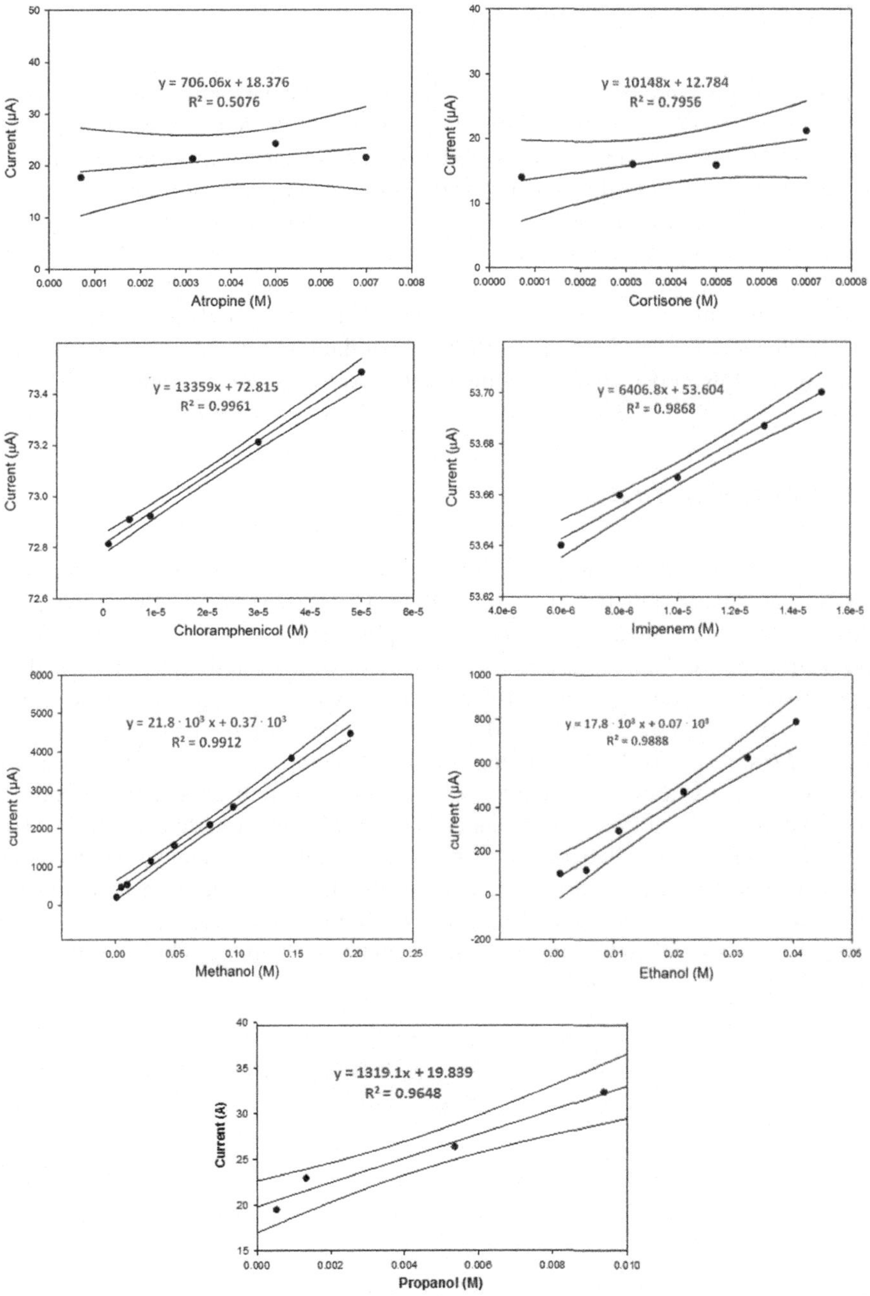


**Fig. 2** Calibration curves for the seven studied organic molecules and univariate calibration equations for various analytes

## 2.2 Principal Component Analysis

At first, principal component analysis was applied to the data, after mean centering, and 2 components (accounting for more than 99.9% of the total variance) appeared to be significant according to cross-validation. The results of PCA are graphically displayed in Fig. 3, in terms of scores and loading plots. The loading plot seems to indicate that there are two kind of exponential decays characterizing the analytes, one (contributing the most to PC1) which is faster, and a second one (more relevant for PC2) which is slower. By comparing these outcomes with the corresponding scores plot, one could infer that in ethanol, to a lesser extent in methanol and chloramphenicol, are characterized by a higher contribution of the faster component is dominant, whereas the slowest one contributes the most in particular for methanol.

## 2.3 Multi-way Analysis

In order to have a better insight into the characteristics of the systems, the data were further analyzed by power-slicing followed by PARAFAC data processing [2]. This approach takes advantage of the holographic nature of the exponential decays and rearranges the data into a three-way structure which can then be processed by appropriate multi-way algorithms (in this case, PARAFAC). By doing this, it was possible to identify four exponential decays (with time constants 7.1 s, 33.0 s, 539.0 s and 9502.6 s, respectively) accounting for practically all the variance present in the data, and to evaluate the contribution of each of these four exponential trends to the behavior of the different substances. The results are reported in Fig. 4. The Figure highlights how in general the current drop is higher for ethanol, methanol and chloramphenicol, but also that, for these molecules, the slower decaying components are more relevant. On the other hand, the other molecules show a lower overall intensity drop and, when looking at the individual contributions, a higher amount of the faster decaying components.

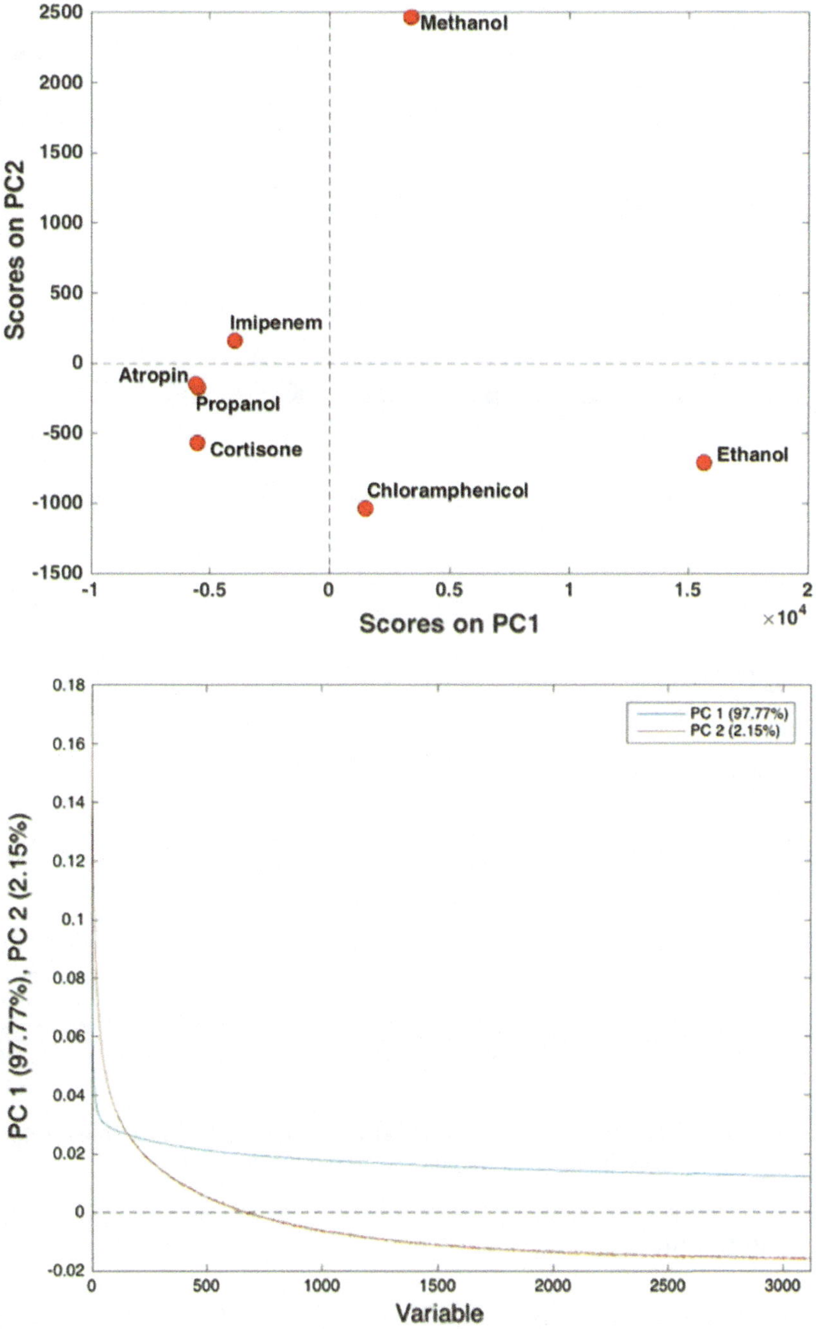


**Fig. 3** Results of PCA on the analyzed data set: scores plot (upper panel); loading plot (lower panel)

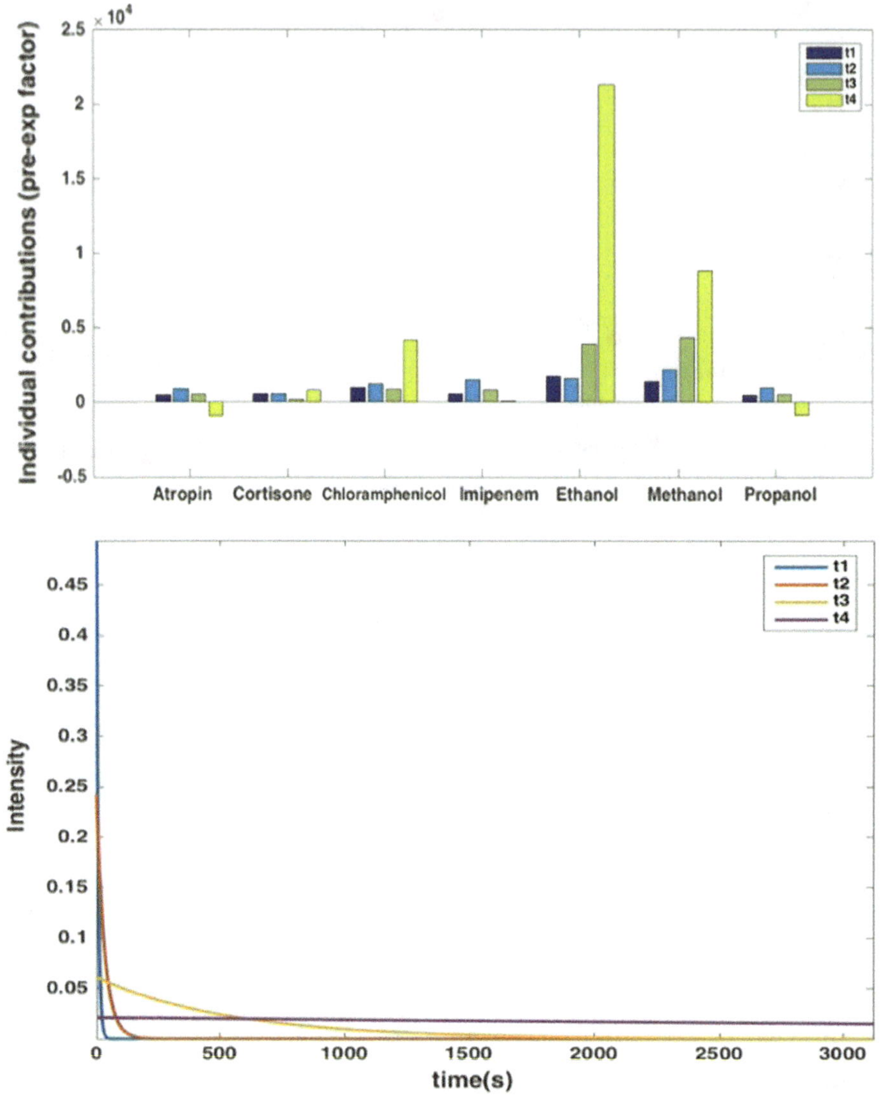


**Fig. 4** Results of Slicing/PARAFAC on the analyzed data set: relative contributions of the different exponential decays (upper panel), whose profiles are reported in the lower panel

## 3 Conclusion

In conclusion, the results obtained from this first approach, based on the use of a DMFC cell type sensor and chemometric methods, seem innovative and promising in the light of further and wider developments in the qualitative and quantitative analysis of several other organic molecules.

**Acknowledgements** This work was funded by the University of Rome "La Sapienza", Center "Prote-zione dell'Ambiente e dei Beni Culturali (CIABC)" and "Istituto per lo Studio dei Materiali Nanostrutturati (ISMN)" of CNR.

## References

1. Tomassetti M, Angeloni R, Merola G, Castrucci M, Campanella L (2016) Catalytic fuel cell used as an analytical tool for methanol and ethanol determination. Application to ethanol determination in alcoholic beverages. Electrochim Acta 191:1001–1009
2. Pedersen HT, Bro R, Engelsen SB (2001) In: Webb GA, Belton PS, Gil AM, Delgadillo I (eds) Magnetic resonance in food science: a view to the future. The Royal Society of Chemistry, London, UK, pp 202–212

# Studies on Silver Nanoparticles Production Mediated by Sugars

**Annalisa Scroccarello, Flavio Della Pelle, Simona Scarano and Dario Compagnone**

**Abstract** In this work, a simple strategy for the formation of silver nanoparticles (AgNPs) mediated by sugars (SGs) has been studied, focusing the attention on the kinetic of the process. In optimized conditions, SGs, together with a proper capping agent, are able to form AgNPs reducing $Ag^{+}$ and stabilizing, at the same time, the resulting AgNPs colloidal suspension. In order to investigate the AgNPs formation, the attention was paid on four SGs (fructose, glucose, sucrose, and xylitol); localized surface plasmon resonance (LSPR) related to the AgNPs formed was monitored over time. Furthermore, the morphology of the AgNPs, was also studied at the different stages of the formation process.

**Keywords** Silver nanoparticles · LSPR · Sugars structure

## 1 Introduction

In recent years, nanomaterials (NMs) have received increasing attention in different fields thanks to their unique optical, electrical, magnetic, catalytic, biological, and mechanical properties. NMs have become the favorite tools to realize sensors and sensing strategies, allowing to achieve new and unique analytical performances. In particular, metal nanoparticles (MNPs), among the several uses in analytical chemistry, have been employed as direct and indirect 'probes' for food quality and safety evaluation [1]. However, the number of studies about the formation of AgNPs still remain very low compared with other metal-based NPs (e.g. gold, copper, platinum, etc.), probably because of the extremely high reactivity of silver [1–4] that results in a difficult handling. The kinetics of the MNPs formation results to be crucial to understand the 'synthetic route' and the conditions to obtain stable MNPs colloidal

A. Scroccarello (✉) · F. Della Pelle · D. Compagnone
Faculty of Bioscience and Technology for Food, Agriculture and Environment,
University of Teramo, 64023 Teramo, Italy
e-mail: ascroccarello@unite.it

S. Scarano
Department of Chemistry 'Ugo Schiff', University of Florence,
Via Della Lastruccia 3-13, 50019 Sesto Fiorentino, Italy

 
G. Di Francia et al. (eds.), *Sensors and Microsystems*, Lecture Notes
in Electrical Engineering 629, https://doi.org/10.1007/978-3-030-37558-4_5

suspensions. According to literature, the formation of colloidal MNPs is classically composed by three main steps: (i) the "nucleation step", where metal reduction occurs and the NPs seeds are formed (this step is fast and challenging to monitor, and starts when a synthetic or natural reducing agent keeps in contact with the metal source); (ii) the seeding step followed by the 'growth step' (growth and stabilization of the nanoparticles occur until the most stable shape and size is reached); (iii) the 'plateau step' (stable and well dispersed/suspended NPs) [3]. In this last step, an overgrowth/aggregation can rapidly occur, and the nano-size domain can be lost, in particular with not fully stabilized nanoparticles. It should be noticed that each MNPs formation step can be spectroscopically followed because of the LSPR feature (Fig. 2b). The plasmonic MNPs absorption band is an intrinsic feature of related to the MNPs size, shape and chemical nature [1, 8]. MNPs have been thus exploited in a wide range of applications in chemistry, biology, and nanotechnology, resulting to be elective materials to realize colorimetric or spectrophotometric sensing strategies [1–7]. AgNPs formation has been proposed by our group in previous works, aimed to assay SGs in food samples [9, 10]; in this work kinetic and morphological studies have been performed.

## 2 Material, Methods, and Apparatus

The AgNPs formation was conducted according to Della Pelle et al. [9], in brief, the AgNPs were obtained directly in cuvette via quantitative reduction of $AgNO_3$, mediated by SGs. The mix was composed by: 5 μL of CTAC ($1.0 \times 10^{-3}$ mol $L^{-1}$), 25 μL of $AgNO_3$ solution ($2.0 \times 10^{-2}$ mol $L^{-1}$), appropriate dilution of SGs standard (fructose, FRU; glucose, GLU; sucrose, SCR; xylitol, XYT), in a final volume of 500 μL of deionized water. The reaction was triggered by the addition of NaOH (10 μL, 5.0 mol $L^{-1}$). All the reagents were used at room temperature. The reaction mix was incubated at room temperature (25 °C), for 10 min in a WineLab instrument (CDR Foodlab s.r.l., Florence, Italy), and the absorbance value was recorded at a fixed wavelength at 430 nm using the kinetic mode, collecting extinction values over time (s). The AgNPs formed upon sugars action were also characterized by UV-Vis spectra (JENWAY 6400 Spectrophotometer, Barlworld Scientific, Staffordshire, UK) and Transmission Electron Microscopy (TEM) (Philips Electronic Instruments, Mahwah, NJ, USA) (Fig. 1).

## 3 Formation of AgNPs by Sugars: Kinetic Study and Morphological Characterization

The kinetic of the MNPs formation results to be crucial to understand and optimize their synthesis, mainly in order to obtain a stable colloidal suspension. Thus, with the

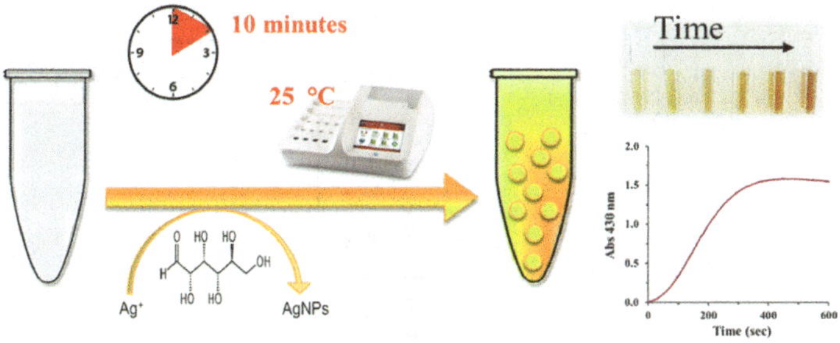


**Fig. 1** Scheme of the AgNPs sugars-mediated formation strategy

aim to investigate the AgNPs formation trend in real time, four standard SGs (GLU, FRU, SCR, and XYT) were studied as models at fixed wavelength (maximum of AgNPs LSPR: 430 nm) by monitoring the increase in absorbance over time until the steady state. The SGs were selected in order to understand how their structure may influence the formation and stabilization of the AgNPs. To this aim, the selected SGs were let to react (using the procedure reported in Sect. 2) at the same concentration ($5.0 \times 10^{-5}$ mol $L^{-1}$) to directly compare their behavior over 600 s. As displayed in Fig. 3a the sugars shown very similar trends, that may be described through three stages: an initial latent starting phase, a linearly increasing phase, followed by the gradual decrease of the growth (i.e. in the curve slope) and finally the plateau phase (Fig. 2a). These three phases correspond to the AgNPs seed-nucleation, growth and final-stabilization (Fig. 2a). During nucleation, silver seeds are formed by SGs (nucleation step, Fig. 2a). During the growth step, the metal seeds core increases (further $Ag^+$ reduction) with a concurrent external shell formation (composed by sugars and capping agent) around the Ag(0) core [9]. After exceeding critical AgNPs dimensions and concentrations, the AgNPs growth is slowed down and finally stops (Fig. 2b purple lines). At this stage, if the reaction is not stopped or the SGs concentration is too high, the AgNPs stabilization is compromised and the colloidal stability decreases. Due to the overgrown, AgNPs tend to interact each other, causing aggregation and collapse/precipitation. This phenomenon is clearly shown in the purple circle of the Fig. 2a inset.

Moreover, studying the kinetics of AgNPs formation mediated by SGs, the resulting time versus $LSPR_{max}$ curves returns information on the different SGs reactivity. It is easy to notice in Fig. 3a, where kinetic studies were conducted with FRU, GLU, SCR, and XYT. Clearly the different SGs structures influence the AgNPs formation rate, particularly in the nucleation step (from 0 to 100 s). This first phase is immediately followed by the silver seeds growth (from 100 to 400 s) where the higher growth rate was obtained by GLU (Fig. 3a, violet line), followed by FRU, XYT, and SCR. In this phase, the dimension of the AgNPs is low and the growth is influenced by both reducing and capacity of the SGs employed [9, 10]. Thus, the slope of the linear portion of the sigmoidal curve returns information on the SGs reactivity, in particular, higher slopes correspond to faster reaction rates. Finally, after exceeding a critical AgNPs dimension, the AgNPs growth is slowed down until it stops, while

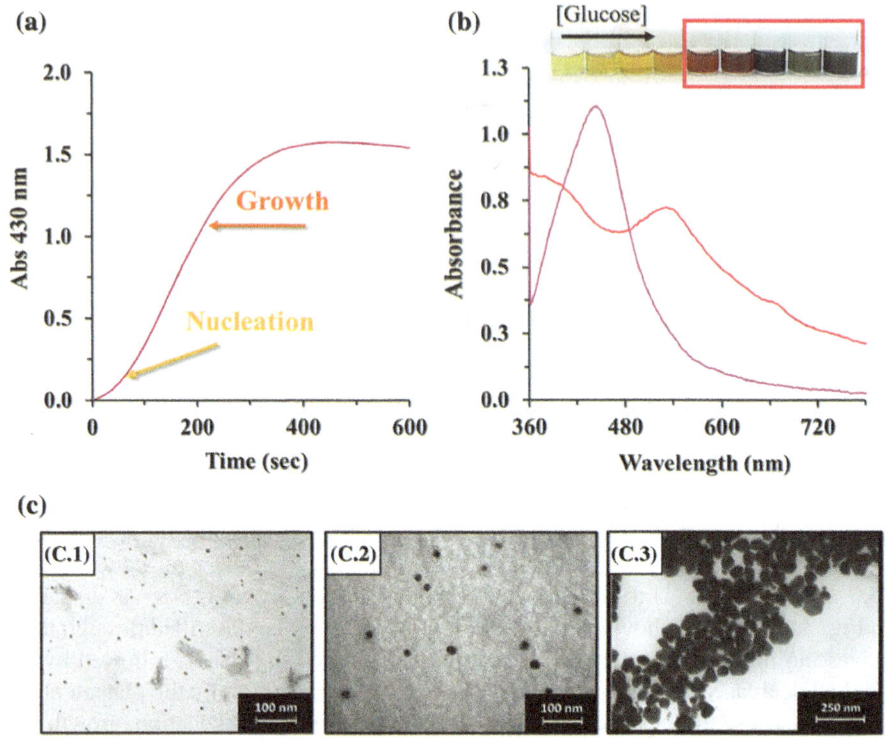


**Fig. 2** **a** AgNPs-GLU mediated formation curve monitored at fixed wavelength (430 nm). **b** LSPR absorbance peak of stable (pink line) and overgrown (purple line) AgNPs synthetized by GLU 2.5 × $10^{-6}$ mol $L^{-1}$ and 2.5 × $10^{-4}$ mol $L^{-1}$, respectively. In the inset, the color scale of AgNPs synthetized by increasing amount of GLU. Circled in purple indicates overgrown AgNPs. **c** TEM magnification of AgNPs formed with GLU 5.0 × $10^{-5}$ mol $L^{-1}$ at different time of reaction: from 0 to 100 s (C.1), from 100 to 600 s (C.2), over 1000 s

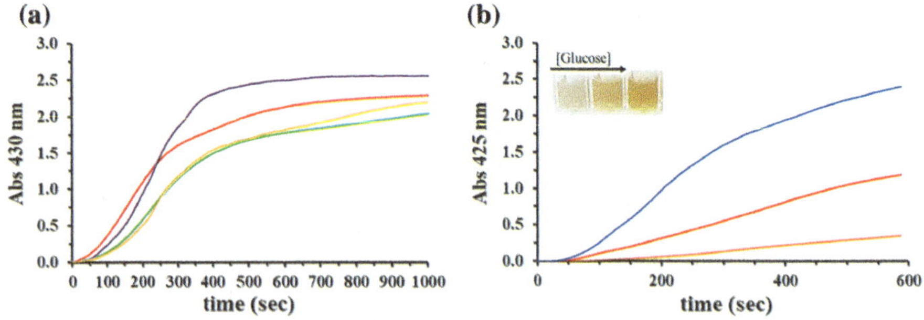


**Fig. 3** **a** AgNPs formation curve obtained using GLU (red line), FRU (violet line), SCR (yellow line) and XYT (green line) at 5.0 × $10^{-5}$ mol $L^{-1}$. **b** AgNPs formation curve obtained with increasing amount of GLU (1.0 × $10^{-5}$ mol $L^{-1}$ orange line, 4.0 × $10^{-5}$ mol $L^{-1}$ red line, and 8.0 × $10^{-5}$ mol $L^{-1}$ blue line)

the external shell, composed by SGs and capping agent, totally surrounds the Ag(0) core [1, 8–10]. As clearly shown in Fig. 3a, for all the SGs tested, the plateau phase is reached around 400 s, and the final absorbance value is strictly related to the intrinsic ability of SGs to quantitatively form AgNPs [9, 10].

Focusing on the SGs structure, among the studied SGs, FRU and GLU are both open structures with ketonic and aldehydic groups, respectively. These SGs return the highest final absorbance values (kinetic trend shown in Fig. 3a, violet and red lines for FRU and GLU respectively). On the other hand, SCR and XYT, a cyclic disaccharide and a polyalcohol, respectively, possess only hydroxyl functions (non-reducing sugars) (Fig. 3a yellow and green lines for SCR and XYT respectively) [9], and, in fact, they show the same kinetic behavior; this feature can be related to the absence of the reducing group. In this case, SCR and XYT are able to allow the formation of AgNPs acting as stabilizers and surface charge carriers in combination with the CTAC (capping agent). Furthermore, through the GLU kinetic curves, performed with increasing SGs concentration (Fig. 3b), taking the absorbance at the end-point of the reaction, a signal proportional to the standard concentration employed was observed. This behavior underlines, again, how in this kind of AgNPs formation strategy the SG is not only required to trigger the AgNPs formation but possesses an active role, that affects the amount, the size/shape, and the dispersion state of the nanoparticles formed.

## 4 Conclusion

In this work, the active role of SGs in the AgNPs formation was confirmed. Moreover, the formation studies allowed to hypothesize the mechanisms behind the SG-mediated AgNPs synthesis. This AgNPs synthesis confirms to be a valuable tool both for the green-production of AgNPs and for sugars content evaluation in real samples.

## References

1. Della Pelle F, Comapagnone D (2018) Nanomaterial-based sensing and biosensing of phenolic compounds and related antioxidant capacity in food. Sensors 18(2):462
2. Özyürek M, Güngör N, Baki S, Güçlü K, Apak R (2012) Development of a silver nanoparticle-based method for the antioxidant capacity measurement of polyphenols. Anal Chem 84(18):8052–8059
3. Simona S, Emanuela P Minunni M (2017) The early nucleation stage of gold nanoparticles formation in solution as powerful tool for the colorimetric determination of reducing agents: the case of xylitol and total polyols in oral fluid. Anal Chim Acta 993:71–78
4. Nishimura S, Mott D, Takagaki A, Maenosono S, Ebitani K (2011) Role of base in the formation of silver nanoparticles synthesized using sodium acrylate as a dual reducing and encapsulating agent. Phys Chem Chem Phys 13(20):9335–9343
5. Aleksandra S-C, Tułodziecka A, Szłyk E (2012) A silver nanoparticle-based method for determination of antioxidant capacity of rapeseed and its products. Analyst 137(16):3750–3759

6. Della Pelle F, Vilela D, González MC, Sterzo CL, Compagnone D, Del Carlo M, Escarpa A (2015) Antioxidant capacity index based on gold nanoparticles formation, Application to extra virgin olive oil samples. Food Chem 178:70–75
7. Della Pelle F, González MC, Sergi M, Del Carlo M, Compagnone D, Escarpa A (2015) Gold nanoparticles-based extraction-free colorimetric assay in organic media: an optical index for determination of total polyphenols in fat-rich samples. Anal Chem 87(13):6905–6911
8. Della Pelle F, Scroccarello A, Sergi M, Mascini M, Del Carlo M, Compagnone D (2018) Simple and rapid silver nanoparticles based antioxidant capacity assays: reactivity study for phenolic compounds. Food Chem 256:342–349
9. Della Pelle F, Scroccarello A, Scarano S, Compagnone D (2019) Silver nanoparticles-based plasmonic assay for the determination of sugar content in food matrices. Anal Chim Acta 1051:129–137
10. Scroccarello A, Della Pelle F, Neri L, Pittia P, Compagnone D (2019) Silver and gold nanoparticles based colorimetric assays for the determination of sugars and polyphenols in apples. Food Res Int 119:359–368

# Low-Power Integrated Circuit for Orientation Independent Acquisitions from Smart Accelerometers

**Antonio De Vita, Gian Domenico Licciardo, Aldo Femia, Luigi Di Benedetto, Alfredo Rubino and Danilo Pau**

**Abstract** In this work, a digital circuit is proposed to eliminate the dependence of measurements of tri-axial accelerometers from the spatial orientation of the sensor. It is suitable for a large number of applications, from wearable system to human activity recognition and prediction. The circuit implements a new, hardware friendly, vector rotation algorithm which contributes to reduce the occupied area and the power dissipation. The design has been targeted to FPGA and std_cells in 65 nm CMOS technology. In the second case, it returns a power dissipation of 1 $\mu$W and an area of about 0.024 mm$^2$. The obtained results show the possibility to integrate the module with the embedded circuitry of inertial sensors.

**Keywords** Inertial sensors · Smart sensors · Low-power · Wearable systems

A. De Vita · G. D. Licciardo (✉) · A. Femia · L. Di Benedetto · A. Rubino
Department of Industrial Engineering, University of Salerno, Via Giovanni Paolo II, 132, 84084 Fisciano, SA, Italy
e-mail: gdlicciardo@unisa.it

A. De Vita
e-mail: andevita@unisa.it

A. Femia
e-mail: afemia@unisa.it

L. Di Benedetto
e-mail: ldibenedetto@unisa.it

A. Rubino
e-mail: arubino@unisa.it

D. Pau
System Research and Application, STMicroelectronics, 20864 Agrate Brianza, MB, Italy
e-mail: danilo.pau@st.com

 
G. Di Francia et al. (eds.), *Sensors and Microsystems*, Lecture Notes in Electrical Engineering 629, https://doi.org/10.1007/978-3-030-37558-4_6

# 1 Introduction

Over the last years, the development of inertial sensors has enabled several applications exploiting the reduced dimensions and low power dissipations of MEMS. Accelerometers, in particular, are largely used in low-power and wearable applications such as motion sensing, fall detection, and the like. Generally, in such applications the sensor is arbitrarily oriented in the space, and the disturbance deriving from the difficulty to establish a homogeneous coordinate reference system for the measured vectors causes significant inaccuracies [1–3]. Therefore, the possibility to relate all the measurements to a unique coordinate frame is highly required [4]. However, the orientation recalculation and the removal or attenuation of the induced errors is computationally intensive, in particular because real-time processing and very high accuracy are required. The use of general purpose processing units (MCUs, CPUs, or GPUs), does not permit an optimal trade-off between dimensions, power consumption and accuracy, being limited to approximated algorithms [5, 6]. With the purpose to overcome the above limitations, in this work, a new custom circuit is proposed, which makes the acquisitions of a tri-axial accelerometer independent from the orientation of the sensor. This is obtained by implementing a new "hardware friendly" algorithm, specifically derived from the Rodrigues' rotation formula [7]. The circuit is capable to rotate simultaneously each coordinate of a tri-axial accelerometer from a generic coordinate system to a reference one. If the reference system is fixed to the Earth, the measured accelerations can be related to the gravity vector, as schematized in Fig. 1a, eliminating in this way the error due to the misalignment between the coordinate systems.

The circuit has been prototyped on the Xilinx xc7z020clg484-1 FPGA and synthesized with the TSMC 65 nm CMOS std_cells. In this second case it returns an occupied area of 0.024 mm$^2$ and a total dissipated power of 1 μW when the clock

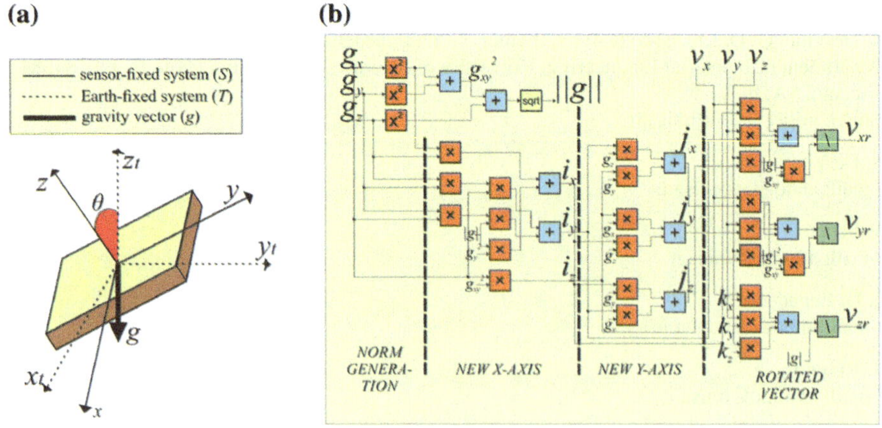


**Fig. 1** **a** Representation of the sensor-fixed reference frame (x, y, z) and the reference frame ($x_t$, $y_t$, $z_t$) with respect to gravity ($\vec{g}$) and the sensor. **b** Proposed calculation scheme

frequency is set to a frequency of 12.3 kHz, compatible with output data-rate of commercial accelerometers for motion sensing. Such results prove the integrability of the proposed circuits in the sensor ASIC.

## 2 The Rotation Algorithm

The vector rotation algorithm is obtained from the Rodrigues' rotation formula. From this an alternative calculation scheme has been derived in order to have the minimum set of arithmetic operators. As in Fig. 1a, the tri-axial accelerometer defines a sensor-fixed system $S = \{x, y, z\}$, while gravity defines an Earth-fixed system $T = \{x_t, y_t, z_t\}$. The proposed circuit calculates the three coordinates in $T$ by rotating $S$ around the rotation axis $u$ by an angle $\theta$, where $u$ and $\theta$ are defined by the gravity vector $\vec{g}$ as:

$$\hat{u} = \frac{\vec{g} \times \hat{z}}{\|\vec{g} \times \hat{z}\|} = \left(\frac{g_y}{\sqrt{g_x^2 + g_y^2}}, -\frac{g_x}{\sqrt{g_x^2 + g_y^2}}, 0\right), \quad \theta = \cos^{-1}\frac{g_z}{\|\vec{g}\|}$$

where $\hat{z}$ is the $z$-axis versor in $S$, $g_z$ the component of $g$ along the $z$-axis in $S$ and $\|\vec{g}\|$ the norm of the gravity vector. If, for example, $z$ must be aligned with $g$, namely if $\hat{z}_t = \frac{\vec{g}}{\|\vec{g}\|} = \left(-\frac{g_x}{\|\vec{g}\|}, -\frac{g_y}{\|\vec{g}\|}, -\frac{g_z}{\|\vec{g}\|}\right)$, the rotated vectors can be calculated as: $v_{xr} = \vec{v} \cdot \hat{x}_t$, $v_{yr} = \vec{v} \cdot \hat{y}_t$, $v_{zr} = \vec{v} \cdot \hat{z}_t$, where:

$$\hat{x}_t = \vec{x}_u + \vec{x}_n = \left(-\frac{|g_x| g_x g_z}{\|\vec{g}\| g_{xy}^2} + \frac{g_y^2}{g_{xy}^2}, -\frac{|g_x| g_y g_z}{\|\vec{g}\| g_{xy}^2} - \frac{g_x g_y}{g_{xy}^2}, \frac{|g_x|}{\|\vec{g}\|}\right)$$

$$\hat{y}_t = \hat{z}_t \times \hat{x}_t = \left(\frac{a_y g_z - a_z g_y}{\|\vec{g}\|}, \frac{a_z g_x - a_x g_z}{\|\vec{g}\|}, \frac{a_x g_y - a_y g_x}{\|\vec{g}\|}\right)$$

$\vec{x}_u = (\hat{x} \cdot \hat{u})\hat{u}$, $\vec{x}_n = \left\|\hat{x} - (\hat{x} \cdot \hat{u})\hat{u}\right\|(\hat{z}_t \times \hat{u})$ and $\hat{x}_t = (a_x, a_y, a_z)$.

The derived calculation scheme is shown in Fig. 1b, from which the huge complexity of the circuit appears evident, as well as the great amount of physical resources required for its straightforward implementation. An accurate reorganization and dimensioning of all the arithmetic operators has reduced the design to a single atomic module, which iteratively operates in order to implement the processing scheme of Fig. 1b. All the partial and final results have been coded with 24-bits fixed point (8.16) with the purpose to simplify the complexity of operators and registers [8, 9].

## 3 The Proposed Architecture

An iterative structure has been implemented by exploiting the modularity of the calculation scheme, in Fig. 1b. The scheme of the circuit is shown in Fig. 2a. The arithmetic core, the Acceleration Rotation Unit (ARU), is detailed in Fig. 2b. It is composed by three 24-bits Booth multipliers operating in parallel, two 24-bits carry-ripple cascaded adders and three multiplexers. The inputs $i1$–$i6$, correspond either to the spatial components of the current $\vec{g}$ and $\vec{v}$ or to feedback intermediate results; three outputs are provided at each clock cycle. The proposed architecture has been tailored for motion sensing with low input data rate, typical of low frequency human activities. Each multiplier is made up of one Booth cell only; therefore 12 cycles are required to carry out the multiplication. A dedicated clock signal (*Mult_clk*) is used, derived by multiplying by 12 the frequency of the global clock (*G_clk*). In order to complete a rotation, 41 cycles are required: 18 cycles are needed for the ARU operations, 8 of which are devoted to the square-root and the remaining 23 cycles are used by the dividers in Fig. 2c. Given the reduced amount of data, the data-stream of input and partially processed data has been managed by registers [10–12]. The iterations have been implemented by the bank of 24 bits registers in Fig. 2a, able to store partial results up to 4 clock cycles. Within each cycle, an input MUX, selected by a 6-bits counter, feeds the ARU with new input data, $(g_x, g_y, g_z, v_x, v_y, v_z)$, or feeds back intermediate data. Three dividers implement the division operations, which are grouped and shifted at the end of the calculation scheme, as shown in Fig. 1b.

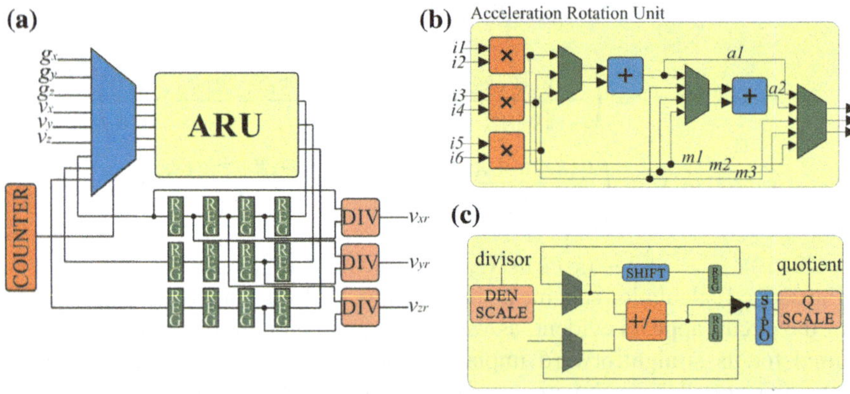


**Fig. 2** **a** Block diagram of the acceleration rotation architecture; **b** scheme of the acceleration rotation unit; **c** scheme of the divisor

### 3.1 Square Root and Division

The square root has been implemented by a Taylor series expansion. The circuit is based on five 30-bytes LUTs (10 words of 24-bit) to store the series coefficients. In total, 8 cycles are required to perform the square root operation. The division operation has been implemented by the conventional restoring algorithm, based on the calculation of the reciprocal of the divisor [13]. Each divider is implemented in an iterative fashion as shown in Fig. 2c. The main arithmetic block is a 24-bit carry-ripple adder/subtractor. Since the divider is used to calculate the reciprocal of a given value, the dividend is always equal to 1. The *DEN SCALE* left-shifts the divisor until its integer part becomes greater than or equal to 1 to properly start the division process. In turn, the *Q SCALE* blocks right-shift the quotient to obtain its final value.

## 4 Results

The proposed design has been targeted to the Xilinx xc7z020clg484-1 FPGA and std_cells in TSMC 65 nm CMOS technology. Although the proposed circuit could be applied to each kind of accelerometer, the proposed setup has been arranged for applications requiring as low power dissipation as possible, like wearable and human activity recognition systems where the obtained results are much more interesting. Results are summarized in Table 1, where also the implementation results of a FP32 coding has been reported to give more context about area and power reduction allowed by the design choices. The sensor output data rate has been set to $f_s = 25$ Hz, which is usually taken as the reference frequency for human activities [14, 15]. Consequently, the minimum allowable value for the global clock frequency has been set to $41\,f_s = 1025$ Hz, and the one of the *Mult_clk* to 12.3 kHz. It is worth to underline that the maximum operating frequencies are much higher: in particular FPGA results returns a critical path delay of 15.4 ns for *Mult_clk* intra-clock paths and of 28.4 ns

**Table 1** FPGA and std_cells synthesis results

| | Floating-point 32-bit | | Fixed-point 24-bit | |
|---|---|---|---|---|
| | FPGA | std_cells | FPGA | std_cells |
| Total power (mW) | 104 | $1.7 \times 10^{-3}$ | 104 | $10^{-3}$ |
| Dynamic power (mW) | <1 | $0.68 \times 10^{-3}$ | <1 | $0.31 \times 10^{-3}$ |
| Static power (mW) | ≅104 | $1.02 \times 10^{-3}$ | 104 | $0.69 \times 10^{-3}$ |
| LUT | 8594 | – | 2760 | – |
| FF | 1285 | – | 1305 | – |
| Area (mm$^2$) | – | 0.05 | – | 0.024 |
| Max op. freq. (MHz) | 10.5 | 31.3 | 64.9 | 109.9 |

for *G_clk*. Therefore, the maximum operation frequencies are 65 and 5.4 MHz for the *Mult_clk* and the *G_clk* respectively. In turn, maximum operation frequency in CMOS technology are 110 and 9.2 MHz. The amount of mapped physical resource has been highly reduced by the choice of a full iterative implementation: 2760 LUTs are required for FPGA implementation and an area occupation of 0.024 $mm^2$ with std_cells. Consider that a large part of the design is occupied by the divisor circuitries, up to 45% of the total number of LUTs in FPGA, since DSP instantiation has been avoided to ensure the portability of the RTL and as much platform independent results as possible. The reduced occupied area is mandatory to limit the power dissipation, which in std_cells is 1 $\mu$W when *G_clk* is set to 1.025 kHz, since about the 70% is due to leakages, although low-power libraries and devices with a high threshold voltage have been used. The corresponding energy per clock cycle is 0.87 nJ, while energy per AR operation is 35.6 nJ. With respect to a FP32 coding, the design choices show a 2× reduction both in area occupation and power consumption, with a negligible impact on the overall accuracy of the system. In this work comparisons with other solutions are absent since, to the best of our knowledge, in the literature a design that offers the same functionality with the same performances does not exist.

## 5 Conclusion

In this work, an ultra-low power fixed-point architecture for AR is proposed, which achieves state-of-art performances in terms of accuracy, power dissipation and occupied area. The ultra-low power consumption makes feasible the integration of the proposed circuit with the sensor in the same die or package. Future improvements will concern the further reduction of the multiplier complexity in the way shown in [16–18] and the introduction of a custom filter stage to improve accuracy.

## References

1. Chatterjee G, Latorre L, Mailly F, Nouet P, Hachelef N, Oudea C (2015) Smart-MEMS based inertial measurement units: gyro-free approach to improve the grade. Microsyst Technol 23:3969–3978
2. Ang WT, Khosla PK, Riviere CN (2003) Design of all-accelerometer inertial measurement unit for tremor sensing in hand-held microsurgical instrument. In: IEEE international conference on robotics and automation (Cat. No. 03CH37422), Taipei, Taiwan, vol 2, pp 1781–1786
3. De Vita A, Licciardo GD, Di Benedetto L, Pau D, Plebani E, Bosco A (2018) Low-power design of a gravity rotation module for HAR systems based on inertial sensors. In: IEEE 29th international conference on application-specific systems, architectures and processors (ASAP), Milan. https://doi.org/10.1109/asap.2018.8445130
4. Jzahn A, Bachmann M, Wenzel P, David K (2017) Focus on the user: a user relative coordinate system for activity detection. In: Brézillon P, Turner R, Penco C (eds) Modeling and using context, CONTEXT 2017. Lecture notes in computer science, vol 10257. LNCS, Heidelberg

5. Mizell D (2003) Using gravity to estimate accelerometer orientation. In: Seventh IEEE international symposium on wearable computers, 2003. Proceedings, White Plains, NY, USA, pp 252–253
6. Bennett TR, Wu J, Kehtarnavaz N, Jafari R (2016) Inertial measurement unit-based wearable computers for assisted living applications: a signal processing perspective. IEEE Signal Process Mag 33:28–35
7. Belongie S (1999) Rodrigues' rotation formula. MathWorld, a wolfram web resource, created by Eric W. Weisstein. http://mathworld.wolfram.com/RodriguesRotationFormula.html
8. Albanese LF, Licciardo GD (2010) An area reduced design of the Context-Adaptive Variable-Length encoder suitable for embedded systems. In: 5th international symposium on I/V communications and mobile network, Rabat. https://doi.org/10.1109/ISVC.2010.5656302
9. Licciardo GD, Albanese LF (2012) Design of a context-adaptive variable length encoder for real-time video compression on reconfigurable platforms. IET Image Process 6:301–308
10. Licciardo GD, Boesch T, Pau D, Di Benedetto L (2016) Frame buffer-less stream processor for accurate real-time interest point detection. Integration 54:10–23
11. Licciardo GD, D'Arienzo A, Rubino A (2015) Stream processor for real-time inverse tone mapping of full-HD images. IEEE Trans Very Large Scale Integr (VLSI) Syst 23:2531–2539
12. Licciardo GD, Cappetta C, Di Benedetto L (2018) Design of a Gabor filter HW accelerator for applications in medical imaging. IEEE Trans Compon Packag Manuf Technol 8:1187–1194
13. Richards RK (1955) Arithmetic operations in digital computers. Van Nostrand, New York, pp 166–172
14. Bulling A, Blanke U, Schiele B (2014) A tutorial on human activity recognition using body-worn inertial sensors. ACM Comput Surv 46
15. Maurer U, Smailagic A, Siewiorek DP, Deisher M (2006) Activity recognition and monitoring using multiple sensors on different body positions. In: International workshop on wearable and implantable body sensor networks (BSN'06), Cambridge, MA
16. Licciardo GD, Cappetta C, Di Benedetto L, Vigliar M (2017) Weighted partitioning for fast multiplierless multiple-constant convolution circuit. IEEE Trans Circuits Syst II Express Briefs 64:66–70
17. Licciardo GD, Cappetta C, Di Benedetto L, Rubino A, Liguori R (2018) Multiplier-less stream processor for 2D filtering in visual search applications. IEEE Trans Circuits Syst Video Technol 28:267–272
18. Licciardo GD, Cappetta C, Di Benedetto L (2016) FPGA optimization of convolution-based 2D filtering processor for image processing. In: 8th computer science and electronic engineering (CEEC), Colchester, pp 180–185. https://doi.org/10.1109/ceec.2016.7835910

# Hand-Held Electronic Nose to Detect Biomarkers of Diseases Through Breath

**Carlos Sanchez-Vicente, J. P. Santos, Jesús Lozano and I. Sayago**

**Abstract** In this communication, we present a home-developed portable electronic nose used in discrimination of biomarkers of different diseases through the breath. We have used samples corresponding with different levels of gravity of illnesses. The electronic nose is equipped with embedded control, instrumentation electronics, rechargeable batteries, pump and electrovalve to control the way of the sampling gas, touch screen and IEEE 802.11 transceiver for wireless communication. Several measurements were made to check the discrimination capability of the prototype in the discrimination of different biomarkers of diseases.

**Keywords** Electronic nose · Gas sensors · E-nose · Biomarkers · Breath · Diseases · Asthma · Diabetes · Volatile organic compounds · Digestive system · Respiratory system

## 1 Introduction

The increased incidence of chronic diseases worldwide, due to lifestyle and environmental conditions, can have a detrimental effect if not diagnosed in time and properly managed. Asthma and diabetes are two of the chronic diseases with the greatest impact on the population of the world. These pathologies affect to 339 million and 422 million of patients respectively. In recent years the number of patients affected by these pathologies has increased significantly, in the case of the diabetes increase 25% in 30 years. Although the mortality from these diseases has been reduced, there is still a considerable percentage undiagnosed or with deficient control of the pathology [1, 2].

C. Sanchez-Vicente · J. P. Santos (✉) · I. Sayago
Institute of Physics Technology and Information, CSIC, Madrid, Spain
e-mail: jp.santos@csic.es

J. Lozano
Industrial Engineering School, University of Extremadura, Badajoz, Spain

C. Sanchez-Vicente
Up Devices and Technologies, Madrid, Spain

 
G. Di Francia et al. (eds.), *Sensors and Microsystems*, Lecture Notes in Electrical Engineering 629, https://doi.org/10.1007/978-3-030-37558-4_7

Incorrect diagnostic and inadequate control of chronic diseases can increase the risk to the patients, aggravating the status of disease and reducing his quality of life. Not only that, late diagnosis or poor control increases in health care cost too. For this reason, it is necessary to have diagnostic equipment that is efficient, portable and economical, so that more health centers, hospitals, and patients have access to them. In addition, it is important for patients that this type of equipment is non-invasive, allowing rapid and risk-free analysis. These diseases modify the composition of the volatile organic compounds (VOCs) present in the breath of the patient, compared to a healthy person. This allows a non-invasive analysis to know the status of these pathologies. The breath is a mixture of nitrogen, oxygen, $CO_2$, $H_2O$, and other gases. It known that the air exhaled of a person has 95% of relative humidity (RH) and around 2500 trace VOCs with concentration in the range of ppm to ppt by volume [3]. Acetone is a volatile closely related to diabetes. Other gaseous compounds present in breath related to diseases are carbon monoxide (CO), and nitric oxide (NO). CO and NO can be detected and evaluated as potential biomarkers of lung diseases. Electronic nose would allow detection of exhaled gases on the breath (VOCs, CO, NO …) and could be used for the diagnosis of diseases.

## 2 Electronic Nose

For this purpose, we present WiNOSE. This is an electronic nose that has been home-designed (CSIC and UEx) to support 8 commercial micro-sensors, which are controlled by DSC Microchip model dsPIC33FJ128GP306. In addition, the e-nose is provided with an electronic pump and an electrovalve to control the way of the sampling gas, embedded control and instrumentation electronics, rechargeable batteries, touch screen and IEEE 802.11 transceiver for wireless communication. The schematic of WiNOSE is shown in Fig. 1. In this case, this e-nose uses 8 commercial micro-sensors which are from Cambridge. Sensors 1–4 correspond to model 801 and sensors 5–8 correspond to model 803 [4, 5].

## 3 Experimental

### *3.1 Biomarkers*

In order to detect these diseases, the majority biomarkers present in the breath CO and NO in the case of asthma and acetone in the case of diabetes are used. Table 1 shows the concentration ranges for healthy and sick people for each biomarker used [6–9].

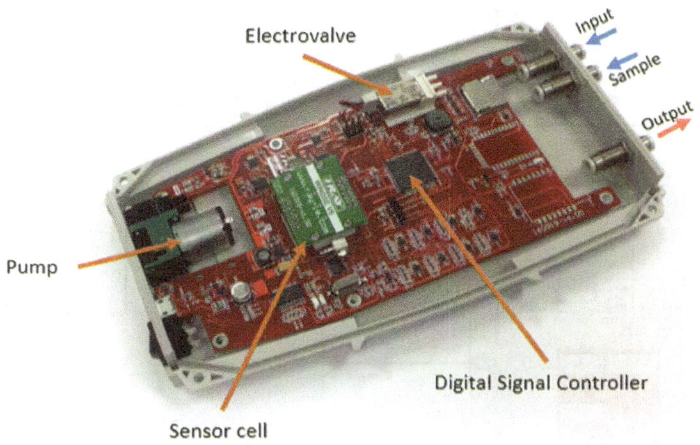


**Fig. 1** Schematic of electronic nose

**Table 1** Concentration range used for the different biomarkers

| | Range for healthy persons | Range for ill persons |
|---|---|---|
| Acetone (ppm) | 0.5–2 | >4 |
| NO (ppb) | 1–20 | >25 |
| CO (ppm) | 1–2 | >5 |

## *3.2 Measurement Setup*

In this work, different samples have been prepared in the laboratory that simulates human breath, varying the concentration of the different biomarkers, to check the discrimination capability of the prototype. The measurement setup shown in Fig. 2 has been used for this purpose.

NO and CO gas bottles have been used for the preparation of asthmatic patient samples. On the other hand, in order to obtain the samples corresponding to patients with diabetes, a permeation tube of acetone has been used. A gas mixing unit was used to obtain the desired concentration of each gas. It should be noted that analyses have been carried out with samples with 50% of RH and without humidity, to study the influence of humidity on the sensors. Because this is a preliminary study, a lower RH value that can be found in a breath of person has been used, with the objective to determine the influence that it has on the sensors and how these behave in the presence of humidity with respect to analyses made without it.

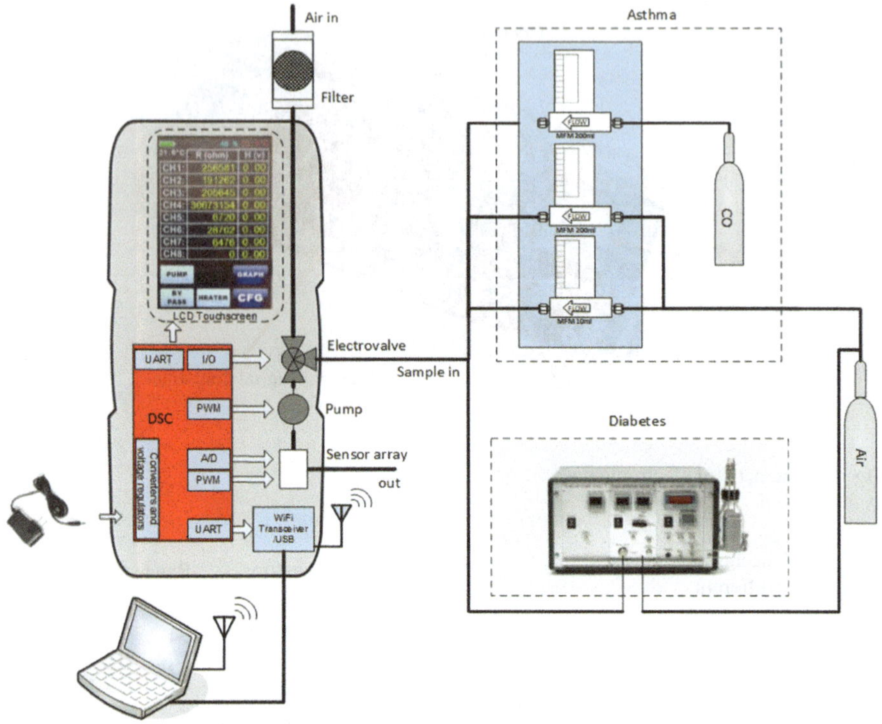


Fig. 2 Measurement setup

# 4 Results

The results obtained can be seen in the graphs included in Fig. 3. The graphs are separated into two columns, the left column shows the analyses carried out without humidity and the right column shows the analyses carried out with a relative humidity of 50%. The concentration range is indicated: the green zone corresponds to healthy persons ant the red zone to ill patients. In Fig. 3, the response is indicated as a percentage of the sensor's resistance variation. Several operation temperatures, from 300 to 400 °C approx. (corresponding heating voltages 1.3–1.8 V) have been used to get different responses. The 4 sensors with the best response are represented.

Table 2 shows the optimal percentage variations that can occur for each of the compounds (both with and without humidity), comparing the responses obtained for the maximum concentration value corresponding to a healthy person and the minimum value corresponding to a sick person.

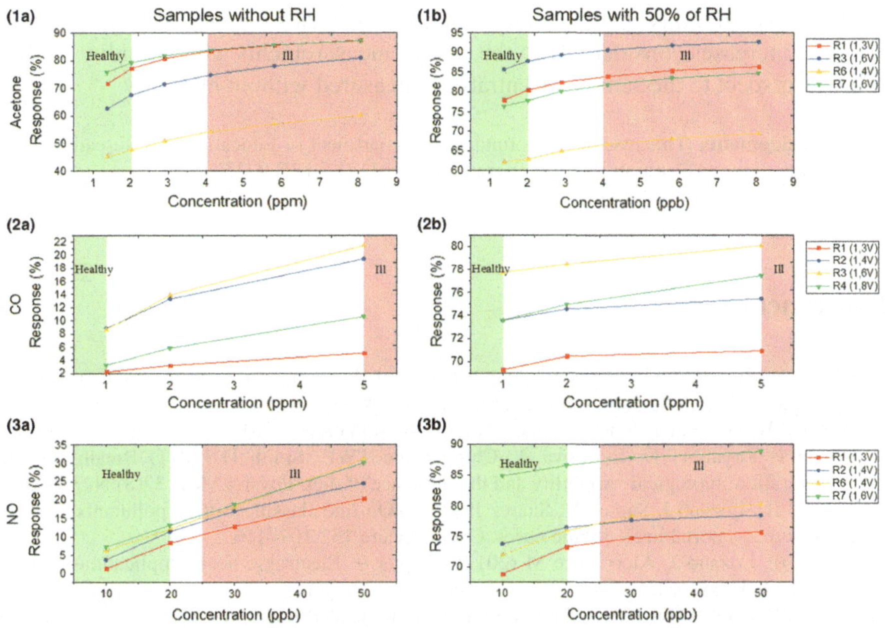


**Fig. 3** Representation of sensor responses for samples prepared in the laboratory. **1a** Acetone samples without humidity. **1b** Acetone samples with 50% RH. **2a** CO samples without humidity. **2b** CO samples with 50% RH. **3a** NO samples without humidity. **3b** NO samples with 50% RH

**Table 2** Percentage variation between the maximum concentration in a healthy person and the minimum that can be found in a sick person, for each biomarker

| | Samples without humidity (%) | Samples with 50% RH (%) |
|---|---|---|
| Acetone | 7.3 | 4.0 |
| NO | 12.8 | 3.8 |
| CO | 6.1 | 2.4 |

# 5 Conclusion

Samples with different concentrations of acetone, CO or NO have been analyzed to evaluate the effectiveness of this equipment against these variations in concentration. The samples have been prepared in the laboratory, using bottles of gases (CO and NO) or permeation tubes of acetone. The electronic nose is able to differentiate between different concentrations of CO and acetone corresponding to patients with diabetes or asthma and healthy persons, as shown in Fig. 3. As can be seen, in samples with humidity there is less variation in response between different concentrations because humidity attenuates the sensor response. Table 2 shows the percentage decrease between samples with 50% relative humidity and those without. With these results, we will evaluate the use of a filtering method to reduce the attenuation of the response produced by humidity. On the other hand, it has been observed that the range of response increases considerably in the case of measurements with humidity for NO and CO, respect to the same concentrations measured without humidity.

**Acknowledgements** This research was funded by Department of Education and Research of the Madrid Autonomous Community (Spain) grant number IND2017/TIC7714.

## References

1. World Health Organization (WHO) (2016) Global report on diabetes. WHO Library Cataloguing-in-Publication Data Global report on diabetes
2. WHO (2018) Global asthma network. The global asthma report 2018
3. Španěl P, Dryahina K, Rejšková A, Chippendale TWE, Smith D (2011) Breath acetone concentration: biological variability and the influence of diet. Physiol Meas 32(8):N23–N31
4. Herrero JL, Lozano J, Santos JP, Suárez JI (2016) On-line classification of pollutants in water using wireless portable electronic noses. Chemosphere 152:107–116
5. Santos JP, Lozano J, Aleixandre M (2017) Chapter 9: Electronic noses applications in beer technology. Brewing technology. SPi Global. InTech, Rijeka
6. Ryter SW, Choi AMK (2013) Carbon monoxide in exhaled breath testing and therapeutics. J Breath Res 7(1):017111
7. Szefler SJ et al (2012) Asthma outcomes: biomarkers. Natl Inst Heal 129:S9–S23
8. Das S, Pal S, Mitra M (2016) Significance of exhaled breath test in clinical diagnosis: a special focus on the detection of diabetes mellitus. J Med Biol Eng 36(5):605–624
9. La esperanza de vida en España se incrementa 2,6 años (online). Available at: https://www.consalud.es/pacientes/la-esperanza-de-vida-en-espana-aumenta-pero-no-su-calidad_45795_102.html. Accessed 21 Oct 2018

# Mid Term Field Validation of the MONICA Air Quality Multisensor

**E. Esposito, S. De Vito, F. Formisano, E. Massera, G. Fattoruso, G. Migliaccio, P. D'Auria, A. Amendola and G. Di Francia**

**Abstract** Accurate sensors validation is a fundamental requirement in air quality monitoring. Validated multisensor could in fact be used as tools to provide indicative measurements to compliment data coming from regulatory monitoring stations. The sparseness of the conventional analyzers, do not allow to have a dense background and prevent the possibility to achieve a high resolution picture of pollutants concentrations in cities and validated Air Quality multisensor could provide a solution to the lack of resolution. Currently, the most accredited validation procedure involves in field data recording in co-location with reference instrumentation. In this work, we show the results of a validation experiment implemented co-locating the ENEA MONICA platform together with an ARPAC (Campania Regional Agency for Environmental Protection) conventional analyzer, during 8 months. The obtained results encourage the possible use of MONICA multisensor platform as a backup tool for reference analyzers.

**Keywords** Air quality multisensor systems · Reference backup data · Field calibration

## 1 Introduction

Air Quality (AQ) multisensor systems have shown to be able to provide, if properly calibrated, high quality data in terms of *indicative* measurements. As such, they could be used as a backup tool for reference analyzers providing a way to obtain meaningful data in the case reference data become unavailable due to failures.

E. Esposito (✉) · S. De Vito · F. Formisano · E. Massera · G. Fattoruso · G. Migliaccio · G. Di Francia
ENEA, DTE-FSD-SAFS, Research Centre Portici, Naples, Italy
e-mail: elena.esposito@enea.it

P. D'Auria
ARPAC—Regional Agency for Environmental Protection, Campania, Italy

A. Amendola
SITE s.r.l.—Sicurezza TErritorio, Centro Direzionale di Napoli, Naples, Italy



G. Di Francia et al. (eds.), *Sensors and Microsystems*, Lecture Notes in Electrical Engineering 629, https://doi.org/10.1007/978-3-030-37558-4_8

Many researchers are interested in Air Quality multisensory systems in order to improve their performances. In the last few years, the achieved results have shown that the performance have gradually reached surprising levels, meeting, for limited periods of time, strict requirements on uncertainty that qualify them for use as *indicative* measurement systems in regulatory air quality monitoring networks [1]. The main issue is the calibration of these systems. It is paramount, actually, to assess and validate their performance so that they could be reliably used for obtaining significant information on AQ in cities. As we know, laboratory based calibration procedures have failed to allow cost effective solutions and their accuracy have shown limited when used in field conditions. At the same time calibration procedures using field data have their robustness challenged by concept and sensor drift appearing at least on a seasonal basis [2]. Their use for long term operations is hence still debated. Adaptation with periodical recalibration is advocated by several studies basis of evidences provided by long term co-location periods [3, 4].

Of course, regulatory monitoring network will continue to represent the backbone for providing the highest quality reference values to be used as the primary data base for remediation and regulation policy design. However, AQ regulatory monitoring stations suffer from periodical failures and are subjected to frequent maintenance operations that put them off-line generating a complete lack of data during several time periods along the year. This situation hampers the validation of short and long term assessments on AQ on urban and regional scale. In this work, we show how AQ multisensor system can be used to provide a backup system for regulatory monitoring achieving 100% availability of high quality AQ data by completing AQ reference stations data with their EU qualified indicative measurements.

A first attempt to obtain a mid-term validation of our AQ systems was carried out using several approaches. First, a simple linear univariate and multivariate calibration procedure was used. Afterwise, we adopted a machine learning methodology, in particular, a shallow neural network estimation algorithm. Validation of their performance is obtained with data coming from a mid-term deployment (6 months, still on-going) in which an electrochemical sensors based intelligent AQ multisensory system is installed inside an AQ regulatory monitoring station, enjoying the access to the very same inlet that serves the regulatory instruments for maximum reproducibility. The setup is actually designed to limiting any interference coming from differences in the real time analyzed air usually found in the common co-location deployments. Results shows that both the analyzed approaches allow for reaching and sustaining good performances during the in place co-location period.

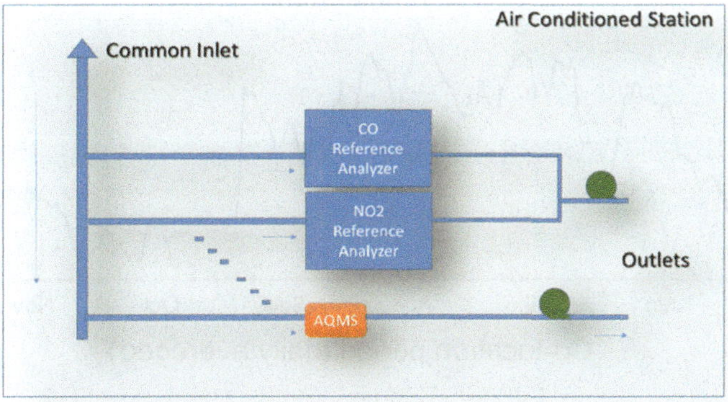


**Fig. 1** Co-location experiment architecture. The reference system common inlet directly feed the AQ multisensory system

## 2 Experimental Setup

### *2.1 Co-location Experiment Architecture*

The co-location experiment started on April 2018 and it's still in place and captured data continuously contribute to the robustness and confidence of the resulting performance figures. Our AQ multisensory platform inlet is directly connected on the main common inlet of the AQ regulatory station, as shown in Fig. 1, actually providing the air flow to all the regulatory instrumentation.

An internal Air conditioning (AC) unit, contributed to keep stable environmental conditions. However, an overlooked wrong set point configuration caused temperature oscillations to peak at 38 °C during June–August 2018 (see Fig. 2).

### *2.2 The MONICA Portable Multisensor Platform*

The AQ multisensory system is based on the Monica device (v 2.0), developed by ENEA. Monica is a chemical multisensory device based on Alphasense A4 electrochemical sensing units and T, RH environmental sensors (sensirion SHT75). Its gas sensors array currently hosts three gas transducers specifically devoted to estimate CO (CO-A4), $NO_2$ ($NO_2$-A43F) and $O_3$ ($O_3$-A431) concentrations, but its modularity allows for the inclusion of a total VOC (PID) as well as to substitute the current sensors with dangerous/toxic gas sensors (e.g. H2S). It is well known that EC sensors are affected by temperature interference as well as non-target gas interference (e.g. $NO_2$ for $O_3$ sensor). Multivariate calibration is hence needed to obtain best performances (see [5, 6]). The entire system is enclosed in a 12 × 10 × 6 cm (500 g)

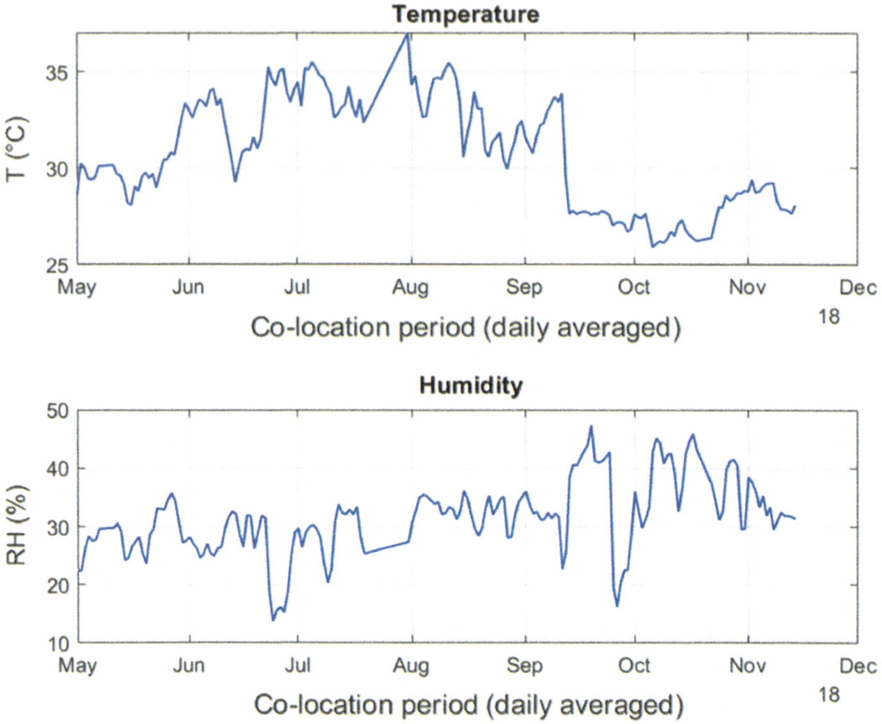


**Fig. 2** Temperature (°C) and relative humidity (%) behavior during the co-location period (daily averaged data are shown)

box with two 3 × 3 cm fans, that helps maintaining a constant air flow near the sensitive edge of the sensor units. The sensors analog front end allows to connect sensors to an ARM microcontroller based STNucleo board (STm32 Nucleo™ 432 MCU board) capturing and digitalizing working and auxiliary electrodes voltages. Raw sensors data are transmitted via Bluetooth 4.0 (BLE) interface to a Raspberry Pi3 based datasink providing for local storage and WAN connectivity services.

Data is captured at 10 samples/min rate and data is averaged on an hourly based so to meet the reference instruments sampling period (60 s).

## 2.3 *ARPAC Reference Monitoring Station*

The AQ regulatory monitoring station involved in the co-location experiment is located in Via Argine (Naples, Italy) along a main road connecting Naples to several towns, around the Mt. Vesuvius area. These are generally interested by significant food and diary product transport. The station is actually operated by ARPAC and

**Table 1** Dataset partition scheme for the two different experiments

| Sample rate | Dataset length | Training set | Validation set | Test set |
|---|---|---|---|---|
| *First experiment* | | | | |
| @1 sample/min | ~30k samples | 60% | 20% | 20% |
| *Second experiment* | | | | |
| @1 sample/h | ~4k samples | 1 month | 1 week | 1 week |

is part of the official regulatory AQ monitoring network, as such it is subjected to regular maintenance activities.

The regulatory instrumentation includes a Teledyne™ 300 CO analyzer, a Thermo Scientific Mod. 42i NOx chemilumin. analyzer, a Chromatotec/Airmotec mod. GC 866 PID airTOXIC BTX analyzer, and a FAI SWAM 5a Dual Channel Monitor Particulate matter (PM2.5 + PM10) analyzer.

The recorded dataset exploited in this work consists of 3856 hourly samples recorded from May to November 2018.

### 2.4 *Data Processing and Field Calibration Methodologies*

Several algorithms have been used in order to obtain a calibration. First, linear univariate and multivariate calibration procedures with Temperature and baseline correction have been used. Then, we applied a nonlinear univariate and multivariate calibration procedures, using shallow neural networks estimations algorithm. In particular, we adopted a three layers Back Propagation Network (BPN) architecture, empirically equipped with 5 standard tansig neurons units in the hidden layer and a linear output layer. Data recorded were downloaded and processed with Matlab™ to obtain a first estimation of raw responses correlations with CO and $NO_2$ as targets.

The first experiment encompassed the first month of the co-location period, in which 1 min data were available. The second experiment was carried out using the hourly averaged data of all the co-location period. In both cases, the dataset partition (training/validation/test sets) procedure is sequential, non-overlapping, using all the available data. Table 1 shows the data partition procedure.

## 3 Results

In this work we focused on CO and $NO_2$ concentrations estimations problem using WE and AE sensors data plus temperature and humidity data as inputs for the adopted methodologies.

In Table 2, the correlation results obtained with the selected calibration procedures, are reported.

**Table 2** Correlation coefficient computed between gas (CO and $NO_2$) reference concentration and (1) raw sensor response, (2) linear concentration estimation, (3) nonlinear concentration estimation, respectively

| | Gas | CC |
|---|---|---|
| 1 | CO raw sensor response | 0.71 |
| 2 | CO linear estimation | 0.76 |
| 3 | CO nonlinear estimation | 0.88 |
| 1 | $NO_2$ raw sensor response | 0.80 |
| 2 | $NO_2$ linear estimation | 0.84 |
| 3 | $NO_2$ nonlinear estimation | 0.94 |

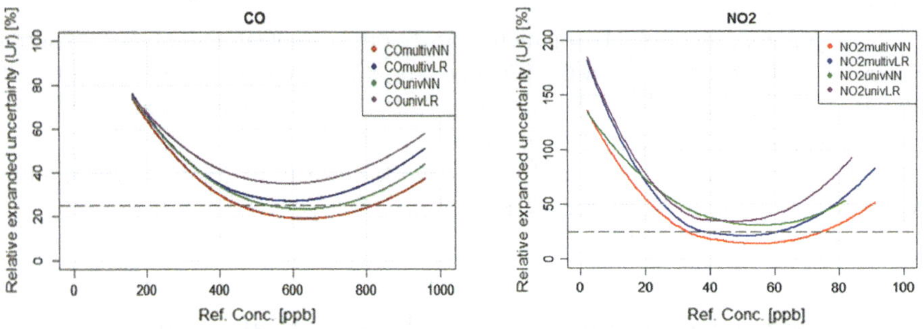


**Fig. 3** Relative expanded uncertainty computed with 4 calibration procedures for CO (left) and $NO_2$ (right). Red: multivariate neural network algorithm; blue: multivariate linear regression algorithm; green: univariate neural network algorithm; purple: univariate linear regression algorithm

Figure 3 shows the measure of the uncertainty (expanded relative uncertainty) for indicative measurements, according to the European Air Quality Directive [7], which defines the Data Quality Objective (DQO) that monitoring methods need to comply with to be used as indicative measurements for regulative purposes. According to the Directive, allowed uncertainties should be less than 50% for PM10 and PM2.5, 30% for $O_3$ and 25% for CO, $NO_x$, $NO_2$ and $SO_2$.

## 4 Conclusions

The obtained results suggest that on-field nonlinear multivariate calibration may allow the achievement of the EU set Data Quality Objectives, both for CO and $NO_2$.

These results encourage the possible use of MONICA multisensor platform to integrate the ARPAC sensors network or, eventually, to replace them while in faulty conditions.

Future works will include the analysis of adaptive ML strategies using a 1-year long dataset, in order to improve AQ multisensory system performances on long-term co-location period and assess the use of AQ multisensory system as a backup tool for reference stations.

## References

1. Borrego C et al (2018) Assessment of air quality microsensors versus reference methods: The EuNetAir Joint Exercise—Part II. Atmos Environ 193:127–142
2. Esposito E et al (2017) Is on field calibration strategy robust to relocation? In: 2017 international symposium on olfaction and electronic nose (ISOEN), Montreal, pp 1–3
3. Castell N et al (2017) Can commercial low cost-sensor platforms contribute to air quality monitoring and exposure estimates? Environ Int 99:293–302
4. Moltchanov S et al (2015) On the feasibility of measuring urban air pollution by wireless distributed sensor networks. Sci Total Environ 502:537–547
5. MONICA Eppela Website: https://www.eppela.com/it/projects/9652-monica-il-tuo-navigatore-personale-antismog. Accessed April 2019
6. Capezzuto L et al (2014) A maker friendly mobile and social sensing approach to urban air quality monitoring. In: IEEE sensors proceedings, pp 12–16
7. Directive 2008/50/EC of the European Parliament, the Council of 21 May 2008 on ambient air quality, and cleaner air for Europe

## References

# A 500 W Microcontroller-Based Powertrain Circuit for Ultrasonic Cleaning Machines

**M. Grassi, A. Liberale, M. Elkhayat and P. Malcovati**

**Abstract** A 500 W microcontroller-based power supply circuit for the piezoelectric transducers used in ultrasonic cleaning machines is presented. The powertrain consists of a boost converter, including AC supply power factor correction, followed by a digitally programmable buck converter delivering the DC bus voltage for a half bridge inverter, driven by a suitable signal from the microcontroller oscillator. Detection of the resonance frequency of the piezoelectric transducer is performed at startup by measuring the load current on the DC bus through a hall sensor.

**Keywords** Ultrasonic cleaning · Powertrain circuit · Piezoelectric transducers

## 1 Introduction

The proposed work reports the study, the design, and the characterization of a microcontroller-based power generator for ultrasonic washing machines [1]. The powertrain circuit is re-configurable and can deliver, in continuous operation, up to 500 W of power to the piezoelectric transducers. Ultrasonic washing machines consist of a tub, filled up with liquid detergents, and a power converter that drives an array of piezoelectric transducers, glued to the bottom of the tub for maximum acoustic power transfer, while the tool to be cleaned is drown into the liquid [2].

In principle, ultrasonic cleaning consists of applying an acoustic carrier to the liquid with adequate power and frequency, in order to cause cavitation. Micro-bubbles are thus generated in the liquid, which clean the immersed tool thanks to their continuous creation and implosion near the surface that has to be cleaned. The effect of the microbubbles on the surface is also called micro-brushing [3]. This kind of cleaning is exploited in several fields, ranging from mechanical factories, to remove oil and grease from precision components or motors, to electronics laboratories, in order to clean up printed and hybrid circuit boards, to hospitals or dentist ambulatories, for surgical tools sterilization.

M. Grassi (✉) · A. Liberale · M. Elkhayat · P. Malcovati
Department of Electrical, Computer, and Biomedical Engineering,
University of Pavia, Pavia, Italy
e-mail: marco.grassi@unipv.it



G. Di Francia et al. (eds.), *Sensors and Microsystems*, Lecture Notes
in Electrical Engineering 629, https://doi.org/10.1007/978-3-030-37558-4_9

The demand for more efficient and reliable ultrasonic cleaning machines implies the need of deeper research on the power converter circuit and its controlling techniques for this specific application. One of the best trade-offs between reliability and cleaning capability in power electronics for ultrasonic washing machines is actually obtained exploiting a rectified AC supply, with or without partial filtering, followed by a half MOS bridge power inverter [4]. With this strategy, the power delivered to the tub may only be defined by varying the frequency of the carrier, given the impedance of the load, since the voltage is imposed by the AC supply.

The powertrain presented in this paper makes use of a microcontroller to control a buck converter, in order to set a regulated DC bus voltage for the power bridge. Moreover, the microcontroller takes care of the generation of the inverter carrier. Other features of the proposed power circuit include the presence of an input Power Factor Corrector (PFC) and the output current measurement, obtained by a hall sensor circuit along the DC bus.

These improvements lead to a larger versatility in terms of compliance with different supply voltages around the world, fewer harmonic kickback to the AC supply even without adding bulky passive filters, a more accurate power regulation towards the tub, and the possibility to have a complete digital control of the powertrain. In particular, the possibility of fine regulation of the DC bus allows delivering a power up to 500 W with different sets of ultrasonic power transducers and tubs, as well as different water levels and immersed tools. The main difficulty for the characterization of the overall system (powertrain+tub full of water and drown tool) is actually the definition of a specific benchmark for the tool cleaning performance, as reported in the preliminary system results.

## 2 The Cleaning Machine Powertrain Circuit

The simplified block diagram of the powertrain circuit, including drivers and actuators, is reported in Fig. 1. It consists of an input rectifying AC-DC boost converter with PFC feature, a DC-DC closed loop programmable buck converter, and a half bridge MOS inverter. Furthermore, an ancillary flyback converter delivers regulated DC voltages of 15–5 V: the first voltage is used to supply the drivers and the controller circuits of the three main power blocks, the second voltage is used to supply the microcontroller and a digital variable resistor in the buck converter feedback. The first power block boosts and stabilizes the rectified voltage input to a chosen and fixed value of 400 V, independently of the country in which the machine is operated, leading to maximum versatility ($V_{AC,rms} = 90$–$250$ V). In addition, its PFC functionality is really important to meet worldwide regulations in terms of harmonic current components and reactive power drawn from the electrical network, while guaranteeing suitable phase alignment between AC current and voltage components at the input.

The second stage is a closed loop buck DC-DC converter with programmable output voltage by means of a digitally-controlled resistor placed in the feedback

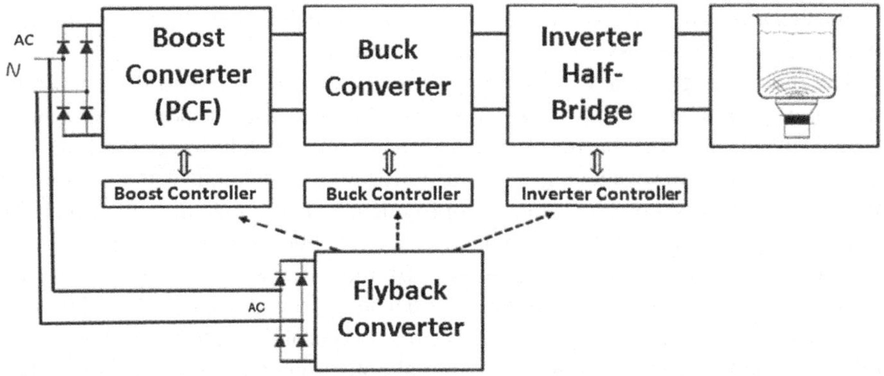


**Fig. 1** Block diagram of the developed ultrasonic cleaning machine

network. By regulating the voltage value of the DC-bus we can set the power delivered to a given load at a given frequency, since the DC bus supplies the half bridge subsequent inverter, which is the last stage of the powertrain. Furthermore, if load tuning near to resonance is operated, the Load Power Factor (LPF) will have a small reactive component (LPF > 0.75).

The half bridge is made up by two power MOS transistors and by a capacitive divider used to balance the inverter. It is driven by a specific gate controller and voltage translator, namely a bootstrap circuit, to work in push-pull avoiding dead-time cross-conduction effects. This driver input is fed by the microcontroller with a square wave carrier of suitable frequency. The inverter delivers then a power square wave to the transducers and thus to the water tub for cleaning.

More in detail, the half bridge is decoupled from the actual transducers by means of a 1:1.6 transformer which has both the function of insulating the tub (connected to ground for safety) from the powertrain and to heighten the load voltage signal for higher piezoelectric transducers performance. Finally, an ad-hoc inductor [5] is present after the transformer to tune the transducer capacitive load for maximum efficiency and reliability, while avoiding excessive current spikes, typical of a capacitive load with fast commutations, which could cause fast ageing of the bridge MOSFETs.

## 3 Cleaning Machine Operation Principle

The powertrain circuit drives the piezoelectric transducers with a square wave at suitable ultrasonic frequency. The optimal frequency is near the resonance value of the load, which is actually composed by the tuning inductor, the transducers, and the mechanical effects of the tub, the water, and the immersed tool to be cleaned. At startup, the microcontroller sets the buck converter output to a moderate voltage value (around 100 V DC) and generates a slow frequency sweep (33–38 kHz). Meanwhile, a hall sensor measures the current on the DC bus, which is then digitally converted by an A/D converter to be read by the microcontroller.

The resonance frequency, in first approximation, corresponds to the point with maximum current on the DC bus for a given voltage. This seeking operation is carried out at machine startup by the microcontroller which drives the bridge at incremental frequency values at reduced power and, at the end of the sweep, the microcontroller sets the system operating frequency near to the value for which the current is maximized, thus obtaining theoretically the maximum power transfer from the powertrain to the load. For the employed 30 L tub full of water, equipped with 10 × 50 W ultrasonic transducers, the optimal carrier frequency is around 36.5 kHz. In the meantime, the DC bus voltage is set by the microcontroller to the desired operating voltage by updating the feedback buck digital potentiometer. Since the carrier duty cycle is fixed to 50% to avoid DC components on the primary side of the transformer, if the load is driven near to the resonance frequency, the power transferred to the cleaning process will be, as stated, easily programmable by regulating the DC bus voltage. Considering N the constant transformer ratio, $R_L$ the resistive value of the load and φ the angular delay of the output current with respect to the exciting voltage at the operating frequency (near the resonance value), we have, neglecting harmonic components except the fundamental:

$$P_{LOAD} \cong \frac{2V_{DC,BUS}^2}{\pi^2 R_L N^2} \cos(\varphi) \tag{1}$$

The square wave carrier spectrum is filtered by the RLC nature of the load itself, which actually, in first approximation, exploits only the fundamental harmonic, avoiding unuseful power dispersion.

## 4 System Characterization

The developed prototype of the powertrain is reported in Fig. 2. From left to right we may find the rectifying boost converter with PFC, the DC-DC buck converter, and the half inverter bridge with their required drivers and controllers. On the right we have instead the transformer and the tuning inductor.

The code loaded into the microcontroller performs an initial frequency sweep at low power whose aim is to seek the resonance frequency of the tuned transducers connected to the tub, which corresponds to the maximum current measured in the DC bus.

We evaluated the electrical efficiency of the powertrain circuit operated at 36.5 kHz with different power levels (up to 500 W of active power on the transducers) with an average load PF of 0.8. The average electrical efficiency is of the order of 89%, while the average power factor on the AC input power is 0.97. The electrical efficiency of the powertrain as a function of the output power, in the described conditions, is reported in Fig. 3.

**Fig. 2** The developed powertrain circuit PCB

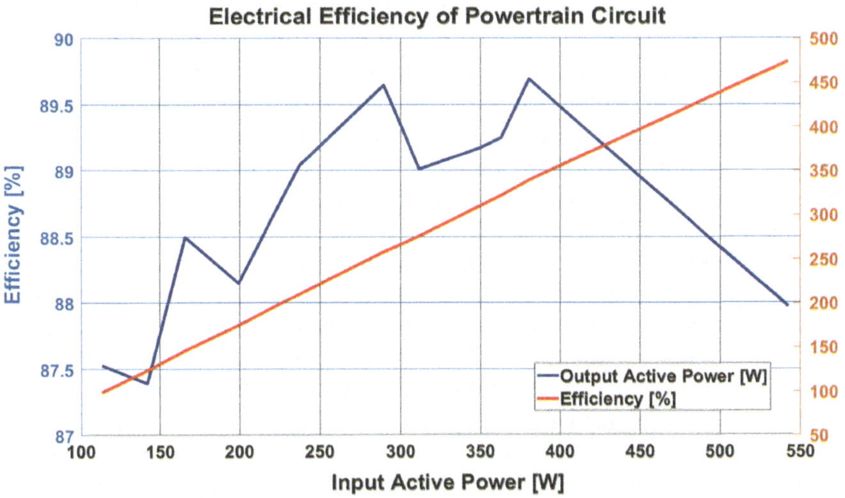


**Fig. 3** Electrical efficiency of the developed powertrain circuit

The actual characterization of the ultrasonic cleaning performed by the developed machine may result difficult because of the absence of a given conventional benchmark. Several solutions have been taken into consideration within the project. A considered technique would be to use a hydrophone or a hydrophone array to measure the rms acoustic pressure in the water. Unfortunately, this technique has been discarded because the cleaning effect is both function of the carrier energy and of how it propagates into different parts of the tub over space and time, while the local measurement of the acoustic pressure is in first approximation only a function of the bare power. Another considered technique has been the use of a transparent tub,

**Fig. 4** Effect of the ultrasonic cleaning in terms of aluminium resist removal at different power

like an aquarium, filled by two detergents of different density and color. During the cleaning process, the envelope of the perturbation of the border line between the two detergents is proportional to the square root of the energy involved in the process, which may be monitored, thanks to this technique, not only on the surface but also at different heights in the tub. Unfortunately, also this second technique has been discarded, because, again, the cleaning effect is not a function of the bare energy involved but also of the probability of micro-bubble generation (cavitation effect).

We decided at the end to exploit, as benchmark, an empiric technique based on the assumption that the cleaning performance is proportional to the effect of physical attack to a thin film resist on a lite mechanical substrate. The best candidate for this evaluation is a standard aluminum film on a flexible plastic sheet, hereinafter target, mounted on a frame immersed in water. For a given operating power value and a fixed operating time, the more the aluminum film is removed from the sheet, the more the cleaning is effective. We put the frame with the target in the center of the long side of the 30 L tub with ten 50 W piezoelectric transducer elements in parallel at different acoustic power values (100, 200, 300, 400 and 500 W) and waited 30 s of time before extracting it. The results are reported in Fig. 4, where we may notice that at higher acoustic power on the transducers corresponds a higher effective aluminum removal effect. Furthermore, we can observe that the effect is not linear as a function of power.

## 5 Conclusions

A laboratory characterization of the powertrain circuit together with microcontroller and piezo-transducers equipped 30 L tub has been carried out exploiting aluminum resist thin film sheets. Results underline that, for a given tub, over a specific critical power value the effectiveness of the cleaning is satisfactory, while under such value is poor. In our case, the minimum power to generate micro-brushing is 200 W, while a satisfactory cleaning effect is obtained at 400 W. For best cleaning-power efficiency trade off, the 30 L tub should be operated in the range 350–450 W. A higher power capability power delivery up to more than 500 W is anyway guaranteed for very high-performance cleaning.

## References


1. Fuchs FJ (1995) Ultrasonic cleaning: fundamental theory and application. NASA-GOV documentation, N95-31783, pp 369–378
2. Wang Y, Draper M, Denley S, Robinson F, Shepherd P (2011) Power converters for power-ultrasonic transducers. In: 46th International universities power engineering conference, Soest, Germany
3. Azhdast MH, Haleh H, Pouladzadeh P, Azhdast A, Soltanabadi M (2009) Theoretical and experimental analysis of beating and cavitation phenomenon on erosion in ultrasonic cleaning process. In: IEEE/ASME international conference on advanced intelligent mechatronics, Singapore, Singapore
4. Buasri C, Jangwanitlert A (2008) Comparison of switching strategies for an ultrasonic cleaner. In: 5th International conference on electrical engineering/electronics, computer, telecommunications and information technology, Krabi, Thailand
5. Alvarez-Arenas TEG, Diez L (2013) Novel impedance matching materials and strategies for air-coupled piezoelectric transducers. In: IEEE Sensors, Baltimore, MD, USA

# Environmental Monitoring Exploiting Optical Fiber Biosensors. The Case of Naphthalene Detection in Water

**Nunzio Cennamo, Luigi Zeni, Ezio Ricca, Rachele Isticato, Vincenzo Manuel Marzullo, Alessandro Capo, Maria Staiano, Sabato D'Auria and Antonio Varriale**

**Abstract** We have developed and characterized a surface plasmon resonance (SPR) biosensor in a plastic optical fiber (POF) to detect traces of naphthalene (NAPTHA) in water. As previously developed for perfluorinated compounds (PFAs), the gold surface of the plasmonic POF sensor has been chemically modified through the formation of specific reactive groups and functionalized with antibodies able to specifically recognize the NAPTHA. The obtained limit of detection value (0.76 ng/mL) is lower than the maximum residue limit value of naphthalene (0.13 μg/mL) referred as the water environmental quality standards.

**Keywords** Naphthalene · Optical biosensors · Plastic optical fibers · Surface plasmon resonance · Water environmental quality standards

## 1 Introduction

Plasmonic platforms combined with several types of biological molecular recognition elements (MREs), including proteins, aptamers, and antibodies (Abs), are reported in literature [1–3]. In particular, Abs are of particular interest as their dissociation constants for antigens can be as low as $10^{-15}$ M, thus offering outstanding affinities.

Surface plasmon resonance (SPR) in optical fibers is widely used as a detection principle for sensors operating in different application fields, including the detection of environmental pollution [4]. In SPR sensing the MRE, coupled with a gold surface

N. Cennamo (✉) · L. Zeni
Department of Engineering, University of Campania "Luigi Vanvitelli", 81031 Aversa, Italy
e-mail: nunzio.cennamo@unicampania.it

E. Ricca · R. Isticato
Department of Biology, University of Naples Federico II, Complesso Universitario Monte Sant'Angelo, Via Cinthia 4, 80126 Naples, Italy

V. M. Marzullo
Institute of Protein Biochemistry, IBP-CNR, Via Castellino 111, 80131 Naples, Italy

A. Capo · M. Staiano · S. D'Auria · A. Varriale
Institute of Food Science, ISA-CNR, Via Roma 64, 83100 Avellino, Italy



G. Di Francia et al. (eds.), *Sensors and Microsystems*, Lecture Notes in Electrical Engineering 629, https://doi.org/10.1007/978-3-030-37558-4_10

by a Self-Assembled Monolayer (SAM), selectively recognizes and captures the analyte present in the sample (aqueous solution), producing a local change in the refractive index (RI) on the gold surface. The extent of RI change depends on the structure of the analyte [1–4]. The most exploited configuration for SPR sensors is based on the classical Kretschmann and Otto configurations, but has the shortcoming of being bulky and requiring expensive optical components, thus it is difficult to be miniaturized and used for remote interrogation and in situ (e.g. point-of-care) application. Jorgenson et al. replaced the prism by a multimode optical fiber [5] with metal deposited on the bare core of the fiber. The use of an optical fiber allows for remote sensing and may reduce the cost and the dimensions of the device. In the optical fiber the propagation of light occurs for angles of incidence exceeding the critical angle, which depends on the RIs of both core and cladding components, thus, SPR only works for surrounding dielectrics whose RI lies in a narrow range. To overcome this drawback, Jorgenson et al. used a polychromatic light source and a spectrometer. This approach results in low cost, easy to implement devices and can offer some attractive advantages such as the remote sensing capabilities. Many different configurations based on SPR in silica optical fiber are described, and, lately, POFs have been introduced for SPR platforms, being advantageous for their excellent flexibility, easy manipulation, great numerical aperture, large diameter, and the fact that plastic is able to withstand smaller bend radii than glass [4]. We have already demonstrated the use of SPR sensors in plastic optical fibers (POFs) combined with specific SAM of Abs (as bio-receptors). In particular, interesting results have been achieved exploiting SPR-POF biosensors in the fields of medical diagnostics for celiac disease antigen monitoring [6], in environmental monitoring with the detection of perfluorinated compounds (PFAs) [4], and in food safety with the detection of butanal [7].

In this work we present an SPR-POF biosensor, based on SAM of a specific Ab deposited on the POF's gold surface, to detect traces of naphthalene (NAPTHA) in water.

## 2 Naphthalene Optical Biosensor

### *2.1 Optical Fiber Platform*

Figure 1a shows a picture of the used SPR-POF platform. This SPR sensor can be made by removing the cladding of POF (along half circumference), spin coating an "optical" buffer layer on the exposed core (this layer is about 1500 nm thick) and finally sputtering a thin gold film (60 nm). The final length of the sensing region is about 10 mm. The selectivity is obtained by SAM of a specific bioreceptor (in this case an Ab) for the considered analyte in tight contact with the gold surface.

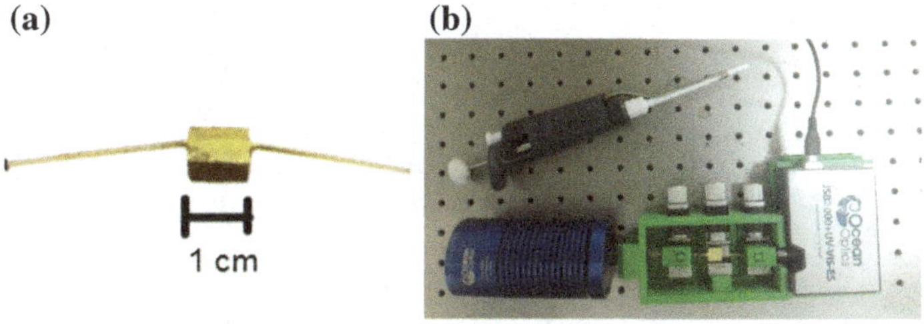


**Fig. 1** Pictures of the SPR-POF sensor (**a**), and of the experimental setup (**b**)

The used POF presents a size of 1 mm in diameter (with a core of 980 μm). The buffer layer is made of photoresist Microposit S1813, with a refractive index greater than that of the POF's core (PMMA), to improve the performances [4].

As reported in Fig. 1b, the experimental setup arranged to measure the SPR spectrum is based on a halogen lamp (with an emission range from 360 nm to 1700 nm), illuminating the SPR sensor system, and a spectrometer (with a detection range from 300 nm to 1000 nm), both produced by Ocean Optic.

## *2.2 Biosensor for NAPHTA*

To obtain the anti-NAPHTA antibodies, a retro-synthetic chemical strategy was applied to modify the NAPHTA structure in a derivative structure. This modified NAPHTA structure was coupled with a protein carrier and used for immunization. As reported in Fig. 2, the gold surface was derivatizated and functionalized with the produced antibody by using the EDC/NHS amino-coupling immobilization protocol. In the functionalization process, the resonance wavelength shifted to higher values step by step.

Figure 3 shows the SPR spectra of the plasmonic biosensor obtained at increasing concentrations of NAPHTA in water solutions, from 0 to 2.5 μg/ml. The resonance wavelength shifted to higher values by increasing the concentration of the analyte. The obtained results indicate that this biosensor is able to sense the presence of naphthalene in water or in sea-water [8]. The limit of detection (LOD) value has been calculated to be 0.76 ng/ml, a value lower than the maximum residue limit value of naphthalene (0.13 μg/ml) referred as the water environmental quality standards (EQS).

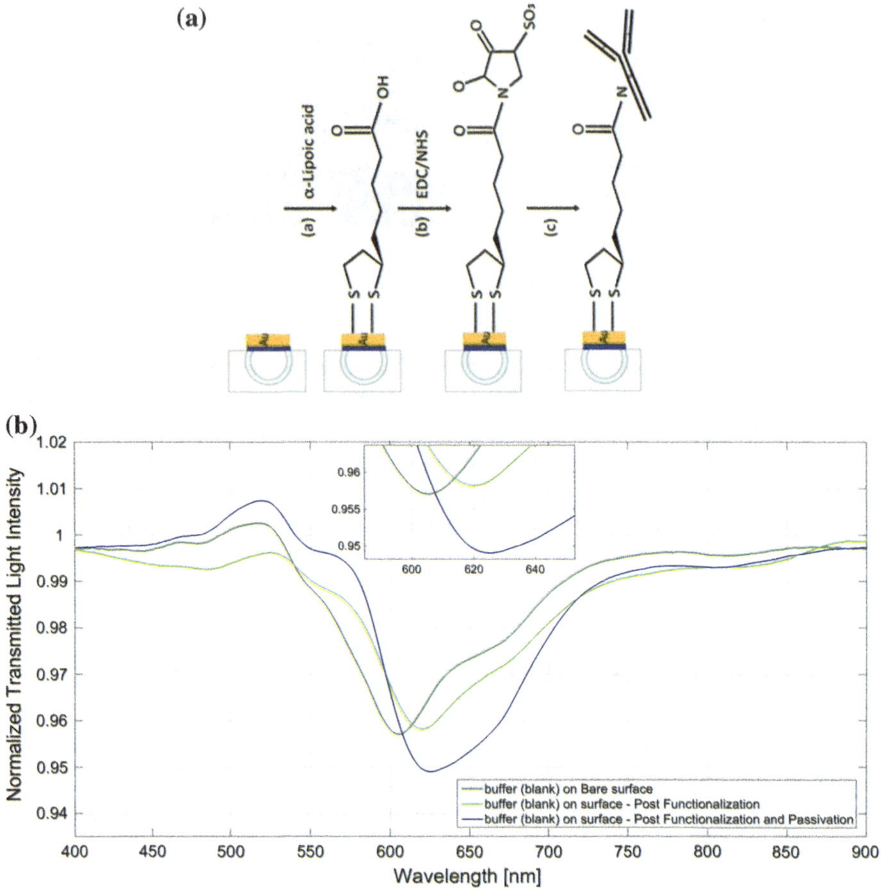


**Fig. 2** **a** Outline and **b** SPR spectra, obtained in the functionalization steps

## 3 Conclusion

In conclusion, we have reported a novel sensor useful to monitor the presence of naphthalene in water solutions, exploiting a simple and low-cost optical biosensor system with a remote sensing capability offered by the POF.

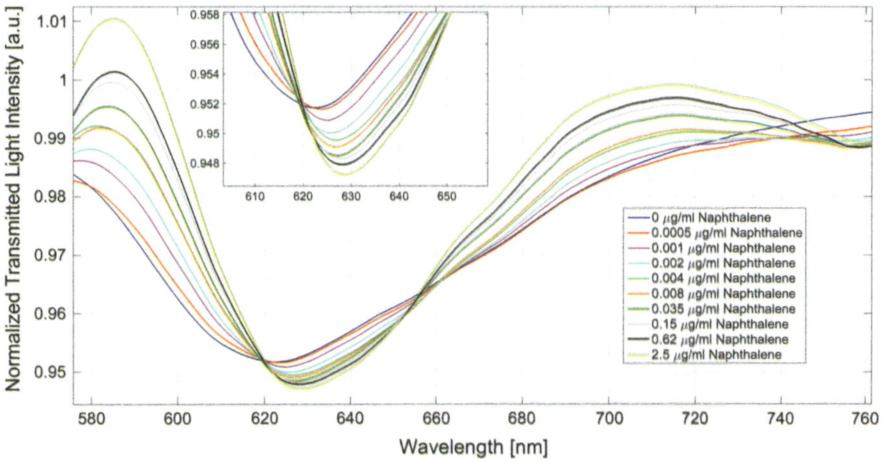


**Fig. 3** SPR spectra obtained for different NAPHTA concentrations in water. Inset: zoom of the wavelengths around the resonances

# References


1. Caucheteur C, Guo T, Albert J (2015) Review of plasmonic fiber optical chemical sensors: improving the limit of detection. Anal Bioanal Chem 407:3883–3897
2. Olaru A, Bala C, Jaffrezic-Renault N, Aboul-Enein HY (2015) Surface plasmon resonance (SPR) biosensors in pharmaceutical analysis. Crit Rev Anal Chem 45:97–105
3. Nguyen HH, Park J, Kang S, Kim M (2015) Surface plasmon resonance: a versatile technique for biosensor applications. Sensors 15:10481–10510
4. Cennamo N, Zeni L, Tortora P, Regonesi ME, Giusti A, Staiano M, D'Auria S, Varriale A (2018) A high sensitivity biosensor to detect the presence of perfluorinated compounds in environment. Talanta 178:955–961
5. Jorgenson RC, Yee SS (1993) A fiber-optic chemical sensor based on surface plasmon resonance. Sens Actuators B Chem 12:213–220
6. Cennamo N, Varriale A, Pennacchio A, Staiano M, Massarotti D, Zeni L, D'Auria S (2013) An innovative plastic optical fiber based biosensor for new bio/applications. the case of celiac disease. Sens Actuators B Chem 176:1008–1014
7. Cennamo N, Di Giovanni S, Varriale A, Staiano M, Di Pietrantonio F, Notargiacomo A, Zeni L, D'Auria S (2015) Easy to use plastic optical fiber-based biosensor for detection of butanal. PLoS ONE 10:1–12
8. Cennamo N, Zeni L, Ricca E, Isticato R, Marzullo VM, Capo A, Staiano M, D'Auria S, Varriale A (2019) Detection of naphthalene in sea-water by a label-free plasmonic optical fiber biosensor, Talanta 194:289–297

# Optical Chemical Sensing Exploiting Inkjet Printing Technology and Molecularly Imprinted Polymers

**Nunzio Cennamo, Luigi Zeni, M. Pesavento, Simone Marchetti, Salvatore Baglio, Salvatore Graziani, Vincenzo Marletta, Antonio Pistorio and Bruno Andò**

**Abstract** We reported a low-cost optical chemical sensor for the detection of furfural (furan-2-carbaldehyde, 2-FAL) in aqueous media. This novel optical sensor platform is based on a PET (polyethylene terephthalate) substrate with a pattern of silver nanoparticles, printed on it by InkJet technology. To obtain the selective detection of the analyte, the slab waveguide has been covered by a specific molecularly imprinted polymer (MIP) for 2-FAL. The sensing device has connected to a light source and a spectrometer by two plastic optical fibers (POFs), to monitor the interaction between the MIP and 2-FAL. The performances reported in this work are comparable to those of plasmonic POF sensors combined with the same MIP.

**Keywords** Inkjet technology · Optical chemical sensors · Furfural · Plastic optical fibers · Molecularly imprinted polymers

## 1 Introduction

The determination of 2-FAL in aqueous matrices of food interest (as for example wine, coffee or beer [1]) is becoming a very crucial task in relation to its toxic and carcinogenic effect on human beings. Besides, 2-FAL is also relevant for flavouring and aroma of food, which is an important factor in quality control and quality assurance. For example, in wines this factor is among the most important ones [2]. During ageing the concentration of some compounds formed during the alcoholic fermentation decreases, but new compounds appear, in part deriving from the evolution of the wine components themselves and in part extracted from the wood containers in which wines are usually stored. Furanic compounds, in particular 2-FAL, which can

N. Cennamo (✉) · L. Zeni
Department of Engineering, University of Campania "Luigi Vanvitelli", via Roma, Aversa, Italy
e-mail: nunzio.cennamo@unicampania.it

M. Pesavento · S. Marchetti
Department of Chemistry, University of Pavia, via Taramelli, Pavia, Italy

S. Baglio · S. Graziani · V. Marletta · A. Pistorio · B. Andò
Dipartimento di Ingegneria Elettrica Elettronica e Informatica, University of Catania, Catania, Italy

 
G. Di Francia et al. (eds.), *Sensors and Microsystems*, Lecture Notes in Electrical Engineering 629, https://doi.org/10.1007/978-3-030-37558-4_11

be formed in wine by sugar dehydration, and can be extracted from wood too, have a high impact on the aroma. Moreover, they can be used as ageing markers, since their concentration increases with ageing [2].

One interesting and different example of 2-FAL used as chemical marker is the detection of 2-FAL in oil filled transformers [3]. In this case, 2-FAL is a by-product of the thermal degradation of cellulose, so of the paper insulation (paper windings, pressboards) in transformers. For these reasons their determination by a fast, easy, low-cost, and immune to electromagnetic interferences method is of interest. The optical chemical sensor here reported appears to be particularly interesting.

An optical approach has previously proposed as potentially suitable for on line detection of 2-FAl (chemical markers) in insulating mineral oil of transformers, which is based on Surface Plasmon Resonance (SPR) in D-shaped POFs. It has been implemented in a plastic optical fiber which acts as optical platform for the sensing element, a Molecular Imprinted Polymer layer in contact with a thin gold film [3].

In this work we will report a very simple optical platform, based on inkJet printing technology, combined with an MIP receptor for the detection of 2-FAL. Compared to traditional silicon-based electronics or nanotechnologies, printed devices are cheaper and can fulfil many applications requiring low-cost and throwaway systems. The most relevant techniques for the production of printed devices are Screen Printing and InkJet Printing.

## 2 Optical Chemical Sensor

### *2.1 Optical Platform*

Figure 1 shows the optical sensor, based on a PET substrate with a pattern of silver nanoparticles covered with MIP, two plastic optical fibers (POFs), a white light source, and a spectrometer. These POFs have 500 μm of total diameter, a core of PMMA and a cladding of fluorinated polymer.

As previously reported [4, 5], the process adopted to print the parallel lines on a PET substrate uses a low-cost WF-2010 piezo inkjet printer (Epson, Suwa, Japan) and a metal ink for the creation of conductive patterns. The latter is the silver nanoparticles solution "Metalon® JS-B15P" by Novacentrix (Austin, TX, USA). The adopted PET substrate is the Novele™ IJ-220 Printed Electronics Substrate by Novacentrix, suited to low-cost and low-temperature applications and specially designed for inkjet-compatible conductive inks. We used as pattern a very simple layout based on parallel lines, as shown in Fig. 1, where the thickness of the silver nanoparticles lines on the PET substrate is about 2 μm, the length is 5 cm, the spacing is 500 μm and the width is 300 μm. The thickness of the PET substrate is 140 μm, and the other dimensions of the PET substrate are 1 cm × 5 cm (see Fig. 1).

The experimental setup arranged to measure the binding between the MIP receptor and the analyte (2-FAL) is based on a halogen lamp (with an emission range from

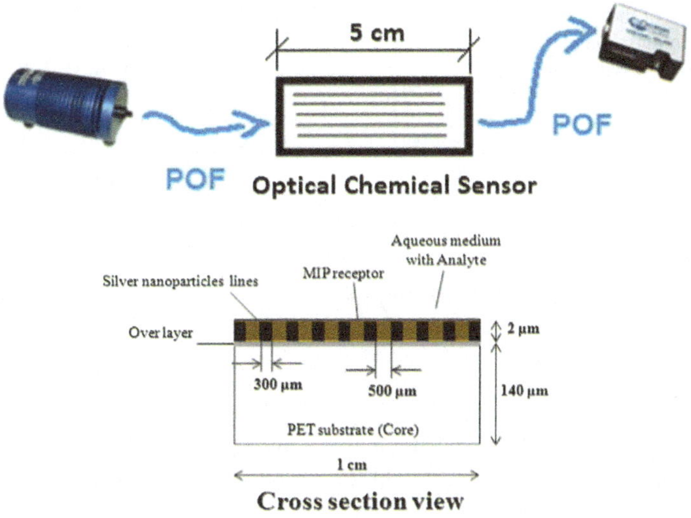


**Fig. 1** Outline of optical chemical sensor with the experimental setup

360 nm to 1700 nm), illuminating the sensor by POF, and a spectrometer (with a detection range from 300 nm to 1000 nm), both produced by Ocean Optic, as reported in Fig. 1.

## *2.2 Chemical Receptor Layer (MIP)*

On the optical platform, based on parallel lines of silver nanoparticles on PET, the MIP prepolymeric mixture (200 μL) have been dropped and spun for 2 min at 1000 rpm. The prepolymeric mixture for MIP was prepared according to the procedure reported in [3] and the thermal polymerization was then carried at 80 °C. The template was extracted by repeated washings with 96% ethanol.

Figure 2 shows the wavelengths of interest relative to the experimentally obtained transmission spectra, for different concentration of 2-FAL in water (from 0 to 58 ppm). As shown in Fig. 2, when the concentration of the 2-FAL increases, the output signal (I) increases. The importance of silver nanoparticles pattern is demonstrated by the fact that the behaviour is the opposite of a similar sensor configuration without parallel lines of silver nanoparticles on PET (in this case the output signal decreases when the 2-FAL concentration increases) [6].

The experimental results show that this novel approach presents a limit of detection of about 0.03 ppm, comparable to that of plasmonic sensors combined with the same MIP [3], but with the advantage of an easier preparation procedure.

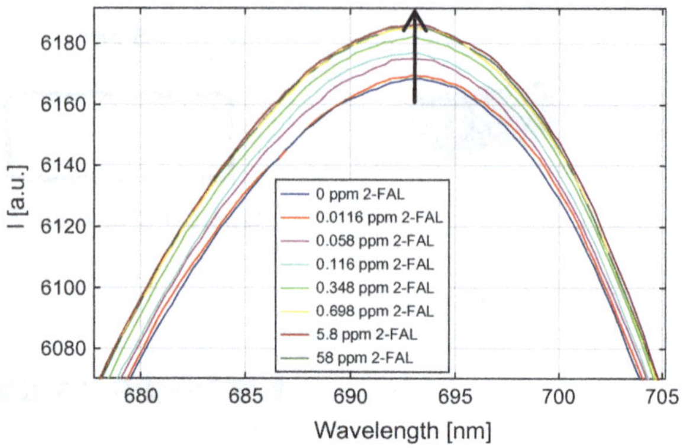


**Fig. 2** Zoom of the wavelength range around 690 nm of transmission spectra, obtained by the sensor for different 2-FAL concentrations in water

The detection range is up to about 0.7 ppm, this too similar to that obtained with the sensors previously developed, based on SPR in D-shaped POF (with an homogeneous gold layer).

## 3 Conclusion

In this work we report a preliminary investigation on the possibility of implementing an optical printed device which could be a good solution when thinness, short developing time and reasonably priced features of printing equipment are required.

In particular, we have demonstrated that the presented optical printed device, based on InkJet Printing technology, can be used to monitor a thin MIP receptor layer for the determination of 2-FAL in aqueous matrices.

## References

1. Rico-Yuste A, Gonzalez-Vallejo V, Benito-Pena E, de las Casas Engel T, Orellana G, Cruz Moreno-Bondi M (2016) Furfural determination with disposable polymer films and smartphone-based colorimetry for beer freshness assessment. Anal Chem 88:3959–3966
2. Camara JS, Alves MA, Marques JC (2006) Changes in volatile composition of Madeira wines during their oxidative ageing. Anal Chim Acta 563:188–197
3. Cennamo N, De Maria L, Chemelli C, Profumo A, Zeni L, Pesavento M (2016) Markers detection in transformer oil by plasmonic chemical sensor system based on POF and MIPs. IEEE Sens J 16:7663–7670
4. Andò B, Baglio S, Bulsara AR, Emery T, Marletta V, Pistorio A (2017) Low-cost inkjet printing technology for the rapid prototyping of transducers. Sensors 17:748
5. Andò B, Baglio S, Lombardo CO, Marletta V, Pistorio A (2015) A low-cost accelerometer developed by inkjet printing technology. IEEE Trans Instrum Meas 65:1242–1248
6. Cennamo N, Zeni L, Pesavento M, Marchetti S, Marletta V, Baglio S, Graziani S, Pistorio A, Andò B (2019) A novel sensing methodology to detect furfural in water, exploiting MIPs, and inkjet-printed optical waveguides. IEEE Trans Instrum Meas 68:1582–1589

# Remote Monitoring of COPD Patients During Non-invasive Mechanical Ventilation by a New Tele-medicine Device

**A. V. Radogna, S. Capone, P. Siciliano, S. Sabina, N. Fiore, G. A. Di Lauro, V. De Luca, L. T. De Paolis, C. G. Leo, P. Mincarone, E. Sabato and F. Satriano**

**Abstract** Chronic obstructive pulmonary (COPD) disease is a progressive illness that will intensify over time. Exacerbation are frequent and likely caused by bacterial and viral respiratory infections. The management of COPD includes home assistance by noninvasive ventilation (NIV), but no remote monitoring system is generally applied. In this work a telemedicine device based on Smart Breath Analyzer (SBA) devoted to the remote-monitoring of parameters in the exhaled air of COPD patients during NIV was developed. System architecture and preliminary pre-clinical test results were here presented.

**Keywords** COPD · Non-invasive mechanical ventilation · Breath analyzer · Gas sensors

---

A. V. Radogna (✉) · S. Capone · P. Siciliano
Institute for Microelectronics and Microsystems, National Research Council (CNR-IMM), pal. A3, Campus Ecotekne, Str. Prov. Lecce-Monteroni Km 1.2, 73100 Lecce, Italy
e-mail: antonio.radogna@le.imm.cnr.it

S. Capone
e-mail: simonetta.capone@cnr.it

S. Sabina · C. G. Leo
Institute of Clinical Physiology, National Research Council (CNR-IFC), pal. A7, Campus Ecotekne, Str. Prov. Lecce-Monteroni Km 1.2, 73100 Lecce, Italy

E. Sabato
"A. Perrino" Hospital, Pulmonology Ward, 72100 Brindisi, Italy

F. Satriano
U.O.S. Dipartimentale di Pneumologia Territoriale, Cittadella Della Salute, ASL Lecce, 73100 Lecce, Italy

N. Fiore · G. A. Di Lauro
Dedalo Solutions srl, 56037 Peccioli, PI, Italy

V. De Luca · L. T. De Paolis
Department of Engineering for Innovation, University of Salento, 73100 Lecce, Italy

P. Mincarone · E. Sabato
Institute for Research on Population and Social Policies, (CNR-IRPPS), 72100 Brindisi, Italy



G. Di Francia et al. (eds.), *Sensors and Microsystems*, Lecture Notes in Electrical Engineering 629, https://doi.org/10.1007/978-3-030-37558-4_12

## 1 Introduction

Chronic Obstructive Pulmonary Disease (COPD) is a major cause of mortality in the world [1, 2] and exacerbations of COPD (ECOPD), during which symptoms acutely worsen, are common [3, 4]. These episodes are often triggered by infection and are the second most frequent cause of emergency hospital admission. Non-invasive ventilation (NIV) is usually implemented to treat COPD. Ventilotherapy improves the blood gas exchange of oxygen ($O_2$) and carbon dioxide ($CO_2$) in the patient's lungs and supports the work of breathing when the patient's ventilatory muscles cannot do this effectively. Short-term NIV is well-established in treating COPD patients hospitalised due to acute exacerbations of COPD, but it's also widely applied to improve quality of life and health condition in COPD patients being treated at home [5].

However, management of domiciliary NIV is a delicate issue in terms of organization and cost of health assistance service; frequent calls for technical intervention are required for adjusting ventilator setting. Basic models of mechanical ventilator for home NIV don't allow memorization and transmission of the monitored data. At present home care programs controlled by telemonitoring are relatively rare and generally based on services as teleconsultation or videoconference [6, 7]. For this reason there is a growing interest in developing telemedicine systems for a remote monitoring of respiratory variables and patient surveillance under NIV.

In this work we introduced a star-network tele-medicine system strongly based on Internet-of-Things paradigm. Leaf nodes of the network are new Smart Breath Analyzers (SBAs) based on gas sensors for the analysis of exhaled air from patients during a NIV therapy. The system aim to be as a useful new non-invasive, instrumental and infrastructure tool for remote patient monitoring supporting respiratory diseases screening practices.

## 2 The ReSPIRO Tele-Medicine System

This work is part of an Italian regional project (acronym project ReSPIRO: Rete dei Servizi Pneumologici: Integration, Research & Open-innovation) (http://respiro.eresult.it). ReSPIRO project proposed a new socio-health model able to overcome what is still in force today in Apulia by Italian Health Service for patients suffering COPD; the model integrates hospital and home care by an assistance protocol involving clinicians, hospitals for nursing home assistance and companies in charge of technical assistance for NIV. In Apulia, in particular, the Brindisi local health agency has already started testing some programs of hospital-community integration and has expressed interest in using the ICT to foster hospital-territory integration. This integration requires data and information from ad hoc sensors for monitoring the patient's health state and anticipating worsening of health condition.

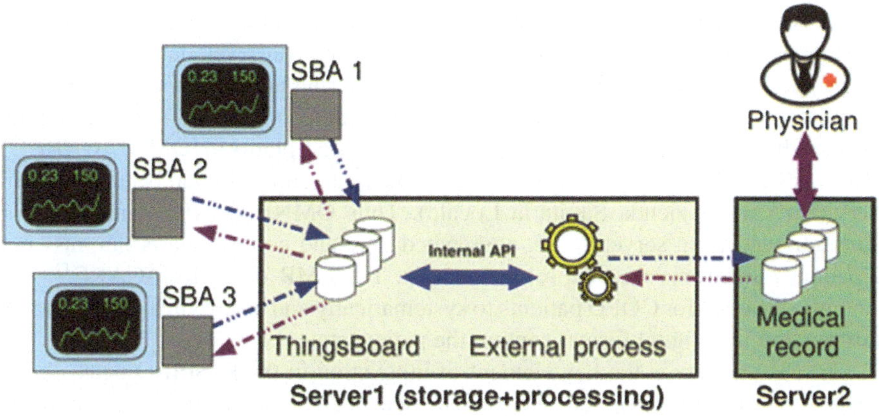


**Fig. 1** ReSPIRO tele-medicine system diagram

Within the project a novel tele-medicine system based on central IoT platform connected to several Smart Breath Analyzers (SBAs) modules were realized. Physicians and healthcare staff can take advantage of this system in order to remotely check the effectiveness of the therapy and assess any events of exacerbation. The system diagram is schematized in Fig. 1.

The system is designed to receive data from more SBA devices, installed at home of patients under non-invasive mechanical ventilo-therapy. The SBA is a module independent from ventilator; it's inserted side-stream to the expiratory line of bi-tube breathing circuit. The coupling of a SBA to breathing circuit makes it a supplementary universal module compatible with any commercial mechanical ventilator arranged for bi-tube breathing circuit. The device acquires sensor data and sends them, via TCP/IP communication, to ICT remote platform. In particular, SBAs devices perform a bi-directional communication through MQTT protocol, a well-known protocol suited for embedded devices, with a first remote server (server 1) for storage and signal processing. Server 1 hosts both an IoT platform for interfacing the domiciliary SBAs devices to the database, and an external application process for data processing and analysis. As regards to the IoT platform, our choice fell on ThingsBoard since it's a 100% open-source IoT plaform enabling real-time data visualization (IoT Dashboards) and device connectivity via industry standard IoT protocols (like MQTT and HTTP); useful technical support with transport encryption is also available on web community. JavaScript Object Notation (JSON) was used as data exchange format.

The analysis on how monitored parameters change over time is performed by means of control charts, implemented as a separated system process on the same server machine of ThingsBoard. Control charts are a useful tool to monitor parameters in a process of whatever nature, determining whether a process is stable or not by detecting variations generated by special causes.

Although ThingsBoard is an excellent storage server, it provides only simple aggregation operators such as minimum, maximum, average value, sum and others and it doesn't allow more complex data analysis or aggregation. For this reason, we implemented an external application process that retrieves data from ThingsBoard, performs some basic computations like filtering and delivers processed data by means of REST services to the electronic medical record (EMR) software located on an additional server (server 2), a multifunctional IoT platform (OMNIACARE$^{TM}$, eResults srl, Italy) managed by the territorial hospital unit of Italian National Health Service (ASL—Azienda Sanitaria Locale). Thus OMNIACARE$^{TM}$ integrates all data retrieved from server 1 (i.e. aggregated data and control charts outputs) into a patient's electronic medical record (EMR). The EMR on OMNIACARE$^{TM}$ was suitably designed for COPD patients to systematically and electronically store health information in a digital format capture the patient state across time.

The SBA device is the key element of innovation in the ReSPIRO network and basically is a custom designed electronic device that includes a bunch of components to monitor various signals from sensors exposed to exhaled air, the electronic circuitry for signal conditioning and the fluidic components (sensor cell and gas lines) allowing the connection in side-stream to the expiratory line of the bi-tube breathing circuit. The SBA device has been carefully designed with the target of minimizing air leaks in device gas lines thus allowing the ventilator's normal operation and ensuring the patient's safety. Details of the device structure is given elsewhere [8–10]. The SBA hardware architecture together with an image of the final device is schematized in Fig. 2.

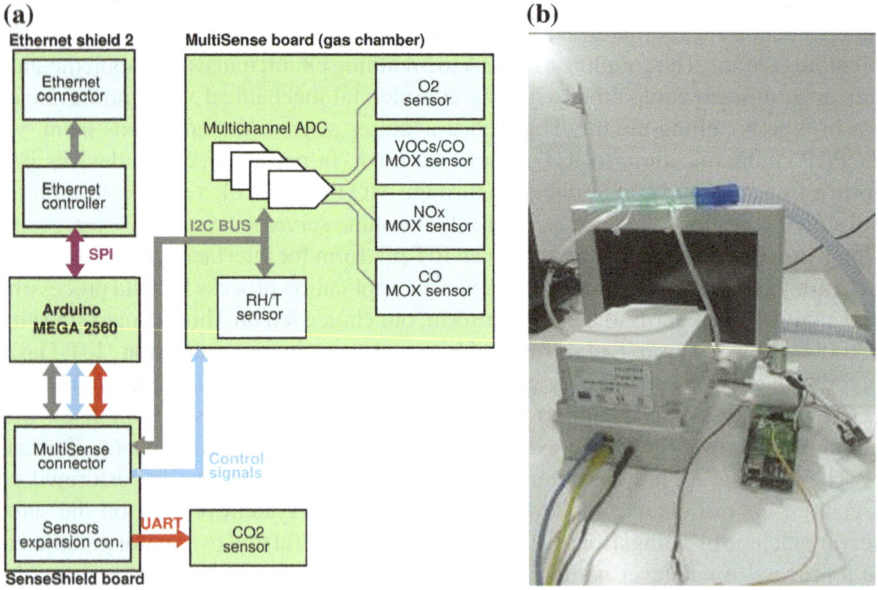


**Fig. 2** **a** Smart Breath Analyzer architecture; **b** Device connected to respiratory circuit of a mechanical ventilator

## 3 Functional Tests

A clinical trial on a small cohort of patients suffering COPD and receiving NIV therapy has been planned by using the ReSPIRO tele-medicine system. The trial will be carried out during patient's ventilo-therapy sessions both in out-patients department under the supervision of a pneumologist and in-home settings. The observational study related to the planned clinical trial was approved by Ethics Committee of Local Health Service Unit of Lecce and Brindisi on patients with GOLD (Global Initiative for chronic Obstructive Lung Diseases) stage IV. Prior to participation in the study, all subjects will provide written, signed, informed consent.

The functioning of the SBA device and the implemented firmware for sensor sampling procedure were verified by functional tests. First functional tests were carried out in local without sending the data to developed remote server architecture, in order to differentiate the control of a correct data acquisition to the possible problems that may occur in data transmission. Hence, the data were sent (via device USB port) to the serial port of a PC. A healthy volunteer, accepting to undergo a noninvasive ventilo-therapy session, was recruited; for volunteer safety, the ventilator parameters in BILEVEL ST mode have been set according to the pneumologist's indications. Figure 3 reports the temporal registration of all the sensors tracings according to the sequential sensor sampling protocol.

As it can be observed, the sampling protocol, according to which the sensors are sampled one at a time, works correctly; step1: signal sampling from 3 MOX-based

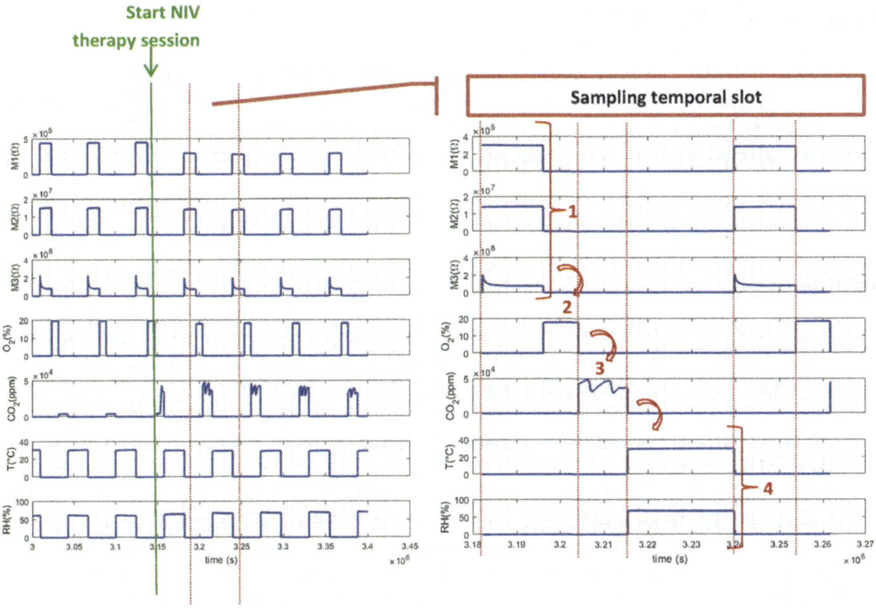


**Fig. 3** Breath signals taken in a functional test in local from an healthy volunteer

sensors (sensor resistance M1, M2 and M3 in Ohm); step 2: signal sampling from $O_2$ sensor ($O_2$ concentration in %); step 3: signal sampling from $CO_2$ sensor ($CO_2$ concentration in ppm ($\equiv 10^{-4}$%)); step 4: signal sampling from R.H. & T sensor (R.H. in %; T in °C). As expected, the $O_2$ concentration decreases and $CO_2$ concentration increases from inspiratory values to physiologic exhaled values for a healthy subject. The signals from MOX-based sensors also changed toward lower resistance values indicating a promising sensitivity to exhaled VOCs. Humidity content in exhaled air is typically high ($\approx$90%) and an increase of R.H. was indeed registered during the NIV session compared to the value before the start. The acquired sensor traces collected from all the sampling windows for each sensor are hence valid for a next data analysis and correlation with routine clinical examinations and clinical data. Of course, an appropriate suitable interpretation of the monitored parameters by a physiological point of view is required.

## 4 Conclusions

A tele-medicine system, based on SBAs was designed, realized and verified in laboratory by functional tests. The SBAs' star-based network system has the aim to fill the gap between healthcare services and COPD patients that relies on home-assisted mechanical therapy to cope with the disease. The preliminary functional tests demonstrated the effect possibility to enter the SBAs in medical practice in home-ventilo-therapy protocols by suitable interpretation of the monitored parameters by a physiological point of view.

**Acknowledgements** The work was founded by project ReSPIRO (Rete dei Servizi Pneumologici: Integration, Research & Open-innovation—Cluster Tecnologici Regionali, Apulia Region, project cod. F29R1T8). We thank eResults s.r.l. for providing the Omniacare platform. We thank SAD s.r.l. for providing ventilators and the necessary technical assistance and support in experimental set up.

## References

1. Rabe KF, Watz H (2017) Chronic obstructive pulmonary disease, seminar. Lancet 389(10082):1931–1940. https://doi.org/10.1016/S0140-6736(17)31222-9
2. Pocket guide to COPD diagnosis, management, and prevention, a guide for health care professionals, 2018 Report, © 2018 Global Initiative for Chronic Obstructive Lung Disease, Inc. www.goldcopd.org
3. Wedzicha JA, Miravitlles M, Hurst JR, Calverley PMA, Albert RK, Anzueto A, Criner GJ, Papi A, Rabe KF, Rigau D, Sliwinski P, Tonia T, Vestbo J, Wilson KC, Krishnan JA (2017) Management of COPD exacerbations: a European respiratory society/American thoracic society guideline. Eur Respir J 49:1600791, 16 pp. https://doi.org/10.1183/13993003.00791-2016
4. Viniol C, Vogelmeier CF (2018) Exacerbations of COPD. Eur Respir Rev 27:170103, 9 pp. https://doi.org/10.1183/16000617.0103-2017

5. Murphy PB, Hart N (2018) Home non-invasive ventilation for COPD: how, who and when? Arch Bronconeumol 54(3):149–154. https://doi.org/10.1016/j.arbres.2017.12.005
6. Ambrosino N, Vagheggini G, Mazzoleni S, Vitacca M (2016) Telemedicine in chronic obstructive pulmonary disease. Breathe 12(4):350–356. https://doi.org/10.1183/20734735.014616
7. Vitacca M, Montini A, Comini L (2018) How will telemedicine change clinical practice in chronic obstructive pulmonary disease? Ther Adv Respir Dis 12:1–19. https://doi.org/10.1177/1753465818754778
8. Radogna AV, Capone S, Di Lauro GA, Fiore N, Francioso L, Casino F, Siciliano P, Sabina S, Sabato E (2018) A smart device for supporting mechanical ventilo-therapy. In: IEEE proceedings of international conference on IC design and technology (ICICDT 2018) pp 177–180. https://doi.org/10.1109/ICICDT.2018.8399785
9. Radogna AV, Capone S, Di Lauro GA, Fiore N, Francioso L, Casino F, Siciliano P, Sabina S, Sabato E (2019) A smart breath analyser for monitoring home mechanical ventilated patients. In: Andò B. et al (eds) Sensors. CNS 2018. Lecture notes in electrical engineering 539, Springer, Cham, pp 465–471. https://doi.org/10.1007/978-3-030-04324-7_56
10. Radogna AV, Capone S, Di Lauro GA, Fiore N, Longo V, Giampetruzzi L, Francioso L, Casino F, Siciliano P, Sabina S, Leo CG, Mincarone P, Sabato E (2019) A novel tele-medicine system to improve therapy monitoring in chronic respiratory diseases. In: Leone A, Caroppo A, Rescio G, Diraco G, Siciliano P (eds) Ambient assisted Living. ForItAAL 2018. Lecture notes in electrical engineering 544. Springer, Cham, pp 201–205. https://doi.org/10.1007/978-3-030-05921-7_16

# Micro-incubator Based on Lab-on-Glass Technology for Nanosatellite Missions

**L. Iannascoli, F. Costantini, N. Lovecchio, A. Buzzin, D. Caputo, G. De Cesare and A. Nascetti**

**Abstract** The study and quantification of the effects of the space environment on human body is a primary task for future manned deep space missions. The risk models for radiation exposures incurred by astronauts beyond low-Earth orbit, have different limitations due to the difficulty to have terrestrial parallels on which to base risk estimates. Indeed, no terrestrial sources fully reproduce the deep space energy spectrum and the multi directional flux of the cosmic radiation. In situ analysis would therefore be fundamental in order to enable reliable studies about the effects of the radiation environment on living organisms as well as to evaluate customized radiological countermeasures for astronauts. A micro-incubator suitable for cubesat missions for studying in situ the effects of the space environment on cellular cultures is presented. The device is based on lab-on-chip technology with integrated thin-film sensors and actuators for the active control of the environmental conditions of the cell culture and for the monitoring of its metabolic status. In particular, the device includes an incubation chamber connected to a microfluidic network for the supply of nutrients and/or pharmaceuticals. A second network is used for the distribution of carbon dioxide through a thin gas-permeable membrane. On-chip on-demand production of carbon dioxide can be eventually achieved from the pyrolysis of sodium bicarbonate stored in a separate reservoir with a dedicated thin film heater. The same network can be used to supply a controlled atmosphere from a pressurized tank. The on-chip hydrogenated amorphous silicon photodiodes are used to measure the light emitted by genetically-modified cell cultures that express a bio-luminescent behavior when subjected to given stress conditions. Accurate temperature control is achieved by means of additional on-chip thin-film diodes and a transparent indium-tin-oxide heater located beneath the incubation chamber. From technological point of view, the system relies on the combination of different thin- and thick-film fabrication

L. Iannascoli · N. Lovecchio · A. Buzzin · D. Caputo · G. De Cesare
Department of Information Engineering, Electronics and Telecommunications, Sapienza University of Rome, Via Eudossiana 18, 00184 Rome, Italy
e-mail: lorenzo.iannascoli@uniroma1.it

F. Costantini · A. Nascetti (✉)
School of Aerospace Engineering, Sapienza University of Rome, Via Salaria 851, 00138 Rome, Italy
e-mail: augusto.nascetti@uniroma1.it

 
G. Di Francia et al. (eds.), *Sensors and Microsystems*, Lecture Notes in Electrical Engineering 629, https://doi.org/10.1007/978-3-030-37558-4_13

technologies jointly used with the aim to achieve a compact, automated and low-power device that represents a viable solution for biological experiments aboard cubesat satellites.

**Keywords** Micro-incubator · Lab-on-chip · Nano-satellite · Space missions

## 1 Introduction

The greatest long-term risk for astronauts exposed to space radiation environment, beyond Low Earth Orbit (LEO), is to develop degenerative pathologies like cancer, experience central nervous system (CNS) decrements, exhibit degenerative tissue effects or develop acute radiation syndrome [1]. The intensity of the ionizing radiation becomes extremely relevant for missions outside the Earth's magnetosphere, which are currently under consideration within the solar system manned exploration missions being planned by the main space agencies as well as by private companies. Its therefore of fundamental importance to study and quantify the effects of the space environment on human body as a primary task for future manned deep space missions. Ground-based studies often suffer the limitations of the available risk models for radiation exposures beyond low-Earth orbit [2] arising from the difficulty to fully reproduce in laboratory the deep space energy spectrum and the multi directional flux of the cosmic radiation. In situ analysis would therefore be fundamental in order to enable reliable studies about the effects of the radiation environment on living organisms as well as to evaluate customized radiological countermeasures for astronauts.

As part of this framework, we propose a scientific payload that allows experiments on cell cultures to be carried out on nano-satellites by satisfying requirements such as compact size, automated operation and low-power consumption. This goal is achieved through the use of lab-on-chip technology, which enables the fabrication of extremely compact devices that integrate several laboratory functions that may range from sample preparation to the analytical outcome. Lab-on-chip devices have been proven to be powerful instruments to accomplish complex chemical or bio-chemical analysis on a single chip with very limited consumption of samples and reagents [3]. Such devices are ideal for space mission where experiments need to be almost totally automated, and volume, weight and power must be kept as small as possible. However, many lab-on-chip devices presented in literature still require external instrumentation for their operation (such as microscopes, cameras or pumping systems) making them less appealing to nano-satellite based space missions. The lab-on-chip device proposed here, instead, consists of a fully integrated microsystem [4], which includes the incubation chamber with the surrounding microfluidic network and on-chip integrated thin-film electronic sensors and actuators that eliminate the need for external desktop instrumentation and to perform experiments in a standalone fashion.

The present paper is arranged as follows. The system overview is reported in the next section with details of the system components reported in subsections. Here are also described the materials and methods employed for the fabrication of the microfluidic and electronic components. Finally, conclusions are drawn.

## 2 Micro-incubator for Cell Culture

With the purpose of studying the radiation effects on cell cultures, a microfluidic incubator has several advantages over conventional equipment, such as the ability to more closely mimic a cell's natural environment by continuous perfusion culture or by creating chemical gradients [5], the conservation of samples and reagents and the minimization of disposables [6], besides low weight, volume and power consumption.

Figure 1 provides a schematic view of the proposed system for the maintenance and growth of cell cultures. The device is made up of two main elements: a polymeric microfluidic unit and a 50 × 50 $mm^2$ glass substrate (smart-glass) on which hydrogenated amorphous silicon (a-Si:H) sensors and thin-film resistive heaters are fabricated.

The smart-glass includes thin-film photodiodes that allow the implementation of luminescence-based analytical protocols commonly used in laboratory. Accurate temperature control is achieved by means of additional on-chip thin film diodes and a

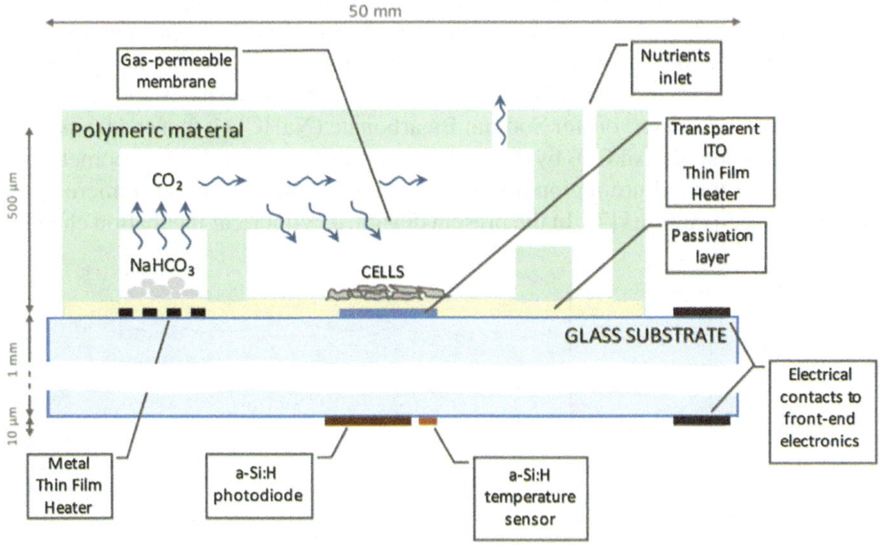


**Fig. 1** Schematic view of the proposed micro-incubator for cell cultures based on lab-on-chip technologies

transparent indium-tin-oxide heater located beneath the culture chamber. Additional thin film resistive heaters are planned for the production in situ of carbon dioxide ($CO_2$) used to regulate the environmental conditions of the cell culture.

The polymeric microfluidic unit, which hosts the cell culture as well as the culture media and the supplies, is permanently bonded on the top face of the smart glass. The thin-film devices are fabricated on the opposite face to prevent any interaction between the 'biological' side and the 'electronics' side, thus making the system more robust and reducing the risk of contamination of the cell culture or of its supplies. The bottom side also hosts the electrical contacts needed to interface the device with the electronic control board.

## 2.1 *Polymeric Incubation Chamber*

One of the key benefits of the use of microfluidics for cell cultures studies is the ability to control parameters of the cell micro-environment at relevant length and time scales. In detail, an incubator system must be able to maintain the pH and osmolality of the culture within physiological levels and guarantee a regular refresh of cell nutrients [7].

According to these requirements, the building blocks of an incubator system are: an incubation chamber, a series of channels and reservoirs for culture media replacement, a thin gas-permeable membrane and a network for the distribution of suitable atmosphere. The device proposed here incorporates all of these elements.

As is shown in Fig. 2 the polymeric microfluidic structure is made up of different levels, each of which has a specific objective. The first layer at the bottom of the microfluidic structure, marked as 1, is the one which is bonded to the smart glass: this layer hosts the incubation chamber, inlet and outlet ports for fresh nutrients and waste respectively, and a reservoir for Sodium Bicarbonate ($NaHCO_3$) that can be used for the on-chip production of $CO_2$ by thermal decomposition of $NaHCO_3$. Geometry and dimensions of the culture region are fundamental design aspects for a microfluidic system with cell cultures [7]. In the present design, a cylindrical incubation chamber

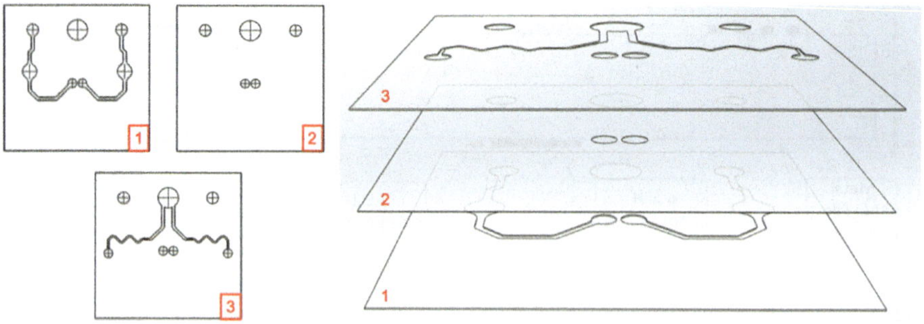


**Fig. 2** Layout of the three layers that make up the microfluidic part

whit a volume of 100 μl has been chosen. The second layer of the structure consists of a 50 μm-thick gas-permeable membrane to ensure gas-exchange with the incubation chamber. The last layer hosts a network for the distribution of carbon dioxide across the chip. Access holes are also made in the top layers in order to allow chip loading. Eventually, a top sealing layer can be used during the experimental phase (not shown).

The polymeric microfluidic structure can be fabricated using different materials or combinations of materials. For this work, Pressure Sensitive Adhesives (PSA) have been chosen. The advantages of these materials rely on low cost, easiness to assemble multilayer structures and rapid prototyping processes; this is suitable especially in the preliminary stages of system implementation. Moreover, experiments with PSA technology were already performed in space environment, mainly on the ISS. Each layer is fabricated separately by patterning the PSA by means of a cutting plotter and then assembling the entire structure layer by layer. The position and geometry of the different microfluidic elements have been affected by the fabrication constraints of the smart-glass chip considering the expected temperature distribution during operation.

## *2.2 On-Chip Sensors and Actuators*

In order to design, test and characterize the active components of the smart-glass a set of intermediate devices has been preliminarily fabricated. In particular, two glass substrates one with sensor-side only and one with heater-side only (Fig. 3) have been produced.

The heaters for the control of the temperature of the incubation chamber are made in Indium Tin Oxide (ITO), which is transparent, in order to allow the optical monitoring of the cell culture by means of the photodiodes that are fabricated on the opposite side of the substrate. The heaters are passivated by a 5 μm-thick SU-8 layer to ensure electrical insulation from the microfluidic part that is assembled on top of them.

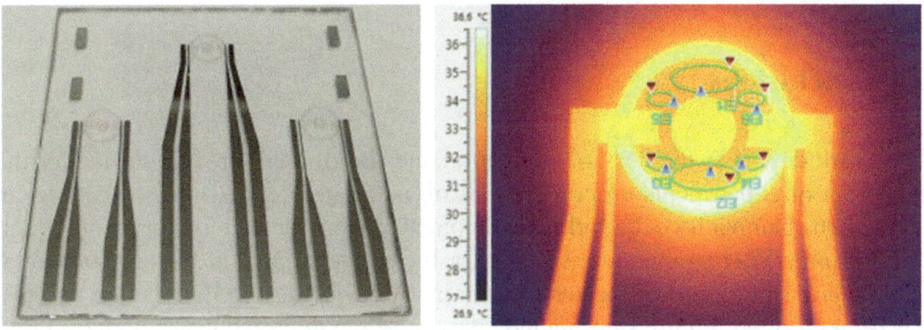


**Fig. 3** Left: fabricated sample of the smart-glass heaters. Right: heater characterization by thermal imaging

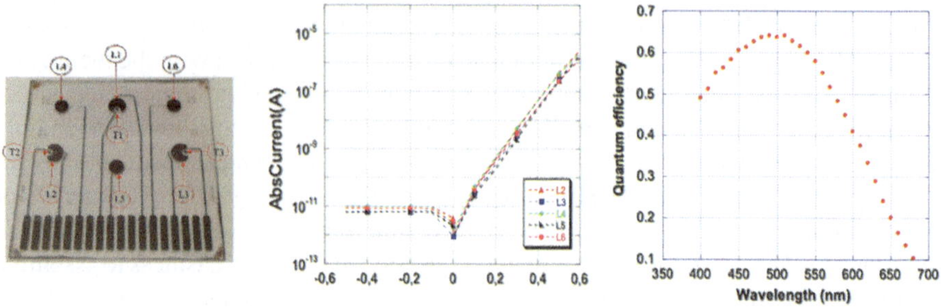


**Fig. 4** Left: sample of the smart-glass sensors (the combined large C-shaped photodiode with the small temperature sensor are visible is T1-L1, T2-L2 and T3-L3). Center: current-voltage characteristics in dark conditions of five out of six on-chip a-Si:H photodiodes. Right: spectral quantum efficiency of one of the a-Si:H photodiodes of the smart-glass device

The a-Si:H diodes are deposited on the bottom glass side and implement both the temperature sensors and the photosensors. The sensors are p-type, intrinsic, n-type a-Si:H stacked structures deposited by Plasma Enhanced Chemical Vapour Deposition (PECVD). The difference between the two types of sensor is just in the front contact layer (on glass surface), which is metallic for the temperature-diodes while it is made of transparent ITO for the photodiodes. In order to allow the monitoring of the temperature of the incubation chamber and ensure optical measurements of the cell culture, a particular shape of the devices has been designed, with a large C-shaped photodiode and small temperature-diode, as reported in Fig. 4.

Current-voltage characteristics have been measured for all the a-Si:H diodes to verify both the dark current and the diode ideality factor which give indication of the overall process quality (center in Fig. 4). Spectral quantum efficiency measurements in reverse bias conditions have been performed on the photodiodes as shown in Fig. 4 (right).

Temperature sensors have been characterized by measuring the variation of the forward voltage drop at constant forward bias current of $10^{-5}$ A/cm$^2$ by varying the temperature on a hot-plate. A sensitivity of 3.3 mV/°C has been found.

## 3 Conclusions

This work describes the design, fabrication and test of a lab-on-chip device for the execution of experiments on cell cultures on nanosatellite missions. It features an incubation chamber for cells with a-Si:H optical sensors for the monitoring of the cell culture and a:Si:H temperature sensors for the active control of the environmental parameters using thin-film heaters. Preliminary device design, fabrication and test has been reported. Future steps include the use of the microincubator with bacterial cultures, the test of the on-chip generation of the carbon dioxide and finally the test with mammal cells cultures.

## References

1. Jeffery JPS, Chancellor C, Scott GBI (2014) Space radiation: the number one risk to astronaut health beyond low earth orbit. Life
2. Barcellos-Hoff MH, Blakely EA, Burma S, Fornace AJ, Gerson S, Hlatky L, Kirsch DG, Luderer U, Shay J, Wang Y et al (2015) Concepts and challenges in cancer risk prediction for the space radiation environment. Life Sci Res 92–103
3. Abragall P, Gue A (2015) Lab-on-chip technologies: making a microfluidic network and coupling it into a complete microsystem a review. J Micromechanics Microengineering 17
4. Caputo D, de Angelis A, Lovecchio N, Nascetti A, Scipinotti R, de Cesar G (2015) Amorphous silicon photosensors integrated in microfluidic structures as a technological demonstrator of a true Lab-on-chip system. Sens Bio-Sens Res 3:98–104
5. Halldórson S, Lucumi Moreno E, Gómez-Sjöberg R, Fleming R (2015) Advantages and challenges of microfluidic cell culture in polymethylsiloxane devices. Biosens Bioelectron
6. DePalma A (2017) Microfluidics for cell culture and analysis
7. Young LEWK, Beebe DJ (2010) Fundamentals of microfluidic cell culture in controlled microenvironments. Chem Soc Rev

# Portable Electronic Nose Device for the Identification of Food Degradation

**G. Zambotti, M. Soprani, E. Gobbi, R. Capuano, V. Pasqualetti, C. Di Natale and A. Ponzoni**

**Abstract** This work is aimed to detect degradation of different food matrices through the use of a miniaturized (USB-stick sized) electronic nose. This electronic nose is based on metal oxide gas sensors and works according to temperature profile protocols. The tested foods were fish or chicken kept at room temperature. The obtained results show that this device is able to discriminate the degradation of both foods at the safety threshold.

**Keywords** Fish · Chicken · Electronic nose · Metal oxides · Microbiological detection

## 1 Introduction

The electronic nose (EN) technology has been recently employed in the field of food quality and safety control of various food and agricultural products. Particularly, ENs have been applied in process monitoring, freshness evaluation, shelf-life investigation, authenticity determination, and product traceability [1–4]. However up to now much less emphasis has been given to microbiology applications of chemical sensor devices. The capability of ENs in the detection of microorganisms is mainly related to the variations of the volatile compounds, the so called olfactory fingerprint,

G. Zambotti (✉) · M. Soprani · E. Gobbi · A. Ponzoni
National Research Council (CNR), National Institute of Optics (INO), Via Branze 45, 25123 Brescia, Italy
e-mail: giulia.zambotti@ino.cnr.it

G. Zambotti · M. Soprani · A. Ponzoni
Department of Information Engineering, University of Brescia, Via Branze 38, 25123 Brescia, Italy

E. Gobbi
Department of Molecular and Translational Medicine—AgriFood and Environmental Microbiology Platform (Pi.Mi.A.A.), University of Brescia, Viale Europa 11, 25123 Brescia, Italy

R. Capuano · V. Pasqualetti · C. Di Natale
Department of Electronic Engineering, University of Tor Vergata, Via del Politecnico 1, 00133 Rome, Italy

 
G. Di Francia et al. (eds.), *Sensors and Microsystems*, Lecture Notes in Electrical Engineering 629, https://doi.org/10.1007/978-3-030-37558-4_14

of food products caused by the microbial metabolism. The fingerprint variation can be either associated with the appearance of new chemical compounds (primary or secondary metabolites) or to the changes in the relative amount of original volatile compounds without changes in the qualitative composition [5, 6]. Routine control of food contamination by EN is one of its most promising industrial applications as sensor systems offer major advantages over current process control analyses: good selectivity, correlation with data from traditional microbiological tests, low cost, portability and usable at working sites, rapidity to be used at-line or on-line, and no need for sample preparation so allowing for easy and fast spoilage detection [7]. For all these reasons we developed a system able to identify food contamination in less time with respect to classical techniques and at reduced costs [8].

## 2 Materials and Methods

The electronic nose used in this study is the commercial JLM Minimox from JLM Innovation (www.jlm-innovation.de). A picture of this device is shown in Fig. 1.

It is implemented with two different types of Metal Oxide (MOX) sensors, a Figaro TGS 852 and a AMS CCS811 sensors. To achieve a discrimination capability with only two sensors, a custom temperature modulation protocol was developed to enhance the discriminative capability of the system [9]. The heater is supplied with a voltage of 70% of the maximum allowed for 10 s, followed by 10 s in which the voltage is set to 50% of its maximum.

A single portion of fish or chicken (10 g) was put in a closed beacker hosting the EN and tested. Several experiments have been carried out at environment temperature (25 °C) allowing the food to degrade for up to 48 h. The EN was exposed to the head-space generated from the food specimen, in order to follow the volatile compounds' change during the spoilage process. We compared the results obtained from the electronic nose with those from classical microbial techniques. Parallel tests have been carried out with empty boxes and boxes containing water to cross- check the sensor behaviour with environmental air and humidity.

**Fig. 1** The commercial electronic nose JLM Minimox from JLM innovation used in this study

Some preliminary measurements were performed using gas chromatography coupled to mass spectrometry (GC-MS) technique, in order to evaluate volatile compound profile evolution during food degradation.

## 3 Results and Discussion

The sensors worked with custom temperature profile protocols and provided real time responses every 20 s (the period of the temperature modulation). It is possible to obtain informations based on the width, height, shape of the curve that represents the features of our sensors [8]. Based on different features it is possible to perform a PCA such as the one shown in Figs. 2 and 3. Figure 2 is obtained by using the

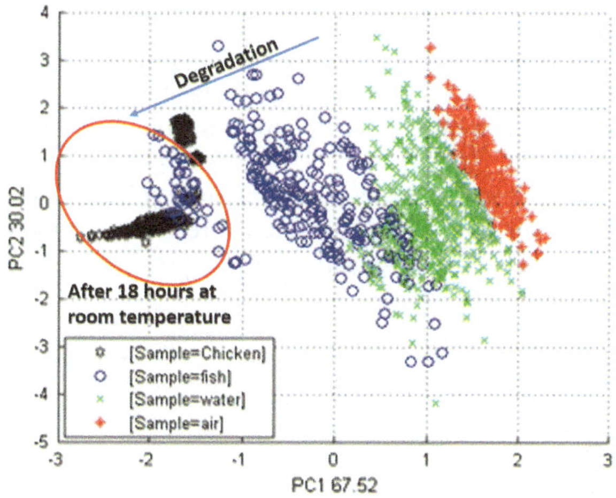


**Fig. 2** Principal component analysis plot (PC1 and PC2). The blue color corresponds to fish kept at room temperature. The black color corresponds at chicken. The blue line identifies increasing time at room temperature. The best discrimination corresponds to the degradation of the fish and chicken starting from CFU/g $> 10^7$ after 18 h how is identified in red circle

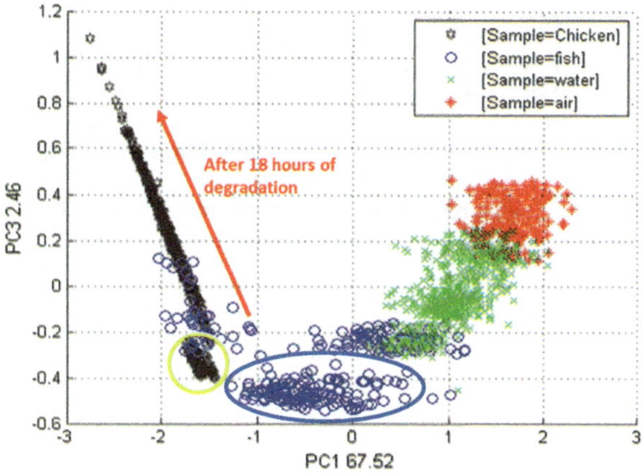


**Fig. 3** Principal component analysis plot (PC1 and PC3). It's possible to see that after 18 h the point corresponding chicken and fish are overlapped

two main components, i.e. the components of the graph that represent the greatest expression of the information. The first thing to note is that there is a separation between water, air, fish and chicken olfactory fingerprints. The fish smell is closer to the water smell compared to the chicken, perhaps because of its higher moisture content. By the time, the fish and chicken olfactory fingerprints overlap. Precisely this occurs after 18/20 h of storage at room temperature. It is possible to note how the sensors detect the physiological degradation of the chicken and fish: their smell changes and intensifies with increasing time, according with microbiological counts (from $10^5$ CFU/g to $>10^9$ CFU/g). Interestingly, these data are in accordance with the legislative requirements for shelf-life; for example fish can be considered edible only when its bacterial load is up to 107 cfu/g [10].

Figure 3 shows the plot of PC1 and PC3 from the previous PCA. From this graph it is evident how the fish and the chicken have two different smells, being perfectly separated even at the same time of their degradation (circle blue and yellow).

Based on these considerations, interestingly it is possible to identify a region typical of the "no more edible" foods, different from that of the safe foods.

Modification of fish headspace composition during degradation process was confirmed also by gas chromatographic analysis (Fig. 4). The quantitative and qualitative modification of released volatile compounds following the degradation of the food is evident.

This result is encouraging because it indicates the potential capability of the electronic nose to discriminate the real degradation regardless of the food kind (at least among fresh fish and chicken).

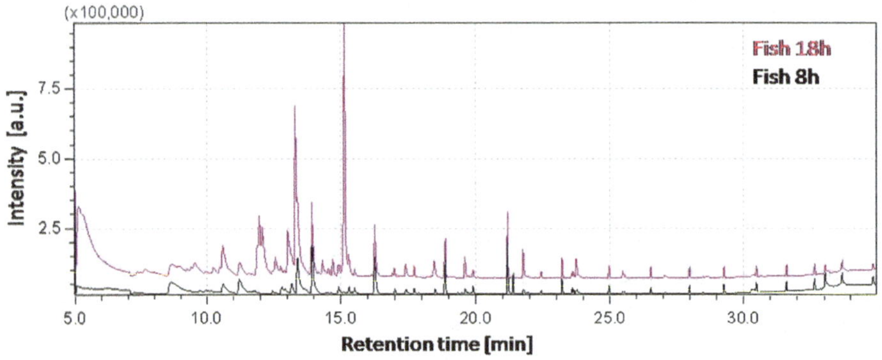


**Fig. 4** Chromatogram comparison between fish headspace samples analyzed after 8 (black curve) and 18 h (pink line) at room temperature storage condition

# 4 Conclusion

This system is able to distinguish types of food matrix and different concentrations of odours correlated to increasing amounts of microbial counts and degradation of food. It is an easy-to-use, versatile tool; for these reasons, it could be very appealing for industrial and domestic applications.

**Acknowledgements** This work has been partially founded by the Lombardia Region and the National Research Council through the project FHfFC and by Lombardia Region and Fondazione Cariplo through the project EMPATIA@LECCO.

## References

1. Wei XB, Shao XF, Wei YY, Cheong LZ, Pan LQ, Tu K (2018) Rapid detection of adulterated peony seed oil by electronic nose. J Food Sci Technol 55:2152–2159
2. Breijo EG, Guarrasi V, Peris RM, Fillol MA, Pinatti CO (2013) Odour sampling system with modifiable parameters applied to fruit classification. J Food Eng 116:277–285
3. Ponzoni A, Depari A, Falasconi M, Comini E, Flammini A, Marioli D, Taroni A, Sberveglieri G (2008) Bread baking aromas detection by low-cost electronic nose. Sens Actuators B Chem 130:100–104
4. Chen QS, Zhao JW, Chen Z, Lin H, Zhao DA (2011) Discrimination of green tea quality using the electronic nose technique and the human panel test, comparison of linear and nonlinear classification tools. Sens Actuators B Chem 159:294–300
5. Rock F, Barsan N, Weimar U (2008) Electronic nose: current status and future trends. Chem Rev 108(2):705–725
6. Gobbi E, Falasconi M, Zambotti G, Sberveglieri V, Pulvirenti A, Sberveglieri G (2015) Rapid diagnosis of Enterobacteriaceae in vegetable soups by a metal oxide sensor based electronic nose. Sens Actuators B Chem 207:1104–1113
7. Perera A, Pardo A, Barrettino D, Hierlermann A, Marco S (2010) Sens Actuators B Chem 146:477–482
8. Sanaeifar A, ZakiDizaji H, Jafari A, de la Guardia M (2017) Early detection of contamination and defect in foodstuffs by electronic nose: a review. Trends Anal Chem 97:257–271
9. Ponzoni A, Depari A, Comini E, Faglia G, Flammini A, Sberveglieri G (2012) Sens Actuators B Chem 175:149–156
10. Jay JM, Loessner MJ, Golden DA (2005) Modern food microbiology, 7th edn. Springer, pp 115–132

# Ultrasonic Transducer for Broadband Applications

A. S. Fiorillo, S. A. Pullano, M. Menniti, M. G. Bianco and C. D. Critello

**Abstract** In-air sonar system takes advantage of transducers based on different materials and technologies for navigation in unstructured and harsh environments. Most technologies suffer from a limited frequency band and a lack of versatility in those applications where a detailed set of information is required. Herein, we propose an ultrasonic sensor inspired by bats' cochlear structure, characterized by broadband (i.e. 30–95 kHz). The transducer is manufactured by folding a thin film of Polyvinylidene Fluoride, according to a logarithmic spiral geometry. Besides the bandwidth, experimental results showed that the ultrasonic transducer working in the bimodal configuration, is characterized by omnidirectional pattern, capable of reproducing the spectrogram of most echolocating bats.

**Keywords** Ferroelectric polymer · Piezoelectricity · Cochlea · Ultrasonic transducers · Biosonar

## 1 Introduction

Electronic locating systems, such as sonar, have been widely investigated, working at various frequencies (in air or water) for the most different tasks (e.g. the localization of shoals of fish, detection of target in military field, etc.) [1–4]. However, sonar is not only used in engineering applications but also found in nature. Most low frequency ultrasound (US) sonar, indeed, were inspired by the study of certain mammals such as dolphins and bats, which employ an echolocation system to perceive the surrounding environment [5, 6]. Some species of bats possess an advanced system of echolocation characterized by constant-frequency (CF) and frequency-modulated (FM) signals in order to detect, localize, classify and track targets in harsh environment. Biosonar of bats can be extremely sensitive up to 200 kHz, and one representative example is that of Pteronotus Parnellii, also known as the mustached bat [7]. It emits four calls,

A. S. Fiorillo · S. A. Pullano (✉) · M. Menniti · M. G. Bianco · C. D. Critello
BATS Laboratory, Department of Health Sciences, University Magna Græcia of Catanzaro, Catanzaro, Italy
e-mail: pullano@unicz.it



G. Di Francia et al. (eds.), *Sensors and Microsystems*, Lecture Notes in Electrical Engineering 629, https://doi.org/10.1007/978-3-030-37558-4_15

and each of them is followed by a shorter downchirp centered at 30.5 kHz, 61 kHz, 92 kHz, and 123 kHz, respectively [7–9].

Many research efforts in the mimesis of the echolocating system have been focused on piezoelectric transducer technology, mainly on acoustic impedance matching materials and the development of devices with a broader working frequency band, which is still a bottle neck in the realization of bioinspired sonar systems [10]. Therefore, the development of broadband US transducers can make a significant progress in the context of mimicking CF–FM bat echolocation, which could also improve the performance of current robotic sonar systems for medical applications [11].

In the present work, we propose an ultrasonic transducer inspired by the spiral shape of the bat's cochlea, which has been designed as an in-air ultrasonic receiver working in the low frequency range. The US sensor is made of Polyvinylidene Fluoride (PVDF), which offers good acoustic matching and sensitivity, as well as excellent design flexibility [12].

## 2 Spiral Sensor Configuration

The proposed logarithmic spiral geometry (Fig. 1) bases its working principle on a clamped hemicylindrical ultrasonic transducer, whose resonant frequency was inversely proportional to the bending radius [13]. Based on this assumption, the proposed design can be assimilated to a summation of contiguous hemicylindrical transducers with different bending radii. This hypothesis has been investigated using COMSOL Multiphysics® software in terms of vibration modes at several resonant

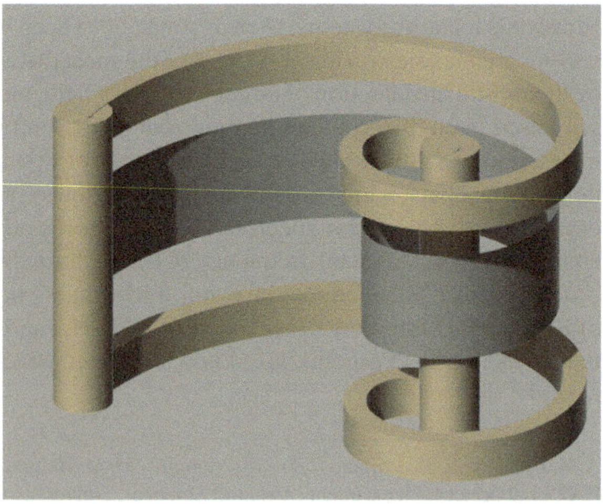

**Fig. 1** The spiral-shaped sensor made of the piezopolymer sheet (grey) anchored at its extremities on a plastic rigid support

conditions [14]. The finite element analysis highlighted pure flexural modes and flexural/radial modes depending on which frequency was evaluated, and this vibrational behavior was likely caused by the intrinsic non-uniformity of the sensor arrangement.

The spiral-shaped transducer has been fabricated from a thin sheet of PVDF (50 mm × 5 mm with a thickness of 28 μm). The film has been left free to resonate in response to the excitation signal, except at the extremities where it has been bonded to an external rigid support (Fig. 1) using epoxy.

## 3 Experimental Results

The piezoelectric sheet has been tested as an ultrasonic receiver, and the echoes at different frequencies have been detected. Figure 2 illustrates the ultrasonic signals at 35 and 95 kHz. Low frequencies echoes showed larger amplitudes than those at higher frequencies because of the predominant vibration of larger arcs. Consequently, the portion of the curved PVDF that generates an electric potential, by means of the direct piezoelectric effect, decreases as the frequency increases. The opposite effect occurs at high frequency, since vibrations affect shorter arcs of the spiral geometry. Therefore, a smaller portion of PVDF is interested, as demonstrated by a higher amplitude of echo signals at lower frequencies. Although the presence of phase distortions, the spiral–shaped sensor was able to "*hear*" high frequencies vibrations evidencing a peculiar characteristic of covering a broad spectrum, which is something unusual for piezoelectric transducers.

Sensitivity measurements of the proposed sensor have been performed using a pulse-echo setup. The sensitivity in dB as a function of the frequency (up to 95 kHz) at two specific angular positions around the sensor, is showed in Fig. 3. It is maintained quite uniform on both horizontal and vertical plane demonstrating the omnidirectional characteristic of the ultrasonic receiver in the frequency range from 30 up to 95 kHz. Sensitivity spanned from −102.11 to −87.89 dB.

## 4 Conclusion

Bats have been always a strong model of bioinspiration for echolocation. This research refers to the area of biomimetic sonar systems proposing a novel ultrasonic piezopolymer transducer using a non–conventional arrangement in the three-dimensional space. The peculiar configuration takes inspiration from the spiral shape of bat's cochlea, demonstrating improved performances in terms of frequency band

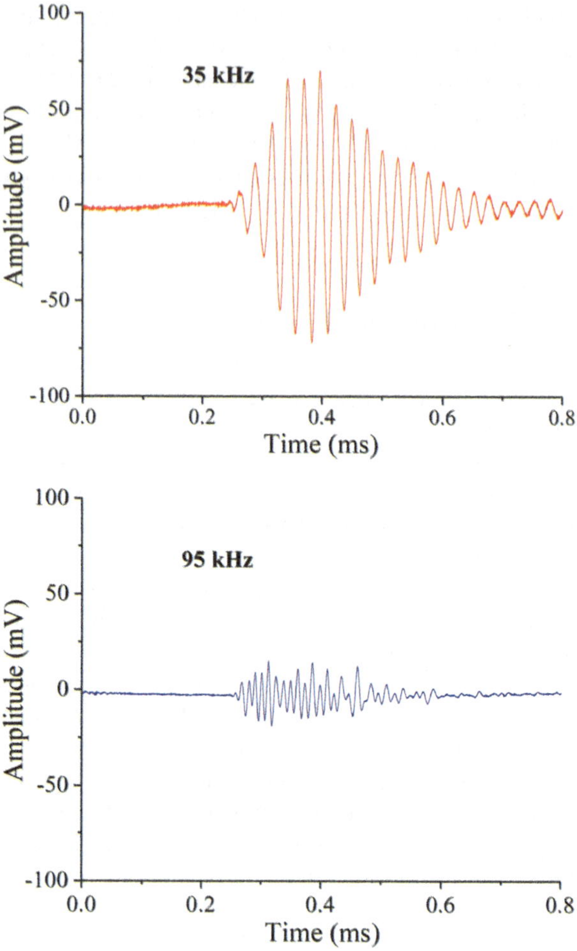


**Fig. 2** Echoes received from the spiral sensor at 35 kHz (top) and 95 kHz (bottom)

and omni-directionality both on the horizontal and vertical planes around the sensor. The spiral–shaped sensor represents a significant advancement for designing bioinspired electronic sonar systems in the area of robotic applications.

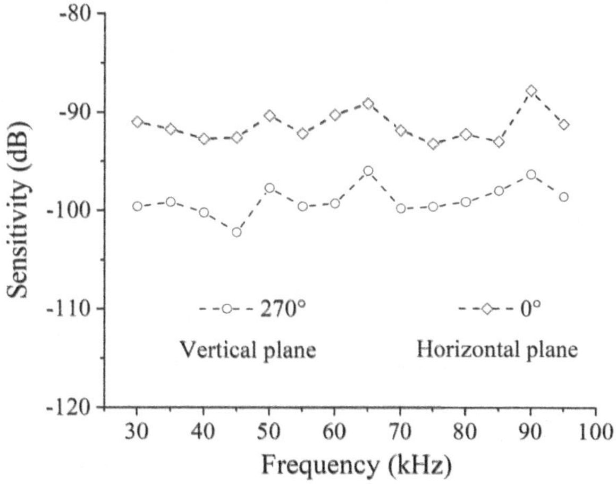


**Fig. 3** Sensitivity of spiral–shaped ultrasonic sensor in the range 30–95 kHz (10 V/Pa as 0 dB scale) at 0° on the horizontal plane, and at 270° on the vertical plane

# References

1. Fiorillo AS, Lamonaca F, Pullano SA (2009) Piezo–polymer based sonar for mobile robots. In: IEEE international workshop on intelligent data acquisition and advanced computing systems: technology and applications, Rende, pp 511–514
2. Steckel J (2015) Sonar system combining an emitter array with a sparse receiver array for air–coupled applications. IEEE Sens J 15:3446–3452
3. Xia W, Jin X, Dou F (2016) Thinned array design with minimum number of transducers for multibeam imaging sonar. IEEE J Oceanic Eng 42:892–900
4. Pullano SA, Fiorillo AS, Vanello N, Landini L (2016) Obstacle detection system based on low quality factor ultrasonic transducers for medical devices. In: IEEE International symposium on medical measurements and applications (MeMeA), Benevento, pp 1–4
5. Corcoran A, Moss CF (2017) Sensing in a noisy world: lessons from auditory specialists, echolocating bats. J Exp Biol 220:4554–4566
6. Fiorillo AS, D'Angelo G (2002) Echo signals processing with neural network in bat-like sonars based on PVDF. In: Proceedings of ultrasonics symposium, Munich, Germany
7. Suga N (2018) Specialization of the auditory system for the processing of bio-sonar information in the frequency domain: mustached bats. Hear Res 361:1–22
8. Heller G, Helversen O (1989) Resource partitioning of sonar frequency bands in rhinolophoid bats. Oecologia 80:178–186
9. Pullano SA, Fiorillo AS (2010) Ferroelectric polymer for bio-sonar replica. In: Lallart Mickael (ed) Ferroelectrics—applications. Rijeka, Intech
10. Chimenti DE (2014) Review of air-coupled ultrasonic materials characterization. Ultrasonics 54:1804–1816
11. Fiorillo AS, Grimaldi D, Paolino D, Pullano SA (2012) Low-frequency ultrasound in medicine: an in vivo evaluation. IEEE Trans Instrum Meas 61:1658–1663
12. Khan H, Ramzmjou A, Warkiani ME, Kottapalli A, Asadnia M (2018) Sensitive and flexible polymeric strain sensor for accurate human motion monitoring. Sensors 18:1–10
13. Fiorillo AS (1996) Ultrasound transducer with low synthetic quality factor. Appl Phys Lett 68:164–166
14. Fiorillo AS, Pullano SA, Bianco MG, Critello CD (2019) Bioinspired US sensor for broadband applications. Sens Actuators A 294:148–153

# Double Approach to Study VOC Composition in Biofluids of Young Men Living in the "Land of Fires" in Campania Region

**Valentina Longo, Angiola Forleo, Alessandra Ferramosca, P. Siciliano, S. Capone, Sebastiana Pappalardo and Luigi Montano**

**Abstract** Air pollution is involved in many pathologies. The pollutants act through several mechanisms that can affect numerous physiological functions, including reproduction. In this work we used a double approach to analyse the Volatile Organic Compound (VOC) component in 4 different biosamples (whole blood, urine, human semen and hairs) of young men (about 18 years old) living in the sadly famous area called "Land of fires" in Campania region (South-Western Italy), known to contain the largest illegal waste dump in Europe, in order to explore hypothesized correlation between VOCs in biofluid and man infertility due to environmental pollution. The double approach is based on two different technologies: (a) SPME-GC/MS (Solid Phase Microextraction—Gas Chromatography and Mass Spectrometry) system and (b) array of gas sensors. The preliminary data seems to be compatible with the EcoFoodFertility Project hypothesis that human semen is a good biological fluid to study volatile component as bioaccumulator of both exogenous and endogenous VOCs.

**Keywords** Air pollution · Volatile Organic Compounds (VOCs) · Male infertility · GC MS · Gas sensors

---

V. Longo (✉) · A. Forleo · P. Siciliano · S. Capone
Institute for Microelectronics and Microsystems, National Research Council (CNR-IMM), pal. A3, Campus Ecotekne, Str. Prov. Lecce-Monteroni km 1.2, 73100 Lecce, Italy
e-mail: valentina.longo@le.imm.cnr.it

A. Ferramosca
Department of Environmental and Biological Sciences and Technologies, University of Salento, Lecce, Italy

S. Pappalardo
Reproduction and Fertility Center, Rome, Italy

L. Montano
Andrology Unit of the "S.Francesco d'Assisi" Hospital, Local Health Authority (ASL) Salerno, EcoFoodFertility Project Coordination Unit, Via M. Clemente, Oliveto Citra, SA, Italy



G. Di Francia et al. (eds.), *Sensors and Microsystems*, Lecture Notes in Electrical Engineering 629, https://doi.org/10.1007/978-3-030-37558-4_16

## 1 Introduction

Air pollution is a major problem of recent decades, which has a serious toxicological impact on human health and environment.

It has a number of different emission sources, but motor vehicles and industrial processes contribute the most to air pollution [1]. Long-term effects of air pollution on the onset of diseases such as respiratory infections and inflammations, cardiovascular dysfunctions, and cancer is widely accepted [2]; hence, air pollution is linked with millions of death globally each year [3]. The Land of Fires gets its name from the high number of fires that break out in the area due to uncontrolled dumping. The triangle of death is an area in the Province of Naples, Campania, Italy, that contains the largest illegal waste dump in Europe due to a waste management crisis. The region has recently experienced increased deaths caused by cancer and other diseases that exceeds the italian national average. The rise in cancer-related mortality is linked to exposure of pollution from illegal waste.

A recent study has revealed the association between male infertility and air pollution [4]. Multicentre EcoFoodFertility project (http://ecofoodfertility.it) is working on such research field; it's specifically a biomonitoring study with multi-disciplinary approaches involving life-style and diet, which has built a nationwide network of actions in several highly polluted living areas of Italy [5]. Moreover, despite improvements in both diagnostic assessment and treatment of infertile couples, management of male-factor infertility remains a challenge, largely because of two factors: (a) the lack of a rapid, noninvasive test to evaluate human semen quality and investigate the cause of male infertility, and (b) the inability to predict gamete quality and embryo viability, which in turn lead to low success rates and a high incidence of multiple births after in vitro fertilization (IVF).

To overcome these limitations, a new science, known as Metabolomics, has been conceived with an expectation that body-fluid analysis can be optimized to create a low-cost, informative and medically relevant methodology for measuring metabolic changes, even when standard clinical chemistry markers are within normal limits [4].

An unexplored subgroup of the entire semen metabolome as well as in other biofluids and in exhaled is its volatile component, known as Volatilome. Volatilome includes all the Volatile Organic Compounds (VOCs) arising from cellular metabolism (endogenous VOCs) and from many different natural and anthropogenic sources with which humans are in continuous contact (exogenous VOCs).

In this work we used a double approach to analyse the VOC component in 4 different biosamples (whole blood, urine, human semen and hairs) of young men (about 18 years old) living in the sadly famous area called "Land of fires" in Campania region (South-Western Italy). The double approach is based on two different technologies: (a) Solid Phase Microextraction—Gas Chromatography and Mass Spectrometry (SPME-GC/MS) system and (b) gas sensor array.

## 2 Methods and Results

The VOCs of different biofluids were extracted by Solid-Phase Microextraction (SPME) technique; SPME was carried out by a Carboxen®/Polydimethylsiloxane (CAR/PDMS) fiber (cod. 57.318, Supelco) which was exposed to each human sample headspace overnight. The gas chromatography/mass spectrometry (GC/MS) analysis of the extracted volatiles was performed using a GC (6890 N series, Agilent Technologies) coupled to a MS (5973 series, Agilent Technologies) equipped with a ZB-624 capillary column (Phenomenex) with the injector temperature set at 250 °C to allow thermally desorption of VOCs. The carrier gas was high purity helium with a flow of 1 ml/min. GC/MS analysis was carried out in full-scan mode with a scan range 30–500 amu at 3.2 scans/s. Chromatograms were analysed by Enhanced Data Analysis software and the identification of the volatile compounds was achieved by comparing mass spectra with those of the data system library (NIST14, $p > 80\%$). Experiments and physiological data, were elaborated by multivariate data analysis techniques using web-based tools available with open access server MetaboAnalyst (version 4) to perform partial least squares-discriminant analysis (PLS-DA).

Each sample (1 ml whole blood, 1 ml urine, 250 μL human semen, some hairs in 5 ml vials with SPME septum) underwent an equilibration phase of about 18 h (overnight), under stirring in a bath at 60 °C. Subsequently, SPME fibers were exposed for 15 min at room temperature (RT) in the sample headspace and the absorbed VOCs have been separated and identified by SPME-GC/MS. As examples, the chromatograms of all the biological samples from a male volunteer recruited within the EcoFoodFertility project are reported in Fig. 1. The chromatograms, related to the extracted VOCs in the human semen headspace, were richer in components than the chromatograms of the other biosamples (urine, blood and hairs).

The second approach is based on a 4-sensor array based on micromachined gas sensors using metal-oxide (MOX) sensing elements operating at a temperature of 250 °C. The sensor responses towards the volatile compounds of the different samples were acquired by applying a constant voltage of 1 V across the electrodes and measuring the electrical current by an electrometer Keithley mod. 6517A equipped with an internal multiplexer module (Keithley mod. 6521).

Concerning the experimental setup for the measurements: the baseline was acquired in dry air in a continuous total flow ($F_{tot}$) of 25 sccm, whereas, for the measurement, the sample headspace, was stripped by means of a deviation of $F_{tot}$ into the vial, kept at RT for 4 min, into a sensor chamber. All fluxes were controlled by mass flow controllers (MFCs) and a multichannel mass flow programmer (MKS mod. 647B). A PC by a LabView software controlled all the gas-mixing protocol and the sensor signal acquiring [6]. The gas-sensing response of the device is defined by the ratio $R_{air}/R_{analyte}$, where $R_{analyte}$ and $R_{air}$ denoted the measured resistance in the presence of the VOCs and in dry air carrier, respectively. In Fig. 2, sensor responses to VOC components of four biosamples of the same man were reported.

The response to hair sample has a different trend, since it's a drier matrix and consumes standard solvent, step by step, at single cycle. Despite we had to use only

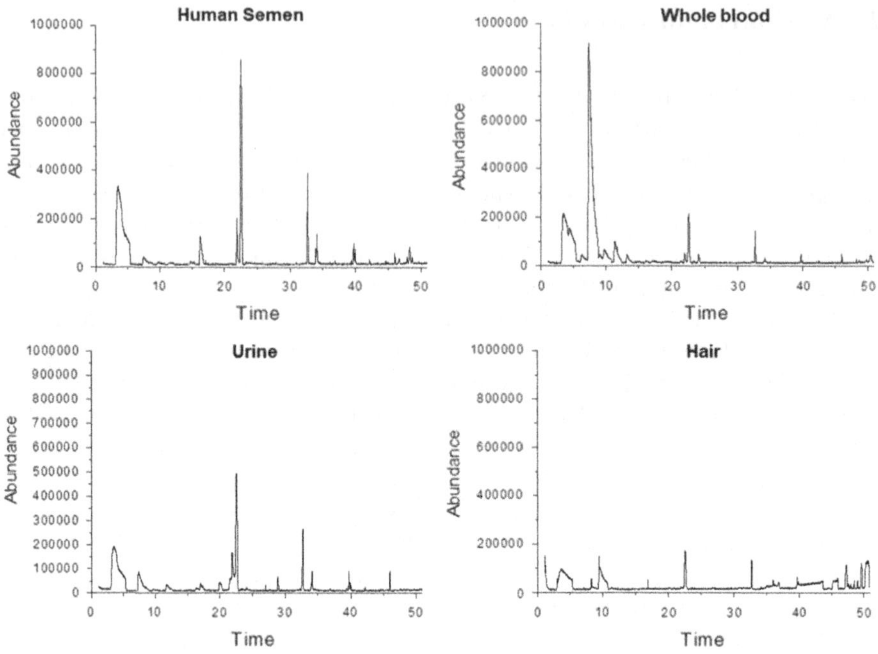


**Fig. 1** Chromatograms of human semen, whole blood, urine and hair sample of a male volunteer recruited within the EcoFoodFertility project

250 μL of human semen compared with 1 mL of blood and urine, it was evident that VOC composition in seminal plasma is much more abundant.

PLS-DA analysis based on 4 groups model showed a good classification between human semen and the other three groups. Also in a 2 groups model the human semen samples was discriminated from each of the other groups of biosamples (hair, urine and whole blood) (Fig. 3).

Also Dendrogram (Fig. 4) which is visualized through Heatmap, demonstrated that the highest and different responses of human semen samples allow to correctly discriminate these samples, while mostly urine and blood ones are completely mixed.

## 3 Conclusion

These preliminary data confirm that human semen is a good biological fluid to study volatile component and, in particular, it can represent a sentinel of fertility alteration due to influence of many external factors reflecting the exposome (diet, environment,

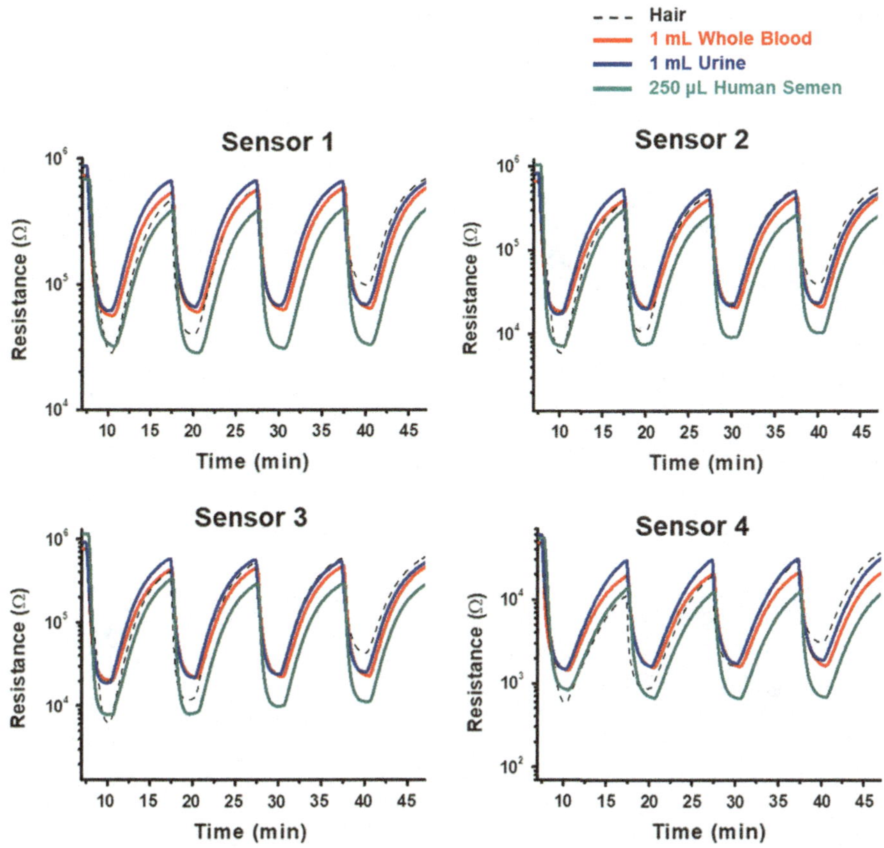


**Fig. 2** Resistance response to the four different analysed biosamples from the same subject for each sensor of the array

life style, tobacco smoke and alcohol consumption, etc.), as highlighted in our previous work [7]. In the framework of EcoFoodFertility project to which this work contribute, it's emerging the hypothesis that human semen is a bioaccumulator of both exogenous and endogenous VOCs.

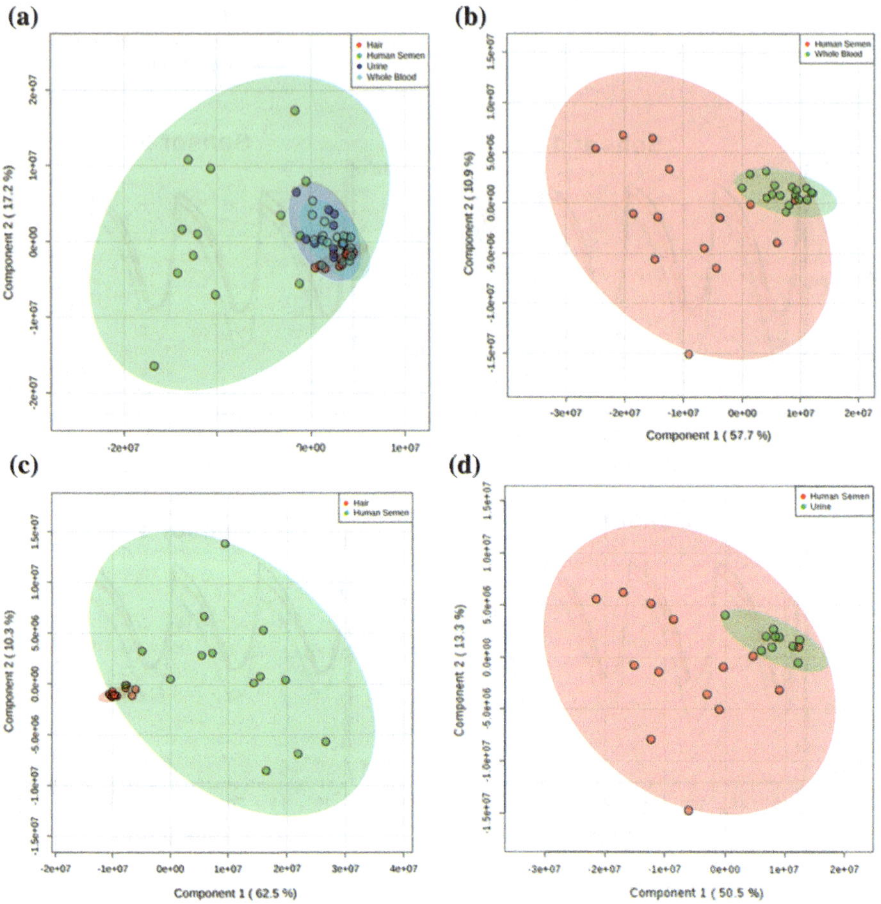


**Fig. 3** PLS-DA analysis of different biological fluids: all biofluid analysis (**a**); human semen versus whole blood (**b**); human semen versus hair (**c**) and human semen versus urine (**d**)

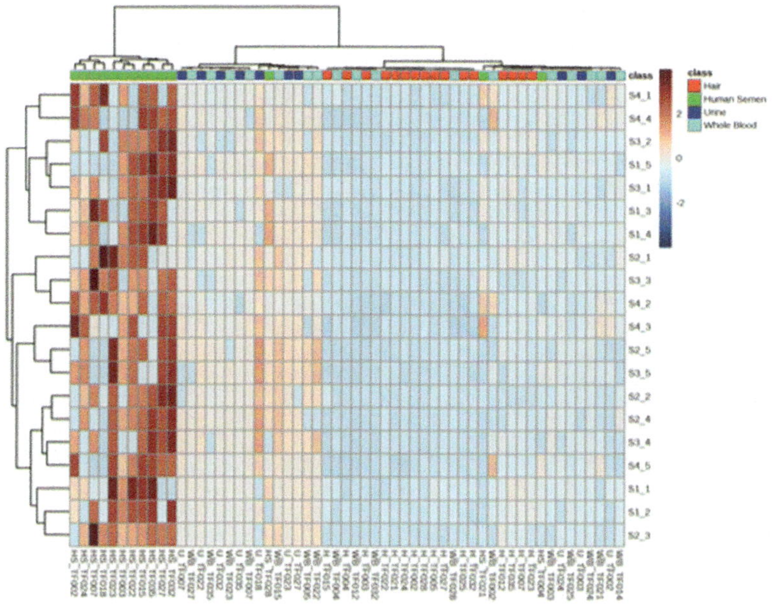


**Fig. 4** Heat map showing differences in the volatile profiles of four biofluids. Dendrogram is also depicted

## References


1. Biggeri A, Bellini P, Terracini B (2004) Epidemiol Prev 28(4–5 Suppl):4–100
2. Habre R, Coull B, Moshier E, Godbold J, Grunin A, Nath A et al (2014) J Expo Sci Environ Epidemiol 24:269–278
3. Carré J, Gatimel N, Moreau J, Parinaud J, Léandri R (2017) Environ Health 16:82
4. Deepinder F, Chowdary HT, Agarwal A (2007) Expert Rev Mol Diagn 7(4):351–358
5. Montano L, Iannuzzi L, Rubes J, Avolio C, Pistos C, Gatti A, Notari T, Raimondo S (2014) Andrology 2:69
6. Forleo A, Taurino AM, Capone S, Epifani M, Francioso L, Spadavecchia J, Siciliano P (2006) Sensor Lett 4:229–234
7. Longo V, Forleo A, Pinto Provenzano S, Coppola L, Zara V, Ferramosca A, Siciliano P, Capone S (2018) Biomed Phys Eng Express 5:015006

# Joule Heating Induced Stabilization of a Resistive Temperature Sensor Based on a Syndiotactic Polystyrene/MWCNT Composite

**Heinz-Christoph Neitzert, Giovanni Landi and Andrea Sorrentino**

**Abstract** Nanocomposite NTC resistor type temperature sensors have been realized by mixing a high concentration of multi-walled carbon nanotubes with Syndiotactic Polystyrene. The initially slightly unstable electrical characteristics have been stabilized by applying gradually higher voltages but limiting the maximum current. After this short burn-in procedure, linear and stable current-voltage and temperature-current characteristics have been obtained and the value of the temperature coefficient could be notably increased.

**Keywords** Temperature sensor · Syndiotactic polystyrene · Carbon nanotubes · Joule heating · Burn-in

## 1 Introduction

The temperature dependence of the electrical resistivity of nanocomposites with conducting fillers in an isolating organic matrix depends on a large variety of parameters. Besides the intrinsic temperature dependent resistivity of the nanofillers, also the interconnection mechanisms [1] and the external electrical contacts are of importance [2]. It has been shown that excellent electrical heaters and temperature sensors with a large operating temperature range and very good stability could be realized using epoxy/CNT composites [3]. In the case of this composites, however, the sample humidity as another parameter modifies also the electrical conductivity. This latter effect can also be used for humidity sensing [4] but is an undesired feature for pure temperature sensors. Syndiotactic Polystyrene (sPS) could be an interesting choice as matrix material for nanocomposite temperature sensors because this thermoplastic material is less susceptible to humidity. Polystyrene/carbon nanotube composites

H.-C. Neitzert (✉) · G. Landi
Dipartimento di Ingegneria Industriale, Università degli Studi di Salerno, Via Giovanni Paolo II 132, 84084 Fisciano, SA, Italy
e-mail: neitzert@unisa.it

H.-C. Neitzert · G. Landi · A. Sorrentino
Institute for Polymers, Composites and Biomaterials (IPCB-CNR), Via Previati 1/C, 23900 Lecco, Italy

 
G. Di Francia et al. (eds.), *Sensors and Microsystems*, Lecture Notes in Electrical Engineering 629, https://doi.org/10.1007/978-3-030-37558-4_17

have been investigated mainly as strain sensors [5] and there is not much information available regarding their application as temperature sensors.

Long-term stability of resistor based sensors is strongly influenced by the stability of the electrical contacts. It has been shown in literature, that the electrical resistivity between a contact metal and carbon nanotubes can be substantially lowered by Joule heating, applying relatively high current pulses with short duration [6, 7]. Here we will show, that also self-heating during a controlled voltage-ramp with current-limitation can be used to improve the electrical characteristics and the temperature performance of a polystyrene/multi-walled carbon nanotube-based temperature sensor.

## 2 Sample Preparation and Experimental

### 2.1 *Sample Preparation*

Composites with polystyrene containing different concentrations of commercial multi-walled carbon nanotubes (MWCNTs) have been realized. Syndiotactic Polystyrene (Questra QA101) was supplied by the Dow Chemical Company. The molecular weight characteristics were: Mw = 320,000 g/mol and Mw/Mn = 3.9, whereas a concentration of syndiotactic triads was greater than 98%.

The carbon nanotubes were obtained from Nanocyl S.A., synthesized by catalytic carbon vapor deposition (ccvd) process. They are multi-walled carbon nanotubes characterized by an average diameter of about 10 nm and the length ranging from 0.1 to 10 μm. The nanotubes were purified to produce a carbon purity >95% and a metal oxide impurity <5% as it results by thermogravimetric analysis (TGA). Neat sPS sample and the composite material containing different concentrations of multi-walled carbon nanotubes (sPS-MWCNT) were prepared using a laboratory-scale conical twin-screw extruder "ThermoHaake Micro-Compounder". Polymers and carbon nanotubes were pre-mixed manually and then loaded into the compounder. All batches, weighing 6 g each, were processed at 295 °C with a screw speed of 100 rpm for 15 min. The samples were then extruded and cooled down to room temperature.

The extruded samples were molded in a hot press (Carver Inc.) at 310 °C, forming 100 mm thick films, which were rapidly quenched in a bath with water at 0 °C. Samples with 1, 3, 6 and 9 wt% MWCNTs have been prepared. The sample with the highest CNT content and hence also the highest value of the electrical conductivity had been investigated in the present study. The crystallinity of the syndiotactic polystyrene, loaded with 9 wt% CNTs had a value of 39%, which is a rather high value when compared to the same polymer material without CNT addition with a crystalline fraction of less than 5%. Lateral electrical contacts have been finally realized by evaporation of gold films.

In Fig. 1a a Scanning Electron_Microscopy (SEM) image of the material prepared with a 9 wt% content of CNTs is shown. Rather homogeneous distribution of

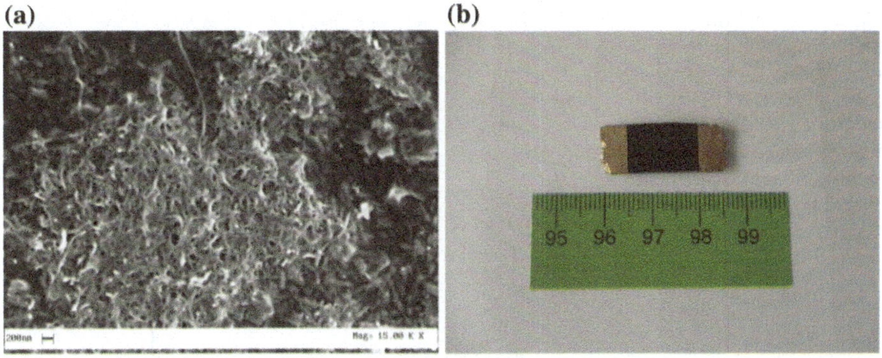


**Fig. 1** **a** SEM image of the syndiotactic polystyrene/MWCNT (9 wt%) composite, **b** photo of the composite sample with coplanar gold electrodes and scale (in cm)

carbon nanotubes within the network can be observed. The photograph of the final temperature sensor with the lateral evaporated gold contacts is shown in Fig. 1b. The contact distance is 14.0 mm, the sample width 11.6 mm and the sample thickness about 1.3 mm.

### *2.2 Measurement Setup for Electrical and Thermal Measurements*

The electrical measurements of Epoxy/MWCNT composites, reported in this paper, have all been performed in a 2-point contact geometry using a Keithley model “2400” Source-Measurement-Unit. The temperature dependent measurements have been done using a HERAEUS type “T5042EK” oven. The oven temperature has been monitored with a small size thermocouple and a FLUKE model “45” Multimeter.

## 3 Results and Discussion

It has been previously shown, that the chosen contact metal has a strong influence on the properties of the nanocomposite electrical characteristics. In the case of epoxy-based temperature sensors, for example, it has been demonstrated, that gold contacts formed a perfect ohmic contact without the need of further annealing procedures [2]. In our case for the syndiotactic polystyrene loaded with the same type of multi-walled carbon nanotubes, however, the as-prepared samples had a rather noisy electrical characteristics with a non-perfectly linear current-voltage dependence.

In order to improve the electrical behavior, a voltage ramp with stepwise increasing voltages from 0 to 30 V (100 mV steps) has been applied and the maximum supplied

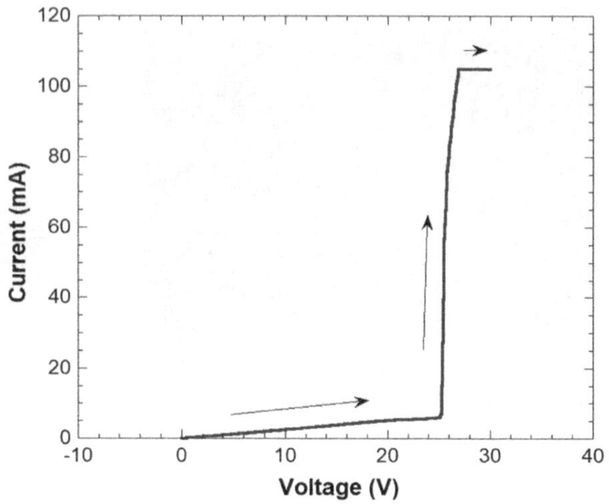


**Fig. 2** Current-voltage characteristic of the syndiotactic polystyrene/MWCNT sample during the Joule heating induced modification. Measurement direction indicated by arrows

current has been set to a value of 100 mA. In this way, only a maximum electrical power of about 3 W had to be dissipated in the device during the step-stress test. In Fig. 2 the results of this step-stress test are shown. Up to a voltage of about 20 V an almost linearly with voltage increasing current is observed, that tends to saturate in the subsequent voltage range up to a value of 25.7 V. For voltages above this threshold value the current is rapidly increasing for more than one order of magnitude and finally at about 26.2 V saturating due to the imposed current limit at a value of about 105 mA.

Comparing the current-voltage characteristics of the sample before and after the step-stress test (see Fig. 3) it can be observed that the sample conductivity in the measured voltage range of ±10 V increased by more than a factor of 18 due to the Joule-heating. The linear fit parameters of both curves can be seen in the insets. The sample resistance value decreased during this burn-in process from a value of 3914 Ω to a value of 208 Ω and the characteristics became almost perfectly linear.

In the following, we show a comparison of the temperature sensing capability of the sample, by monitoring the current for a given voltage during multiple temperature cycles between 30 and 55 °C. The cooling periods—induced by just switching off the regulated oven—were rather slow, so that it was ensured that the temperatures of the sample and of the thermocouple, used to measure the oven temperature had the same values. The results are shown in Fig. 4, where oven temperatures and resulting sample currents are plotted as a function of time. In the case of the 5 temperature cycles done before the step-stress test, an applied voltage of 5 V has been used. Clearly, an NTC-type behavior is observed, but a rather noisy and not very reproducible characteristic is observed. On the other hand, in the case of the 4 cycles done after the burn-in procedure and with an applied voltage of 1 V, a very stable current monitoring trace with a noise-free NTC-behavior has been found.

It should be noted, that the different applied voltages values during temperature cycling have been chosen, in order to maintain an almost constant and small value

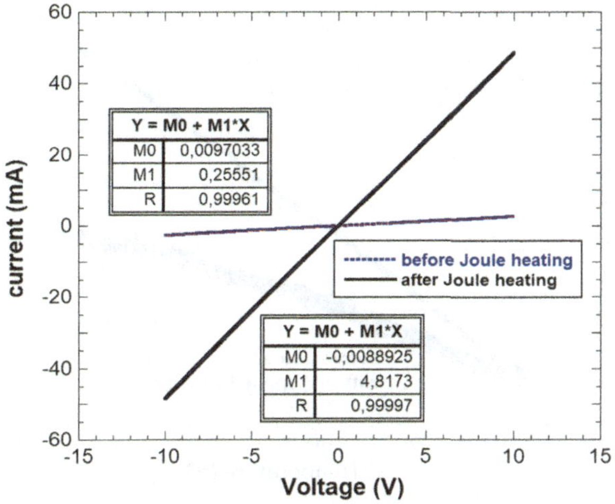


**Fig. 3** Current-voltage characteristics of the syndiotactic polystyrene/MWCNT based temperature sensor before (blue trace) and after (black trace) the Joule heating induced modification with linear fitting parameters (insets)

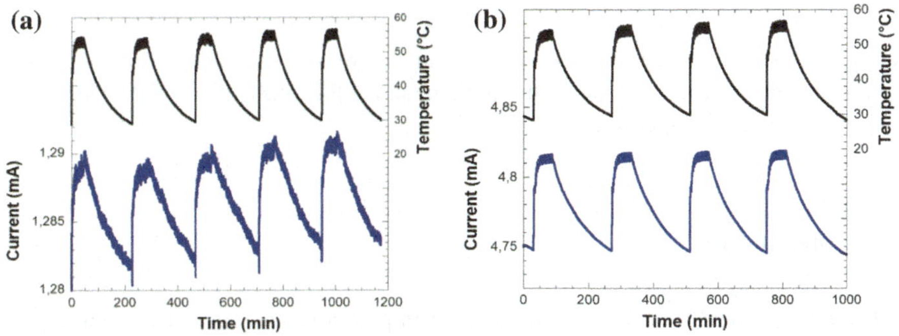


**Fig. 4** Monitoring of the temperatures (upper traces) and of the currents (lower traces) of the nanocomposite sensor, **a** before burn-in with an applied voltage of 5 V and **b** after burn-in with an applied voltage of 1 V

of the electrical power dissipation in the sensors before and after Joule-heating and therefore avoid self-heating during the operation as sensor.

In order to compare the temperature-current characteristics before and after Joule-heating directly, the electrical conductivity of the sample, normalized to its value at 30 °C, has been plotted as a function of the oven temperature just for each last cooling cycle (see Fig. 5). Again a noise-free and almost linear relation is found for the characteristics, measured after Joule-heating as compared to noisy and non-linear characteristics before Joule-heating. It can also be seen, that the sensitivity of the

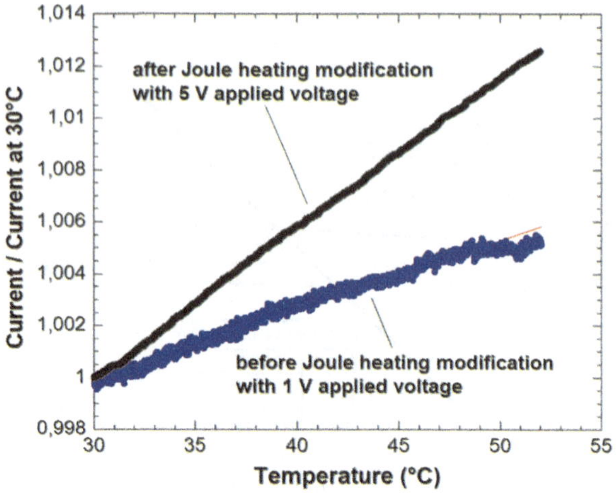


**Fig. 5** Currents of the nanocomposite sensor, measured for a fixed applied voltage (as indicated), normalized to the current at 30 °C as a function of temperature, as measured during the last cooling period during the temperature cycles, shown in Fig. 4a before and shown in Fig. 4b after the burn-in procedure. The lower blue trace is referred to the measurements before and the upper black trace to the measurements after the burn-in procedure

sample resistance to the ambient temperature increased notably due to the Joule-heating procedure, which is a further advantage for the application of the sample as a simple resistive temperature sensor.

## 4 Conclusions

It has been shown, how with a simple burn-in procedure the electrical characteristics of a syndiotactic polystyrene sample, loaded with 9 wt% of multi-walled carbon nanotubes can be modified due to the Joule-heating. In particular, the current-voltage characteristics could be linearized and the sample conductivity increased by more than a factor of 18. Furthermore also the temperature-conductivity characteristics in the temperature range 30–55 °C could be linearized and stabilized and the NTC coefficient increased by about a factor of 3. In this way, the nanocomposite sample with coplanar gold contacts became a linear, very stable and reliable resistive temperature sensor.

**Acknowledgements** Authors acknowledge financial support from the "VINMAC" project, ID [139455], D. R. Lombardia n. 9559 del 02/08/2017, CUP: E67H16000980009.

## References


1. Barone C, Pagano S, Neitzert H-C (2011) Transport and noise spectroscopy of MWCNT/HDPE composites with different nanotube. J Appl Phys 110:113716
2. Neitzert H-C, Landi G (2015) Influence of the contact metallization on the characteristics of resistive temperature sensors based on EPOXY/MWCNT composites (Chap 58). In: Sensors, Proceedings of second national conference on sensors, Rome (Italy) 19–21 Feb 2014. Series: Lecture Notes in Electrical Engineering, vol 319. Springer, Berlin, pp 333–337
3. Neitzert H-C, Vertuccio L, Sorrentino A (2011) Epoxy/MWCNT composite as temperature sensor and electrical heating elements. IEEE Trans Nanotechnol 9:688–693
4. Neitzert H-C, Sorrentino A, Vertuccio L (2013) Humidity sensing of an epoxy/MWCNT composite by electrical conductivity measurements. In: SPIE proceedings 8766 (6yy2013) 87660D
5. Srivastava RK, Vemuru VSM, Zeng Y, Vajtai R, Nagarajaiah S, Ajayan PM, Srivastava A (2011) The strain sensing and thermal-mechanical behavior of flexible multi-walled carbon nanotube/polystyrene composite films. Carbon 49:3928–3936
6. Dong L, Youkey S, Bush J, Jiao J, Dubin VM, Chebiam RV (2007) Effects of local Joule heating on the reduction of contact resistance between carbon nanotubes and metal electrodes. J Appl Phys 101:024300
7. Woo Y, Duesberg GS, Roth S (2007) Reduced contact resistance between an individual single-walled carbon nanotube and a metal electrode by a local point annealing. Nanotechnology 18:095203

# Taming the Techno Leviathan: Why We Should Adopt a Society-in-the-Loop Model Inside IoT Utilities

**Alfredo Adamo**

**Abstract** While the Internet of Things (IoT) has made significant progress along the lines of supporting individual applications, it is only recently that the importance of people as an integral component of the overall IoT infrastructure has started to be fully recognized. Several powerful concepts have emerged to facilitate this vision, whether involving the human context whenever required or directly impacting user behavior and decisions. As these become the stepping stones to develop the IoT into a people-centric utility, this paper outlines how to include the "Society-in-the-loop" approach to govern a lot of ethical, moral concerns. Rapid advances in IoT, Artificial Intelligence and Machine Learning have raised many questions about the regulatory and governance mechanisms for autonomous sensors, machines and infrastructures. We discuss about, in the context of the IoT utility, a lot of concerns raised about algorithms governing our lives, and how the adoption of "Society-in-the-loop" paradigm could be a solution.

**Keywords** IoT (Internet of Things) · Artificial intelligence · Machine learning · HITL (human in the loop) · SITL (society in the loop) · Humanities

## 1 Introduction

In our modern society, science and humanism are often considered two distinguished, often perceived as opposite, entities. The figure of the scientist is regarded as a rational individual, which pursues the research of the truth through experiments and factual plans; while the humanist is a creative individual, with compelling arguments at best, but whose research of truth falls merely in an epistemological discourse.

However, this hasn't always been true. Before the introduction of the scientific method, science and philosophy walked hand in hand in the academic universe. The Italian renaissance, particularly, witnessed the birth of polyvalent intellectuals, who mastered both the scientific, mathematic side of arts and the creative, theoretical facet of natural sciences. As instance, Galileo Galilei was indeed one of the most

A. Adamo (✉)
Alan Advantage, Rome, Italy
e-mail: alfredo.adamo@alanadvantage.com



G. Di Francia et al. (eds.), *Sensors and Microsystems*, Lecture Notes in Electrical Engineering 629, https://doi.org/10.1007/978-3-030-37558-4_18

relevant men of science, but he also contributed to the European philosophic scene with his *The Assayer*, where he poetically states that mathematics is the language with which the entire Universe is coded and created. Leonardo Da Vinci created masterpieces as the Gioconda and The Last Supper, but he was also a pioneer of revolutionary engineering breakthroughs as the helicopter, the tank, or the parachute. It is obvious how technology (although understood in a different way) and humanism were not perceived as mutually independent. The recent, gargantuan developments in the various fields of technology could not help but exacerbate such rupture.

However, I argue that this is not just an intellectual, academic conundrum. The loss of humanism in the recent, techno-scientific progress has led to actual ethical dilemmas, caused by such increasing, uncontrolled social responsibility computers (one should only think of Facebook's data breach by Cambridge Analytica or the, for instance). Dismissing the relevance of the human component in something so influential as technology—and focusing only on the progress for its own sake—is distracting from the main reason technology exists, that is being of support and improvement to the human life.

Nonetheless, it is safe to say that, in recent years, the need to reintegrate the human component in the hi-tech universe has resurfaced in the academic world, with many prominent scholars sharing their publications on such an important matter. Borrowing the words of Luciano Floridi, professor of Philosophy and Ethics of Information at the University of Oxford, it is exactly now, in this age of "*ubiquitous computing*" [1: 43] the Information and Communication Technologies (ICTs) provide, that we should pursue the need to reintegrate humans, and society at large, in the technological progress.

In his *The Onlife Manifesto*, Floridi explores how ICTs' power to permeate every fiber of the social context, together with the newest Artificial Intelligence technologies (as well as apps, robots, and devices of different kinds) allowed the creation of a human-computer hybridization, changing humankind's vision. The result is the creation of a new social framework in which men, and society itself, are struggling to adjust its pre-existing norms, values, and behavioral codes [1: 43]. Such is the impact of ICTs on society that Floridi argues their advent can be seen as "a fourth revolution in our political anthropology". That is after the innovations which changed our understanding of the world, as the ones by Copernicus, Darwin, and Freud [2: 21]. Hence, humanity ended up in a new phase of the information age, in which it almost became an appendix, a stranger who needs to learn how to live in such a new, disruptive reality.

It is obvious, then, that there is a compelling necessity to make the human component central to technology again, to reunite the humanistic side of progress with the hard, scientific facet of the developmental process. Or better: there is the necessity *to reintegrate society in the loop of supervisory control*, to avoid many of the ethical, moral dilemmas we face in our age.

In the next paper I will argue how a wider supervision by society can be achieved by the so-called "Society-in-the-Loop" model (SITL) [3]. Briefly, it is an evolution of the pre-existing Human-in-the-loop system (HITL), which implied supervision from an individual being. SITL, however, does not stop at the individual supervision, but calls into action the wider social context, providing a more inclusive, democratic supervision, avoiding discriminatory algorithms as well.

Following the study of many scholars, my intention is to present a conceptual framework in which the adoption of a SITL system could provide a solution to our conundrum. A reconnection between humanism and technology is possible, and it should start with the inclusion of society in the technological, developmental process.

## 2 Modern IoT and Humans

Beyond legacy embedded systems with constrained applicability, the emerging IoT solutions are becoming more open and integrated by adaptively combining sensors and actuators with actionable intelligence for automatic monitoring and control. However, as a multitude of interconnected and intelligent machines communicate with each other and autonomously adapt to changing contexts without user involvement, the fact that present technology is made by humans and for humans is often overlooked. Indeed, modern IoT systems are still widely unaware of the human context and instead consider people to be an external and unpredictable element in their control loop, because of their unpredictable behavior both as user of the IoT scenarios and as a person who is present in the environment. Therefore, future IoT applications will need to intimately involve humans, so that people and machines could operate synergistically. To this end, human intentions, actions, psychological and physiological states, and even emotions could be detected, inferred through sensory data, and utilized as control feedback.

Many steps in the fields of supervisory control have been achieved in the last years, feeling the compelling necessity to regulate such complex, fast-developing world, but it is only recently, with the evolution of human-computer interaction, that scholars and theorizers feel the necessity of a more relevant human presence in the AI learning processes. An important milestone in the field has been the conceptualization of the Human-in-the-Loop model, as we saw in the introduction to this paper above. Taking part in the loop of reciprocal learning, the human supervisor takes active part in AI Machine Learning process, providing not only a better performance, but also controlling possible computer misbehavior and serving as a legally accountable subject, minimizing the probability of misbehavior at the expense of third parties [3: 7] (Fig. 1).[1]

However, being the essence of progress an unrestrained run towards the future, computers and bots are having more and more computing power and influencing

[1] As fictional as it could seem, however, the European Union, in 2016, has approved a motion aimed at the creation of a body of civil laws to make computers and bots legally accountable for their "actions" [4]. The motion also provides a framework for the establishment of a European agency for the regulation of hi-tech production and several, ethical considerations on the use of machines.

The documents the EU voted upon can be found at:

http://www.europarl.europa.eu/sides/getDoc.do?pubRef=-//EP//TEXT+REPORT+A8-2017-0005+0+DOC+XML+V0//EN#title10.

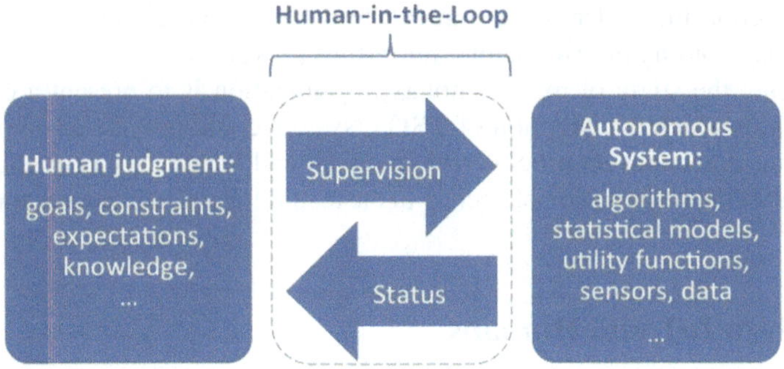


**Fig. 1** Human in the loop

forces on our everyday life. Machines are entrusted with tasks increasingly affecting our actions and decisions, let alone the ways we connect to other humans.

Human behavior may be impacted in either space (for example, the users are encouraged to move to a less congested location) or time (for example, the users are convinced to reduce their current data demand in case the network is overloaded); this is known as the "user-in-the-loop" (UIL).

With UIL, often referred to as "layer 8", the space-time user traffic demand may be shaped opportunistically and better matched with the actual resource supply from the people-centric wireless system. While HITL involves the user whenever human participation is desired or required and UIL extends the user's role beyond a traffic-generating and traffic consuming black box, these trends must account for the fact that people are, in essence, walking sensor networks. Indeed, a wide diversity of user-owned companion devices, such as mobile phones, wearables, connected vehicles, and even drones may become an integral part of the IoT infrastructure.

Hence, they can augment a broad range of applications, in which human context is useful, including traffic planning, environmental monitoring, mobile social recommendation, and public safety, among others. Therefore, we envision that—in contrast to past concepts where the user only assists the network to receive better individual service—future user equipment will truly merge with the IoT architecture to form a deep-fused human–machine system that efficiently utilizes the complementary nature of human and machine intelligence.

## 3 Our Dilemma in a Nutshell

To start with, we should first ask ourselves why it is so important to (re)integrate humans in the supervisory loop. Deep down, isn't progress existing to improve human livelihood, to remove every day's hassles for users and consumers? Shouldn't we instead, through a classic, heuristic trial and error, arrive at the point in history where we could lie down, relax, and let machines wipe out the sweat of our historical fatigues?

Sure, it would be a great life thinking about it now. But, as already demonstrated, the world of hi-tech works in ways much similar to those of a hierarchical, political pyramid, where the top of the system rules and governs the ways algorithms are coded and implemented in our everyday life. That is to say: who decides and programs computer algorithms will end up having a massive influence on other people's lives, in the same fashion as autocratic politicians. Without a more involving, democratic regulation of hi-tech "politics" we could be going towards a "new feudal order", in which a restricted circle of people decides for the vast majority [5: 19].

Although the conceptualization of the HITL model has been a most relevant achievement in the solution of our dilemma, what happens when a computer is entrusted with a task which implies a broader social impact (as an algorithm which could influence mass political preferences or mediate resources and labor within a given country)? Can a single supervisor act on behalf of the entire social context?

It is for this new challenge that Iyad Rahwan, professor at the MIT Media Lab, has proposed an extension of the HITL model in order to integrate the wider social context into fundamental decisions and supervision of digital behavior: The Society-in-the-Loop system. With the integration of society in the loop we could meet the needs of a larger majority of the population, serving as a supervisor not only for the computer performance, but also for programmers and experts behind. In short, such a model would protect the rights of the various societal actors and allow them to enjoy a more sustainable and democratic use of AI and algorithms in their life (Fig. 2).

Nonetheless, if we accept the compelling necessity to integrate what Rousseau called the general will of the people, how do we define, then, what is best for *all* the social actors? How can we safely state that $x$ is better than $y$ in the full respect

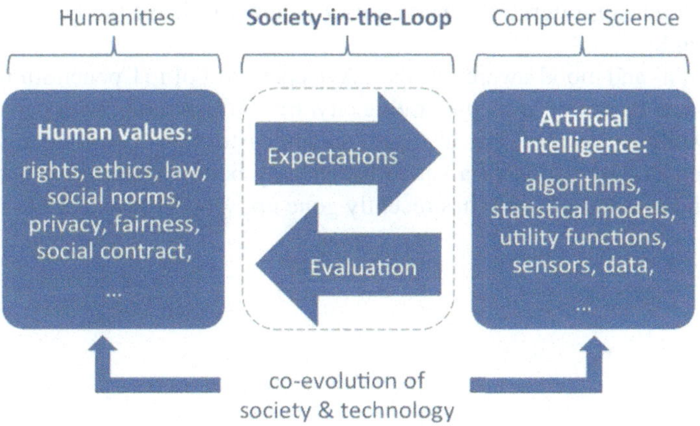


**Fig. 2** It is worth noting how while a HITL system involved individual judgment on the computer's performance, a SITL involves a further consideration on human values, expected to be implemented in the algorithmic applications

of everyone's human rights? Let's have a look to this equation, with which Rahwan summed its SITL model:

$$SITL = HITL + SocialContract$$

Having in mind what HITL and SITL stand for, the only remaining variable to define is exactly the "social glue" for the interests of every individual within the social context: The Social Contract.

## 4 Many Categories of People-Centric IoT Services Emerge and Are Expected to Be Deployed Over the Following Years

Intention- and mission-aware services. These services primarily reflect user's current intention or desire and assist by enabling, for example, situation-aware smart commuting for pedestrians, cyclists, and drivers of scooters, trucks, and other vehicles. This group of applications can help people in a variety of use cases, from highlighting the nearest available parking space on a vehicle's head-up display in urban areas to status reporting on a display or augmented reality (AR) glasses in challenging environments, such as mines, construction sites, etc.

Location- and context-aware services. Another group of services is formed by location- and context-aware applications, such as those communicating alerts from environmental sensors (for example, "put on/take off your mask" when entering/leaving a polluted area). Many more of these services are envisioned to be deployed in the coming years, such as identifying slippery floors and low ceilings, notifying about forgotten trash when a user is about to leave the house, and many other examples.

Condition- and mood-aware services. A deeper level of IoT penetration into people's lives can be achieved by integrating city/area infrastructure with personal medical and wellness devices. For instance, dietary restrictions could be applied on a menu when ordering food or a squad leader may be advised to give a break to a worker whose blood pressure has recently gone up.

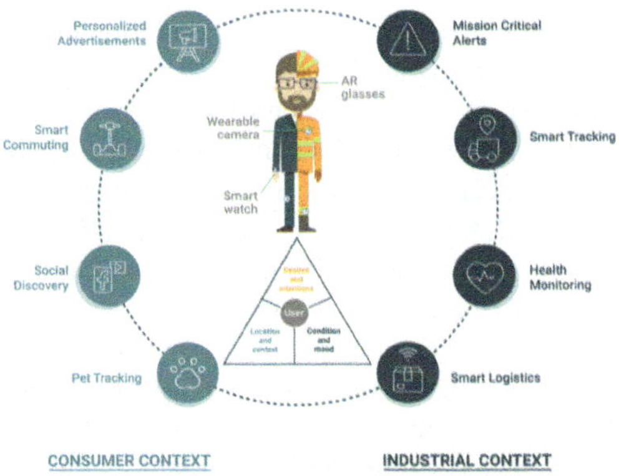


Summarizing the above examples of services, we note that depending on the environment the set of requirements and challenges to implement a particular application may vary considerably.

To further offer a challenges-based grouping, we propose to differentiate between two major contexts: consumer and industrial.

The former is characterized by the presence of numerous devices that are heterogeneous in terms of their communication means and ownership. Therefore, the major challenge in this context is to provide sufficient scalability of the deployed connectivity solution.

On the contrary, the latter context is more challenging in terms of maintaining communication reliability due to more difficult propagation environments. At the same time, the system operator has more control over device population in such areas.

We continue by addressing how people-centric IoT applications are to be engineered, that is, which radio technologies need to be employed in particular scenarios and how to ensure their suitability for the target operating conditions.

## 5 Definition Problems: How Do We Determine What's Best?

Imagine a country devastated by a civil war. The government splits in two factions, those loyal to the former power and those willing to get rid of the old, antiquated political order. Armed conflict is at the door and, eventually, the two factions meet on the battlefield. National confusion and political unrest lead to economic crisis, deaths, and destruction. Finally, the former ruler is beheaded and a new political system is established, changing the governmental landscape of the country. This was Great Britain in the 17th century, when Oliver Cromwell and the Parliament rebelled against the king and established the "republican" Commonwealth.

In hindsight, it is understandable why the Social Contract Theory made its first appearance in England, where the lack of a stable, central power before—and the sudden change in the political system after—the revolution inspired Thomas Hobbes to write his famous *Leviathan*. Briefly, the English philosopher suggested that that the political order (granted by the existence of a superior power, embodied in the state and the government) is the result of a compromise, a *contract*, men agreed upon to enjoy social order and security. Before such an agreement, men lived in a pre-political, pre-moral State of Nature, which he calls the *bellum omnia contra omnes*, the war of all against all. To escape the ferial stage men had to give up part of their freedom and accept a *Pact Unionis*, where they undertook to respect each other and live in harmony, and a *Pact Subjiectonis*, where they agreed upon the establishment of a superior, ruling power which could defend and guarantee social order.

The Social Contract Theory, then, was reexamined by other philosophers, as John Locke and Jean-Jacques Rousseau, who re-elaborated its theoretical foundations—the former moving toward a constitutional contract, where the ruling power is not uncontested; the latter beginning a democratic, fundamental discourse on people's *volonté générale*. However, although their arguments are of the utmost philosophical importance, in all its developments the Social Contract kept the fundamental characteristic of being an agreement all the individuals must agree upon to secure their wellbeing in a given social context.

But how does this pertain to our conundrum? Simple: *if we want to move towards a more involving, democratic use of AI and technology at large, we have to understand what can best serve the general will of all the social actors*, integrating the wider social context into the supervisory loop, as argued above. The SITL system extends decisional power to a larger group, involving two important features a HITL model does not:

(1) Better treatment of common-sense choices an AI cannot face nor process, since it has not the morale a human mind has (as choosing between efficiency and safety or favoring fair and just options although not mathematically perfect).
(2) Better understanding of the social costs and benefits in the implementation of a new innovation. For instance, whether a self-driving car (a much controversial technology, especially after a fatal crash in Arizona, where a pedestrian was killed in the collision) has to prioritize the safety of the driver or of the pedestrians it cannot be decided by the machine itself. It has to be programmed beforehand.

In brief, then, the "Social Contract" is what society agrees upon to deal with many of the problems affecting the individuals' wellbeing and security. As humankind needed a political order to preserve their selves in the pre-political State of Nature, we need, today, a protection for—and more involvement in—our digital selves.

Helbing and Pournaras [6], in an article on *Nature*, express thoroughly how a "top-down control has various flaws". First off, it can be subject to corruption and hacked by extremists and criminals. Second, it fails to address local needs, being bound to limitations in data-transmission rates and processing power. Third, forcibly intervening in individual choices undermines collective intelligence, a most important factor in the creation of a diverse, inclusive society. Fourth, "filter bubbles" consequent to personalized information cause people to be less exposed to other opinions, increasing polarization and conflict. The reduction in pluralism and diversity is very dangerous for ecosystems relying on interdependencies as society and economy. Finally, top-down control can alter people's decisions, disrupting everyday decision-making skills and undermining stability and order, rather than providing a more secure environment.

## 6 Easier Said Than Done

However, although such a model could be a milestone in the achievement of a more democratic hi-tech development, there are still many doubts on how to successfully carry it out. For instance, a fundamental problem is the disciplinary gap existing between machine programming and the legal, ethical values of social sciences. Although professionals and scholars in the various social and legal subjects are able to identify possible computer misbehaviors, it is not that simple to mathematically articulate how a machine should behave lest a computer inflict moral and ethical damages. Furthermore, the human-computer relation is characterized by a constant, reciprocal learning. Human users' IT skills are in constant evolution, thus changing what the wider social context deems acceptable.

Modern demography, moreover, does not help our efforts in implementing such a revolutionary solution, since we should be dealing with a huge amount of people, each with their own opinions and ideas. In this way, we should address every individuals' perception of what is "fair", "correct", and "acceptable", which is almost impossible or, at least, very time consuming. Hence, how could we reunite all people's awareness and have an arithmetic mean of all their feelings? How can we quantitatively define their desires and point of views on ethical matters?[2]

Rahwan [3: 11–12] tries to put forward some solutions to the conundrum, considering some of the solutions put forward by scholars in the field. One of these is the use of *crowdsourcing* techniques and tools to produce a database which could store the general preferences of a society.[3] Together with his colleagues at the Massachusetts Institute of Technology, he developed a public-facing survey tool exhorting the participants to answer ethical dilemmas as: "If this self-driving car is doomed to crash, is it better that it kills $x$ number of pedestrians (including a pregnant woman) or

[2]One should only think that *computational science* is still trying to give an answer to this, researching algorithms that could calculate the optimal results in trade-offs decisions involving several stakeholders. In our case this can be seen through the lenses of a Steinhaus' "fair cake-cutting" problem [7]: we have a number $N$ of stakeholders (all the individuals society is consisted of) and a good (algorithmic computer behavior) they are interested in (or, in this case, they demand for security and wellbeing). Thus, we have a good which should be *fairly divided* among all the stakeholders, without discrimination and in the full respect of everyone's opinion. Steinhaus expressed this conundrum in the form of a cake which should be divided among various gluttonous children. The problem arises in the moment that the cake has assorted toppings, meaning that the good offered is heterogeneous, complicating our research of an *optimal, fair division* among the children/stakeholders.

[3]Crowdsourcing as a solution to deal with trade-offs is also well explored by Conitzer et al. [8].

the passengers (including a family with children, for instance)?" [9]. The perk here is that collecting and analyzing their answers can tell us a lot about how the wider social context deals with such ethical, moral dilemmas.

Moreover, the possibility of having a human audit of algorithms and programs is explored in the paper, but Rahwan believes that such kind of supervision could benefit from automation, entrusting other algorithms with the task to audit computer behavior. This is the overall point we find in Etzioni and Etzioni when they state that: "To ensure proper conduct by AI instruments, people will need to employ other AI systems" [10: 155], foreseeing the use of "oversight programs" to achieve a successful automation of the auditing process.

From a technical-scientific point of view, the progress of GANs seems to indicate a viable path to have an automatic treatment at the base of the SITL-based approach. A generative adversarial network (GAN) is a class of machine learning systems. Two neural networks contest with each other in a zero-sum game framework. This technique can generate photographs that look at least superficially authentic to human observers, having many realistic characteristics. It is a form of unsupervised learning.

However, I find that employing algorithms or GANs to check for possible computer misbehavior is rather tautological. Is defying today's habit to put technology at the center not our main dilemma? The technochauvinist morale to zeroing in on technology for its own sake, excluding the importance of the human component, is what I am trying to point out in this paper: outsourcing the auditing responsibility to other machines would not help (re)integrating humans in the loop, but would confine them again at the borders of the loop, placing at the center of progress the mere will of unrestrained technological development.

I am not, by all means, trying to develop a luddite argument, in favor of the complete destruction of technology to achieve an Amish-like society. I do know that our society will be—if it is not already—data-driven and enmeshed more and more in the massive use of innovative technologies. However, as I already argued above, the use of technology for the sole mean to launch new innovations on the market is creating more problems than improving our lives.

What we need is not *only* the control of machines and algorithms (whether automated or by humans), but the control of those behind the machines, namely the entrepreneurs, the programmers, and the experts coding the algorithms affecting our lives and society. The supervisory loop needs to be extended to the use humans themselves make of our data and our digital selves, aside from the supervision of the pragmatic "actions" of human-programmed computers.

Fortunately, we might see some of these improvements. Back in 2013, Alex "Sandy" Pentland, entrepreneur and professor at the MIT, wrote that "to achieve a data-driven society, we need what I have called the New Deal on Data", where data would be seen as an asset and "individuals would have ownership rights in data that are about them". This means that an individual has full control over his/her personal data, including possession, crystal-clear terms of use, and the right to dispose of or distribute your data [11: 83]. Does this sound familiar? The General Data Protection Regulation (commonly known as GDPR) issued by the European Union in 2016 foresees exactly many of the points explored by Pentland. After the scandal of Cambridge Analytica breaching the data of Facebook's users, this has become even more relevant to the good use of Internet tools as socials and websites.

## 7 Conclusion

Computers and machines are miracles of technology. They have had the power to reshape our understanding of the world and of the relations we build with other humans. Socials have changed the way people know and perceive each other, aside from having changed the way politics in conveyed to—and conceived by—the general public. They are amazing tools for our needs and whims. But that's what they should remain: *a tool.*

In this paper I tried to put forward all the perks, as well as the *current* limitations, of a SITL model. I do not aim at concluding the debate with a finale, sole possible solution. Quite the contrary, I believe it is time to consider the possibility of such a system and to go back to a human-centered vision of technology. As I started this paper, with a title inspired by the disruptive paper by Iyad Rahwan, I would like to conclude quoting his words: "We spent centuries taming Hobbes's Leviathan, the all-powerful sovereign. We must now create and tame the new Techno-Leviathan."

## References


1. Floridi L (2015) Background document: rethinking public spaces in the digital transition: the Onlife Manifesto. In: Floridi L (ed) The Onlife Manifesto. Springer, Cham. https://doi.org/10.1007/978-3-319-04093-6_4
2. Floridi L (2015) Commentary on the Onlife Manifesto. In: Floridi L (ed) The Onlife Manifesto. Springer, Cham. https://doi.org/10.1007/978-3-319-04093-6_4
3. Rahwan I (2017) Society-in-the-loop: programming the algorithmic social contract. DOI, Ethics Inf Technol. https://doi.org/10.1007/s10676-017-9430-8
4. Delvaux M (2016) Motion for a European Parliament resolution: with recommendations to the commission on civil law rules on robotics. Technical Report (2015/2103(INL)), European Commission
5. Citron DK, Pasquale FA (2014) The scored society: due process for automated predictions. Wash Law Rev 89:1–33
6. Helbing D, Pournaras E (2015) Society: build digital democracy. Nature 527:33–34
7. Procaccia AD (2016) Cake cutting algorithms. In: Brandt et al (eds) Handbook of computational social choice. Cambridge University Press, New York, pp 311–329
8. Conitzer V, Brill M, Freeman R (2015) Crowdsourcing societal tradeoffs. In: Proceedings of the 2015 international conference on autonomous agents and multiagent systems. International Foundation for Autonomous Agents and Multiagent Systems, pp 1213–1217
9. MIT (2017) The moral machine. Retrieved 01 Jan 2017, from http://moralmachine.mit.edu
10. Etzioni A, Etzioni O (2016) AI assisted ethics. Ethics Inf Technol 18(2):149–156
11. Pentland A (2013) The data-driven society. Sci Am 309(4):78–83. https://doi.org/10.1038/scientificamerican1013-78

## 7 Conclusion

# Automatic Differential Capacitive Sensing by Means of Linear Interface

**G. Barile, G. Ferri, A. Depari, A. Flammini and E. Sisinni**

**Abstract** In this work we present the development of an integrated CMOS analog interface able to convert differential capacitive sensors variations into a DC voltage. The presented circuit is based on autobalancing bridge techniques improving its performances through the linearization of the input/output characteristic and the achievement of the full-range sensor variations capability. Comparison between theoretical and measured interface static behaviour results are reported.

**Keywords** Sensor interface · Differential capacitive sensors

## 1 Introduction

Differential capacitive sensors are used in several sensing systems such as accelerometers, gyroscopes pressure sensors, position sensors and force sensors [1–8]. In a differential configuration, capacitive sensing principles offer the effective reduction of common-mode noise and parasitic effects. For these reasons, this approach has become progressively popular due to its flexibility and suitability to perform simple and low-cost measurements [9–13]. A differential capacitive sensor can be represented as the series of two capacitors which, under the action of the measurand, change their value in a differential way [14–22]. The proposed interface is shown in Fig. 1: $C_1$ and $C_2$ form the differential capacitive sensor that, excited by a sinusoidal signal Vsin, occupies the left branch of the auto-balanced De Sauty bridge structure, whilst the right one is composed by two reference resistors. Variations in the position of the sensor movable plate alters the balancing condition, so that the differential amplifier error signal $\Delta V = V_a - V_b \neq 0$. The feedback loop aims to restore the balance condition $\Delta V = 0$ by acting on the modulator stage, designed around a multiplier fed by a reference sinusoidal voltage, Vax and a variable DC voltage,

G. Barile (✉) · G. Ferri
Department of Industrial and Information Engineering and Economics,
University of L'Aquila, 67100 L'Aquila, Italy
e-mail: gianluca.barile@univaq.it

A. Depari · A. Flammini · E. Sisinni
Department of Information Engineering, University of Brescia, 25123 Brescia, Italy

 
G. Di Francia et al. (eds.), *Sensors and Microsystems*, Lecture Notes
in Electrical Engineering 629, https://doi.org/10.1007/978-3-030-37558-4_19

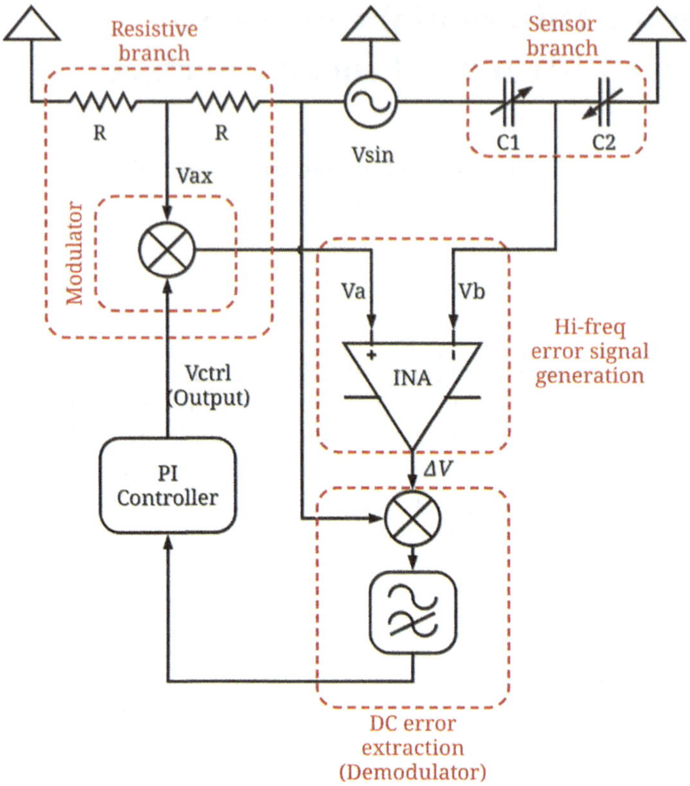


**Fig. 1** Block scheme of the proposed interface

$V_{ctrl}$, which is also the output signal of the interface, so the performed measurement consists of a capacitance to voltage conversion.

## 2 Interface Description and Test Results

In the proposed solution, the feedback circuitry performs the bridge balance by changing the first multiplier (*modulator*) output so to produce a signal $V_a$ which tends to follow $V_b$, hence forcing the error signal $\Delta V$ to zero. The feedback signal $V_{mult}$ is eventually produced by another multiplier (*demodulator*) and finally the useful information, that is the DC voltage $V_{ctrl}$, is extracted by a second order Sallen-Key filter which ensures a good dynamic response. The measurand $x = (C_1 - C_2)/(C_1 + C_2)$ is related to $V_{ctrl}$ through a linear relation: $x = G_{mult} k V_{ctrl} - 1$, being $G_{mult}$ the total multipliers gain that can be manipulated by changing two resistors $R_{g1}$ and $R_{g2}$ while $k$ is related to the multipliers design and is a constant. In order to test the presented solution preliminarily, a discrete board with $\pm 15$ V power supply was designed and fabricated (Fig. 2). The bridge excitation signal is sinusoidal and generated by a Wien oscillator, with a frequency of 9 kHz and an amplitude of 4 V. In Fig. 3 the comparison between theoretical and measured interface static behavior results are reported. In this case also a sensor parasitic compensation has been implemented. Figure 4 depicts the measured interface full-scale error. Finally, each block has been also designed at transistor level, in a standard CMOS technology showing the complete interface integrability.

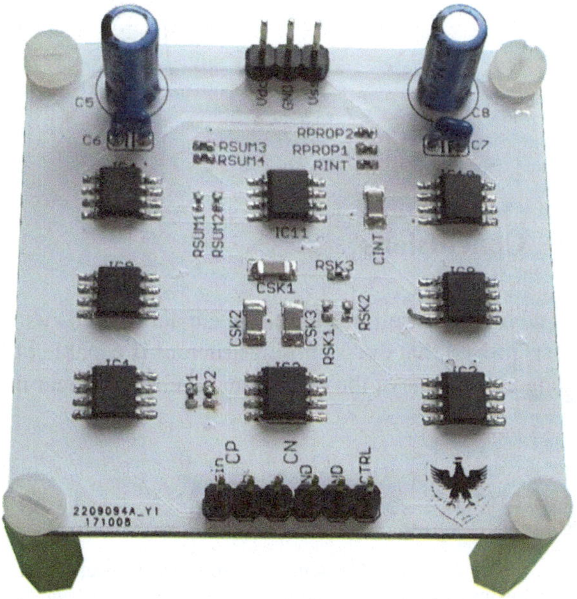


**Fig. 2** Designed prototype board

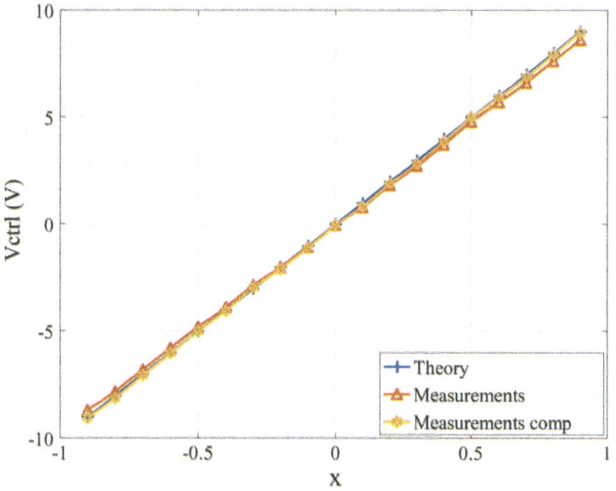


**Fig. 3** Comparison between theoretical and measured interface static behavior

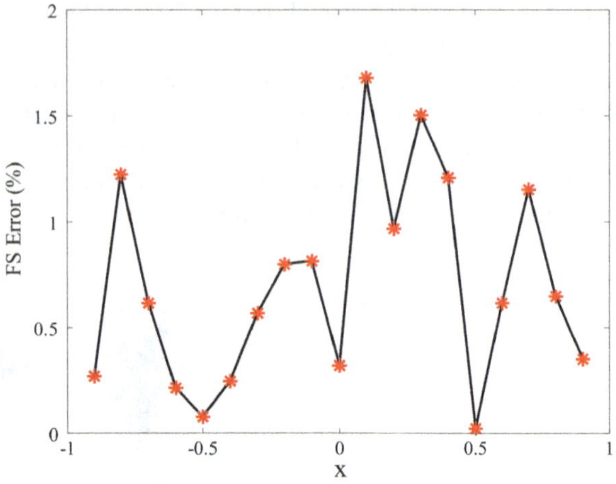


**Fig. 4** Theory versus measurements full scale error

## 3 Conclusions

A novel interface for differential capacitive sensors was presented. Simulations and tests proved that the proposed structure is capable of detecting and quantifying full-range variations of the sensor, maintaining a linear input output static characteristic.

## References


1. Baxter L (1997) Capacitive sensors. In: Capacitive sensors: design and applications. The Institute of Electrical and Electronics Engineers, New York, pp 1–46
2. Puers R (1993) Capacitive sensors: when and how to use them. Sens Actuators, A 37–38:93–105. https://doi.org/10.1016/0924-4247(93)80019-d
3. Lotters J, Olthuis W, Veltink P, Bergveld P (1999) A sensitive differential capacitance to voltage converter for sensor applications. IEEE Trans Instrum Meas 48:89–96. https://doi.org/10.1109/19.755066
4. Xie H, Fedder G (2002) Vertical comb-finger capacitive actuation and sensing for CMOS-MEMS. Sens Actuators, A 95:212–221. https://doi.org/10.1016/s0924-4247(01)00740-3
5. Pedersen T, Fragiacomo G, Hansen O, Thomsen E (2009) Highly sensitive micromachined capacitive pressure sensor with reduced hysteresis and low parasitic capacitance. Sens Actuators, A 154:35–41. https://doi.org/10.1016/j.sna.2009.07.013
6. Zeng T, Lu Y, Liu Y, Yang H, Bai Y, Hu P et al (2016) A capacitive sensor for the measurement of departure from the vertical movement. IEEE Trans Instrum Meas 65:458–466. https://doi.org/10.1109/tim.2015.2490806
7. Tsuchiya T, Funabashi H (2004) A z-axis differential capacitive SOI accelerometer with vertical comb electrodes. Sens Actuators, A 116:378–383. https://doi.org/10.1016/j.sna.2004.05.008
8. Horenstein M, Perreault J, Bifano T (2000) Differential capacitive position sensor for planar MEMS structures with vertical motion. Sens Actuators, A 80:53–61. https://doi.org/10.1016/s0924-4247(99)00251-4
9. Zadeh E, Sawan M (2005) High accuracy differential capacitive circuit for bioparticles sensing applications. In: 48th Midwest symposium on circuits and systems. IEEE

10. Ghafar-Zadeh E, Sawan M, Chodavarapu V, Hosseini-Nia T (2010) Bacteria growth monitoring through a differential CMOS capacitive sensor. IEEE Trans Biomed Circuits Syst 4:232–238. https://doi.org/10.1109/tbcas.2010.2048430
11. Ferri G, Stornelli V, Parente F, Barile G (2016) Full range analog Wheatstone bridge-based automatic circuit for differential capacitance sensor evaluation. Int J Circuit Theory Appl 45:2149–2156. https://doi.org/10.1002/cta.2298
12. Singh T, Saether T, Ytterdal T (2009) Current-mode capacitive sensor interface circuit with single-ended to differential output capability. IEEE Trans Instrum Meas 58:3914–3920. https://doi.org/10.1109/tim.2009.2021241
13. Tan S, Liu C, Yeh L, Chiu Y, Hsu K (2011) A new process for CMOS MEMS capacitive sensors with high sensitivity and thermal stability. J Micromech Microeng 21:035005. https://doi.org/10.1088/0960-1317/21/3/035005
14. Mochizuki K, Watanabe K, Masuda T (1998) A high-accuracy high-speed signal processing circuit of differential-capacitance transducers. IEEE Trans Instrum Meas 47:1244–1247. https://doi.org/10.1109/19.746591
15. Barile G et al (2018) A CMOS full-range linear integrated interface for differential capacitive sensor readout. Sens Actuators A: Phys 281:130–140. Available: https://doi.org/10.1016/j.sna.2018.08.033
16. Vejdani P, Allidina K, Nabki F (2017) Analysis of sensitivity and power consumption of chopping techniques for integrated capacitive sensor interface circuits. J Low Power Electron Appl 7:31. https://doi.org/10.3390/jlpea7040031
17. Brookhuis R, Lammerink T, Wiegerink R (2015) Differential capacitive sensing circuit for a multi-electrode capacitive force sensor. Sens Actuators, A 234:168–179. https://doi.org/10.1016/j.sna.2015.08.020
18. Flammini A, Marioli D, Sisinni E, Taroni A (2005) A multichannel DSP-based instrument for displacement measurement using differential variable reluctance transducer. IEEE Trans Instrum Meas 54:178–183. https://doi.org/10.1109/tim.2004.834052
19. Kar S, Chatterjee P, Mukherjee B, Swamy K, Sen S (2018) A differential output interfacing ASIC for integrated capacitive sensors. IEEE Trans Instrum Meas 67:196–203. https://doi.org/10.1109/tim.2017.2761238
20. Depari A, Sisinni E, Flammini A, Ferri G, Stornelli V, Barile G et al (2018) Autobalancing analog front end for full-range differential capacitive sensing. IEEE Trans Instrum Meas 67:885–893. https://doi.org/10.1109/tim.2017.2785160
21. Barile G, Ferri G, Parente F, Stornelli V, Depari A, Flammini A et al (2017) A standard CMOS bridge-based analog interface for differential capacitive sensors. In: 2017 13th conference on Ph.D. research in microelectronics and electronics (PRIME). IEEE, pp 281–284
22. Barile G, Ferri G, Parente F, Stornelli V, Depari A, Flammini A et al (2017) Linear integrated interface for automatic differential capacitive sensing. In: Eurosensors 2017, MDPI, p 592. https://doi.org/10.3390/proceedings1040592. Accessed 24 July 2018

# Spherical Anemometer for Novel Portable and Fixed-Point Wind Measurement Devices

A. Leoni, G. Barile, M. Muttillo and G. Ferri

**Abstract** In this work, we propose a fully 3D printable spherical directional anemometer of 7 cm diameter, which contains two orthogonal, not intersected inner cylindrical pipes of 6 mm diameter. Measurements results are here presented: test has been performed in a wind tube where the reference air velocity has been monitored by means of a calibrated, high precision hot wire anemometer and compared to the magnitude of the airspeed measured with the proposed spherical anemometer.

**Keywords** Wind measurement · Sensor interface · Anemometer

## 1 Introduction

Nowadays, anemometric systems mainly consist of electromechanical sensors like cup anemometers, which constitute the most preferred solution for fixed measurement stations for weather monitoring [1–6]. Different solutions have been investigated to overcome this technology, like pressure sensor-based or heat transfer-based anemometry, which are more adequate for portable application. However, such solutions cannot guarantee good performance in terms of wind direction estimation, as demonstrated in [7]. Some improvements about this issue have been achieved by the same authors with the implementation of a Constant Temperature Anemometric (CTA) system, which showed better results for what concerns the wind direction measurement but still suffered from a high power consumption [8]. We here propose a fully 3D printable [9–14] compact spherical directional anemometer composed by two orthogonal, not intersected inner cylindrical pipes. Tests results show a precision of about 0.6 m/s for lower wind speed and 0.1 m/s for higher wind speed, in the 0–20 m/s range, where the different sensitivity is due to the nonlinear relationship between airspeed and pressure difference.

A. Leoni (✉) · G. Barile · M. Muttillo · G. Ferri
Department of Industrial and Information Engineering and Economics,
University of L'Aquila, L'Aquila, Italy
e-mail: alfiero.leoni@univaq.it



G. Di Francia et al. (eds.), *Sensors and Microsystems*, Lecture Notes
in Electrical Engineering 629, https://doi.org/10.1007/978-3-030-37558-4_20

## 2 The Proposed Solution

The designed anemometer has a 7 cm diameter, which contains two orthogonal, not intersected inner cylindrical pipes of 6 mm diameter (see Fig. 1). The two ducts are directed along x and y-axis of the local coordinates and a pair of differential pressure sensors are employed to detect the pressure difference that occurs inside the tubes because of the air stream. For a laminar or transition air flow, the Poiseuille law can be exploited to have a good approximation computing the relationship between the pressure difference in Pascal unit of measure, considering two distinct points of each pipe and the average velocity of the airflow $V_{av}$ [m/s], as defined:

$$\Delta P = \frac{32 \cdot \mu \cdot V_{av} \cdot l}{D^2} \tag{1}$$

where $l$ is the distance [m] between the two points of the duct where the pressure difference is evaluated, $D$ represents the diameter [m] of the air duct and $\mu$ is the dynamic viscosity of the air [kg/(m * s)]. On the contrary, for a turbulent regime, it is trivial to produce a suitable analytical study because of the complexity of the physical behavior, therefore an empirical model has to be adopted. According to different studies reported in the literature [15], the Darcy-Weisbach equation can be considered to model the anemometric system and it is possible to derive the following relation for the pressure difference:

$$\Delta P = f \frac{l}{D} \frac{\rho \cdot V_{av}}{2} \quad where\ f = \phi\left(Re, \frac{\varepsilon}{D}\right) \tag{2}$$

Measurements have been performed in a wind tube where the reference air velocity has been monitored by means of a calibrated, high precision hot wire anemometer,

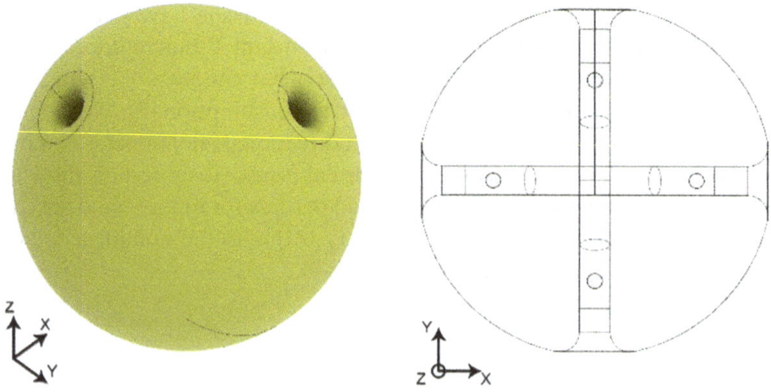


**Fig. 1** Spherical structure design

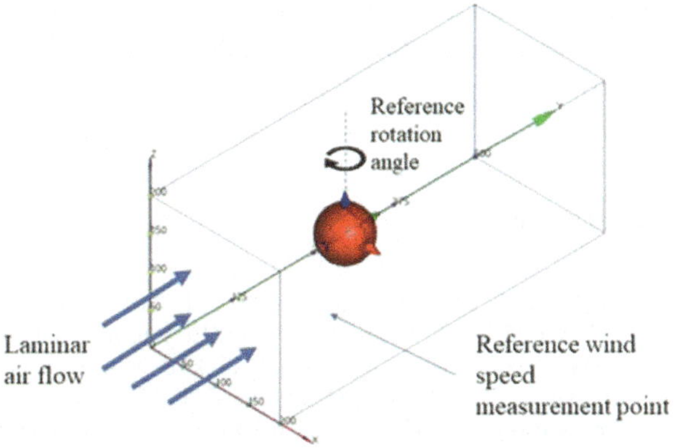


**Fig. 2** Representation of measuring methodology

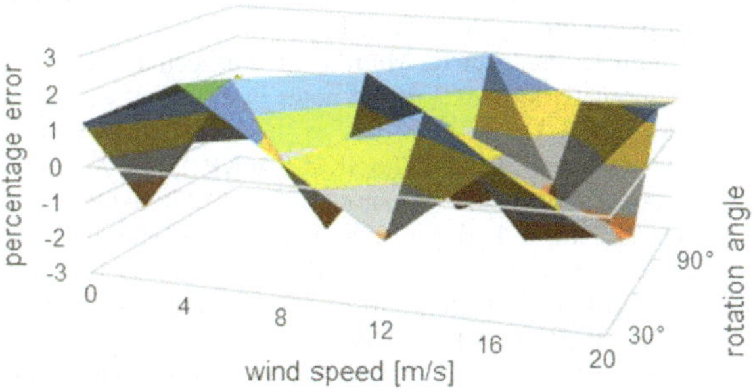


**Fig. 3** Percentage error for wind speed and direction

using high sensitivity measurements techniques [16–20], and compared to the magnitude of the airspeed measured with the proposed spherical anemometer. In addition, to simulate wind direction changing, the sphere has been rotated and the measured wind direction has been compared to the rotation angle of anemometric structure with respect to the wind tunnel local coordinates (Fig. 2) [21]. Thanks to the spherical symmetry of the proposed anemometer, wind angle measurements have been conducted only in the 0° to 90° range, which is representative of the whole space for angle measurement. Measurement results have shown a precision of about 0.6 m/s for lower wind speed and 0.1 m/s for higher wind speed, in the 0–20 m/s range. The sensitivity has a nonlinear relationship between airspeed and pressure difference. For the wind speed detection, we achieved a percentage error less then ±2.5%, while for the wind angle the percentage error is ±3.5%, as shown in Fig. 3.

## 3 Conclusions

We have presented a 3D spherical directional anemometer that can be printed with additive 3D technology. We have performed measurements results that are in good agreement with simulation showing the feasibility of the proposed structure for portable low voltage low power applications also by means of harvesting techniques [22–26].

## References


1. Liu C, Du L, Zhao Z, Fang Z, Liu C, Li L (2014) A directional anemometer based on MEMS differential pressure sensors. In: 9th IEEE international conference on nano/micro engineered and molecular systems (NEMS), Waikiki Beach, vol 1, pp 517–520
2. Hernandez BAJ, Hernandez EM, Perez JYM, Lopez MRC (2006) Arrangement of temperature sensors as a wind sensor. In: 3rd international conference on electrical and electronics engineering, vol 1, pp 1–4, 6–8 Sept 2006
3. Fusacchia P et al (2016) A low cost fully integrable in a standard CMOS technology portable system for the assessment of wind conditions. Procedia Eng 168:1024–1027. https://doi.org/10.1016/j.proeng.2016.11.331
4. Pantoli L, Paolucci R, Muttillo M, Fusacchia P, Leoni A (2017) A multisensorial thermal anemometer system. Lecture notes in electrical engineering, pp 330–337. https://doi.org/10.1007/978-3-319-55077-0_42
5. Jewart C, McMillen B, Cho SK, Chen KP (2006) X-probe flow using self-powered active fiber Bragg gratings. Sens Actuators A: Phys 127:63–68
6. Han D, Kim S, Park S (2008) Two-dimensional ultrasonic anemometer using the directivity angle of an ultrasonic sensor. Microelectron J 39:1195–1199
7. Bruschi P, Dei M, Piotto M (2009) A low-power 2-D wind sensor based on integrated flow meters. IEEE Sens J 9:1688–1696
8. Stornelli V, Ferri G, Leoni A, Pantoli L (2017) The assessment of wind conditions by means of hot wire sensors and a modified Wheatstone bridge architecture. Sens Actuators A: Phys 262:130–139
9. Sahana VW, Thampi GT (2018) 3D printing technology in industry. In: 2018 2nd international conference on inventive systems and control (ICISC), Coimbatore, pp 528–533. https://doi.org/10.1109/icisc.2018.8399128
10. Kadhim NMSM (2018) New technologies and their impact on the development of architectural education. In: 2018 1st international scientific conference of engineering sciences—3rd scientific conference of engineering science (ISCES), Diyala, pp 231–236. https://doi.org/10.1109/isces.2018.8340559
11. Zhang H, Moon SK, Ngo TH, Tou J, Mohamed ABMY (2018) A hybrid machine learning approach for the quality optimization of a 3D printed sensor. In: 2018 International conference on intelligent rail transportation (ICIRT), Singapore, pp 1–5. https://doi.org/10.1109/icirt.2018.8641641
12. Salem MB, Aiche G, Rubbert L, Renaud P, Haddab Y (2018) Design of a microbiota sampling capsule using 3D-printed bistable mechanism. In: 2018 40th annual international conference of the IEEE engineering in medicine and biology society (EMBC), Honolulu, HI, pp 4868–4871
13. Kim H, Jang J, Kim H, Kim KH, Cho D (2018) 3D cell printed corneal stromal analogues for corneal tissue engineering. In: 2018 IEEE international conference on cyborg and bionic systems (CBS), Shenzhen, pp 191–194. https://doi.org/10.1109/cbs.2018.8612218
14. Ortegon T, Vargas M, Uribe-Quevedo A, Perez-Gutierrez B, Rojas D, Kapralos B (2017) Development of a 3D printed stethoscope for virtual cardiac auscultation examination training. In: 2017 IEEE healthcare innovations and point of care technologies (HI-POCT), Bethesda, MD, pp 125–128. https://doi.org/10.1109/hic.2017.8227600
15. Munson BR, Rothmayer AP, Okiishi TH, Huebsch WW (2012) Fundamentals of fluid mechanics, 7th edn., May 2012. ISBN: 978-1-118-11613-5

16. Han FT, Gao ZY, Wang YL (1999) A differential capacitance to voltage converter for electrostatic levitation applications. Sens Actuators A: Phys 99:249–255
17. Ferri G, Stornelli V, Parente FR, Barile G (2016) Full range analog Wheatstone bridge-based automatic circuit for differential capacitance sensor evaluation. Int J Circuit Theory Appl. https://doi.org/10.1002/cta.2298
18. Depari A et al (2018) Autobalancing analog front end for full-range differential capacitive sensing. IEEE Trans Instrum Meas 67(4):885–893. https://doi.org/10.1109/tim.2017.2785160
19. Barile G et al (2018) A CMOS full-range linear integrated interface for differential capacitive sensor readout. Sens Actuators A: Phys 281:130–140. https://doi.org/10.1016/j.sna.2018.08.033
20. Barile G, Ferri G, Parente F, Stornelli V, Depari A, Flammini A et al (2017) A standard CMOS bridge-based analog interface for differential capacitive sensors. In: 2017 13th conference on Ph.D. research in microelectronics and electronics (PRIME). IEEE, pp 281–284
21. Leoni A, Stornelli V, Pantoli L (2018) A low-cost portable spherical directional anemometer for fixed points measurement. Sens Actuators, A 280:543–551
22. Leoni A, Pantoli L, Stornelli V, Ferri G, Russo M, Solic P (2017) 90/900 MHz IC architecture for autonomous systems. In: 2017 2nd international multidisciplinary conference on computer and energy science (SpliTech), Split, pp 1–4
23. Leoni A, Stornelli V, Ferri G, Errico V, Ricci M, Pallotti A, Saggio G (2018) A human body powered sensory glove system based on multisource energy harvester. In: 2018 14th conference on Ph.D. research in microelectronics and electronics (PRIME), Prague, pp 113–116. https://doi.org/10.1109/prime.2018.8430362
24. Stomelli V, Leoni A, Ferri G, Errico V, Ricci M, Pallotti A, Saggio G (2018) A multi-source energy harvesting sensory glove electronic architecture. In: 2018 3rd international conference on smart and sustainable technologies (SpliTech), Split, pp 1–4
25. Leoni A, Pantoli L, Stornelli V, Ferri G, Russo M, Solic P (2018) A combined 90/900 MHz IC architecture for power-assisting in IoT applications. J Commun Softw Syst (JCOMSS) 14(1)
26. Stornelli V, Leoni A, Ferri G, Orengo G, Errico V, Pallotti A, Saggio G (2019) A 10-17 DOF sensory gloves with harvesting capability for smart healthcare. J Commun Softw Syst 15(2). https://doi.org/10.24138/jcomss.v15i2.693

# Landslides Monitoring by Means of Low Cost Wired Sensor Networks

**M. Muttillo, A. Colagiovanni, L. Pantoli and G. Ferri**

**Abstract** In this paper we provide a feasible and functional solution for the automatic monitoring of landslide slopes parameters, allowing to control the dynamic behaviour of the target application, so adopting the necessary actions for any criticism. The here proposed system is a low cost infrastructure wired sensor network composed by sensing nodes and a GSM datalogger in charge of data management and transmission on a remote FTP web server.

**Keywords** WSN · Sensor interface · Environmental monitoring

## 1 Introduction

Landslides problems and rock fall events are even more common in many areas, especially in mountainous regions, and due to geological and climate changes. These phenomena represent a continuous risk for the population, threatening the roads and railways transit and not rarely also the human infrastructure. Some solutions like physical barriers (drape nets, rock gabions, rock fall catchment fences) or indirect remedies based on reforestation or terracing works are trying to mitigate the effects of these serious risks, but these is a continues request for novel prevention activities systems in order to ensure safety [1–3].

For these reasons, the definition and analysis of a complete electronic system and sensor interface [4, 5] for the automatic monitoring [5–12] of landslide slopes parameters, allowing to control the dynamic behavior is a target application for modern applications. Even though, generally these systems are wireless sensors network based architectures with harvesting capability [13–15] in some applications wired solution are more efficient and at low cost.

M. Muttillo (✉) · L. Pantoli · G. Ferri
Department of Industrial and Information Engineering and Economics, University of L'Aquila, L'Aquila, Italy
e-mail: mirco.muttillo@graduate.univaq.it

A. Colagiovanni
SENSing Srl, Spin off Company of the University of L'Aquila, L'Aquila, Italy



G. Di Francia et al. (eds.), *Sensors and Microsystems*, Lecture Notes in Electrical Engineering 629, https://doi.org/10.1007/978-3-030-37558-4_21

We here propose an electronic system for an infrastructure wired sensor network composed by sensing nodes and a datalogger in charge of data management and transmission on a remote FTP web server.

## 2 The Proposed Solution

The proposed system has been developed both at hardware and firmware level. At hardware level, it is formed by three main building blocks: the sensing node, the power supply block and the data logger. The node makes use of a low-power microcontroller (Atmel ATMega328p), which is a 8-bits AVR RISC-based operating at 8 MHz. All the nodes communicate with the data logger by means of a wired link on a RS485 protocol; this control unit, beyond managing the sensors network, transmits the real-time data on a web server by using a GSM on board module. In Fig. 1 the data logger architecture implementing an ATMega2560 microcontroller is shown, while Fig. 2 shows the standalone sensor node architecture. In addition, a GSM module is included for remoting the collected data. The main sensor on the node is the accelerometer employed to evaluate both inclination and eventual unexpected 3-dimensional acceleration. The MMA8451Q from NXP Semiconductors is used to this purpose. It is a small, thin, ultralow power, 3-axis accelerometer with high resolution (14-bits) measurement at up to $\pm 8$ g. Its high resolution (4096 counts/g) enables measurement of inclination changes less than 0.1° making it suitable also for low-effects events. All the slave circuits are programmed to send an asynchronous interrupt output signal forcing the microcontroller to execute a specific routine when an event or interrogation happens. The sensing nodes have been installed above ground on a dedicated rigid rod anchored to a small plinth located at a minimum depth. The nodes can be freely arranged on the site under analysis and the system can

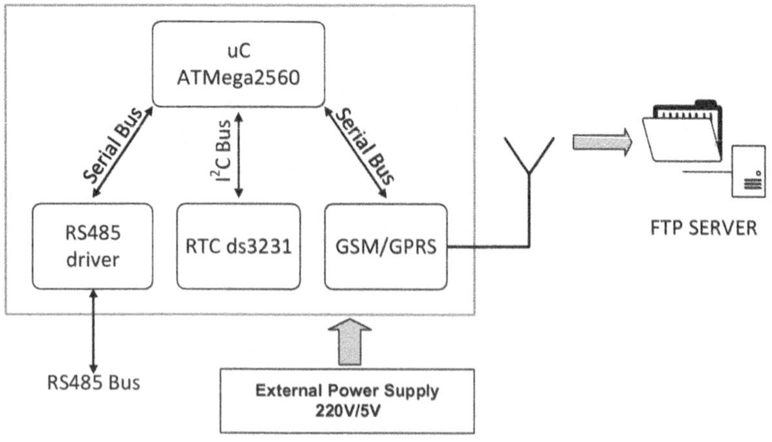


**Fig. 1** Simplified block scheme of the designed data logger block scheme

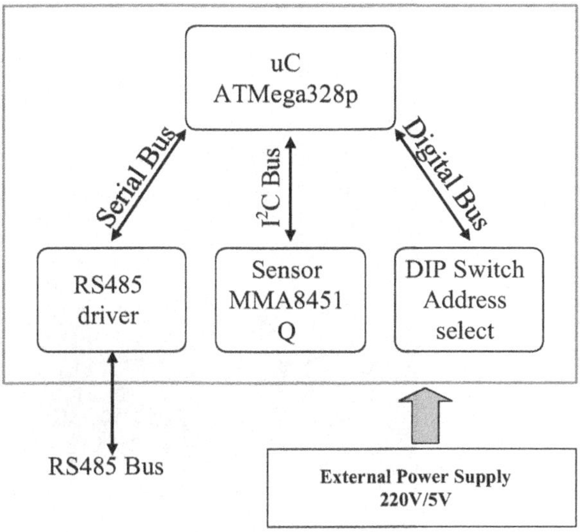


**Fig. 2** Simplified block scheme of the designed node

be used both in a completely autonomous network, thanks to solar panel harvesting architecture mounted on each node and on the datalogger, supplied by a suitable power network. All these characteristics allow to have a complete control of the dynamic of the site. The final prototype has been implemented and tested (and is currently operating) in a real environmental located in the central Italy and applied to a landslide. The monitoring is checked on a dedicated web server with remote login. For confidentiality it is not possible to identify the exact location but, in any case, a photo of an installed sensor node tested in a real environmental is reported in Fig. 3 while in Fig. 4 the Ftp web server data output results are reported for three sensor nodes showing the variation of the inclination in the window period.

## 3 Conclusions

A low cost and ready to market electronic system for an infrastructure wired sensor network has been here proposed. The system is composed by sensing nodes and a datalogger in charge of data management and transmission on a remote ftp web server.

Laboratory test and on the field measured results have confirmed the system feasibility for landslides monitoring applications.

Fig. 3 The proposed sensor node installed in a real environmental Ftp web server data output example

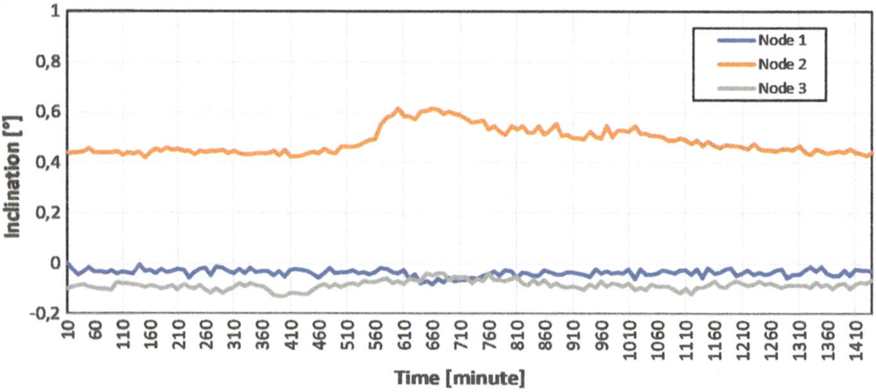


Fig. 4 The implemented Ftp web server data output example

## References

1. Chehri A, Fortier PP, Tardif M (2007) Security monitoring using wireless sensor networks. In: Proceedings of the 5th annual conference on communication networks and services research (CNSR 2007), Fredericton, NB, Canada, 14–17 May 2007, pp 13–17
2. Alippi C, Camplani R, Galperti C, Marullo A, Roveri M (2010) An hybrid wireless-wired monitoring system for real-time rock collapse forecasting. In: Proceedings of the 7th international conference on mobile ad hoc and sensor system (MASS), San Francisco, CA, USA, 8–12 Nov 2010, pp 224–231
3. Okamoto S (2011) Vibration and scattering monitoring of Japanese roofing tile by accelerometer. In: Proceedings of the international conference on fluid power and mechatronics, Beijing, China, 17–20 Aug 2011, pp 30–35
4. Barile G et al (2018) A CMOS full-range linear integrated interface for differential capacitive sensor readout. Sens Actuators, A: Phys 281:130–140. https://doi.org/10.1016/j.sna.2018.08.033
5. Depari A et al (2018) Autobalancing analog front end for full-range differential capacitive sensing. IEEE Trans Instrum Meas 67(4):885–893. https://doi.org/10.1109/tim.2017.2785160
6. Barile G, Leoni A, Pantoli L, Stornelli V (2018) Real-time autonomous system for structural and environmental monitoring of dynamic events. Electronics (Switzerland) 7(12). https://doi.org/10.3390/electronics7120420
7. Cuadra CH, Shimoi N, Nishida T, Saijo M (2013) Estimation of dynamic properties of traditional wooden structures using new bolt sensor. In: Proceedings of the 13th international conference on control, automation and systems (ICCAS), Gwangju, Korea, 20–23 Oct 2013, pp 1593–1598
8. Rotilio M, Cucchiella F, De Berardinis P, Stornelli V (2018) Thermal transmittance measurements of the historical masonries: some case studies. Energies 11(11). https://doi.org/10.3390/en11112987
9. Moghavvemi M, Ng KE, Soo CY, Tan SY (2005) A reliable and economically feasible remote sensing system for temperature and relative humidity measurement. Sens Actuators, A: Phys 117:181–185
10. Gabriele T et al (2015) Smart power management system for home appliances and wellness based on wireless sensors network and mobile technology. In: Proceedings of the 2015 18th AISEM annual conference, AISEM 2015. https://doi.org/10.1109/aisem.2015.7066808
11. de Rubeis T et al (2017) A first approach to universal daylight and occupancy control system for any lamps: simulated case in an academic classroom. Energy Build 152:24–39. https://doi.org/10.1016/j.enbuild.2017.07.025
12. Pantoli L, Paolucci R, Muttillo M, Fusacchia P, Leoni A (2016) A multisensorial thermal anemometer system. In: 3rd National conference on sensors, vol 431, pp 330–337. https://doi.org/10.1007/978-3-319-55077-0_42
13. Leoni A, Pantoli L, Stornelli V, Ferri G, Solic P, Russo M (2018) A combined 90/900 MHz IC architecture for smart tag application. J Commun Softw Syst 14(1). https://doi.org/10.24138/jcomss.v14i1.451
14. Pantoli L, Leoni A, Stornelli V, Ferri G (2016) Energy harvester for remote sensors systems. In: 2016 International multidisciplinary conference on computer and energy science (SpliTech), Split, pp 1–3. https://doi.org/10.1109/splitech.2016.7555933
15. Di Marco et al (2016) Dual band harvester architecture for autonomous remote sensors. Sens Actuators, A: Phys 247:598–603

# Lock-in Thermography for Non-destructive Testing of 3D Printed PLA Items

**Simone Boccardi, Giovanni M. Carlomagno, Giuseppe Del Core and Carosena Meola**

**Abstract** The intention of this paper is to perform a preliminary investigation in order to ascertain if lock-in thermography can be used for non-destructive inspection of samples manufactured with 3D printing. This because of the ever more increasing interest that 3D printing and additive manufacturing are gaining from both the industrial and academic communities. To this end, some samples have been produced by means of the fuse deposition modeling technique and with inclusion of some artificial defects at a certain depth through the thickness. Such samples are inspected with lock-in thermography and the obtained results seem promising.

**Keywords** 3D printing · PLA items · Non-destructive testing · Lock-in thermography

## 1 Introduction

Lock-in thermography (LT) is a well-known technique of non-destructive testing (NDT) that can be applied to inspect several kinds of materials in order to detect both manufacturing defects and degradation occurring during the material operating life [1]. LT allows for contactless NDT and so without any alteration to the part under inspection as well without risks for the operator. The output is a 2D image which represents the material internal conditions. Currently LT is effective to perform NDT on composite materials involving both thermoset and thermoplastic matrices. In particular, it is a valid means to obtain information about production problems like: fibers misalignment, excess or lack in matrix distribution, or impact damages and other material degradations which may occur during the material operating life and that cannot be detected through visual inspection [1–3].

S. Boccardi (✉) · G. Del Core
Dipartimento di Scienze e Tecnologie, Università degli Studi di Napoli Parthenope, Centro Direzionale Isola C4, 80143 Naples, Italy
e-mail: simone.boccardi@uniparthenope.it

S. Boccardi · G. M. Carlomagno · C. Meola
Dipartimento di Ingegneria Industriale, Università degli Studi di Napoli Federico II, Via Claudio 21, 80125 Naples, Italy

 
G. Di Francia et al. (eds.), *Sensors and Microsystems*, Lecture Notes in Electrical Engineering 629, https://doi.org/10.1007/978-3-030-37558-4_22

The idea behind this work is to verify if the advantages offered by LT could be somehow usefully exploited to perform NDT on plastic components which are obtained by means of additive manufacturing with fuse deposition modeling (FDM). This need is justified by the fact that the production of objects by means of 3D-printing is increasing and with even more new applications fields. It is expected an increasing investment in 3D printing in the next future [4, 5]. The driving force resides mostly in the fast production with less waste of material and lower energy consumption. However, alongside their many advantages there is an unexplored world of products defects and performance [6].

## 2 Materials and Samples

Two types of square specimens A and B, whose nominal characteristics are shown in Table 1, were manufactured using Polylactic Acid (PLA+) by SUNLU and FDM technique with a layer thickness of 0.3 mm and an internal filling section of 25% to create the internal honeycomb-like structure. Of course, the external (skin) layers were manufactured with the optimal conditions of 100% infill and 210 °C extrusion temperature to prevent defects formation. To simulate a manufacturing defect, the machine was forced to create a circular zone of 20 mm in diameter with a 100% infill over one PLA layer at a certain depth that is 1.2 mm and 4.6 mm for specimens A-1 and A-2 respectively (Table 1). A photo showing the geometric appearance of a generic specimen is reported in Fig. 1a; instead Fig. 1b, which was taken during the manufacturing process displays the position of the defect inside a type A specimen.

Specimens belonging to type B have been produced by choosing three different temperature values for the fuse deposition modeling process (from 195 up to 210 °C) in order to generate a different gradation of internal defectiveness due to the production temperature lower than the optimal one (Fig. 2).

This because some preliminary 3D printing tests showed that, for the used material and 3D printing system, the lowest defectiveness level was achieved by an extrusion

**Table 1** Details of investigated samples

| Sample | Thickness (mm) | Side of the square sample (mm) | Defect nominal diameter and depth (mm) | Production temperature (°C) |
|---|---|---|---|---|
| A-1 | 3 | 100 | 20, 1.2 | 210 |
| A-2 | 6 | 100 | 20, 4.2 | 210 |
| B-1 | 6 | 80 | – | 210 |
| B-2 | 6 | 80 | Distributed in the middle | 200 |
| B-3 | 6 | 80 | Distributed in the middle | 195 |

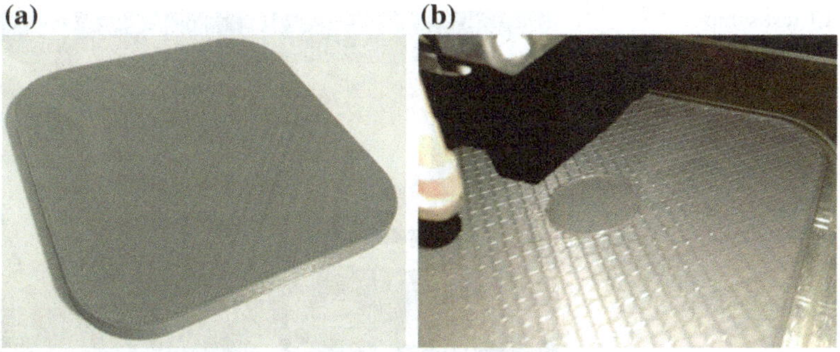

**Fig. 1** **a** Specimens appearance; **b** photo taken during the 3D printing process showing the defect position in a type A specimen

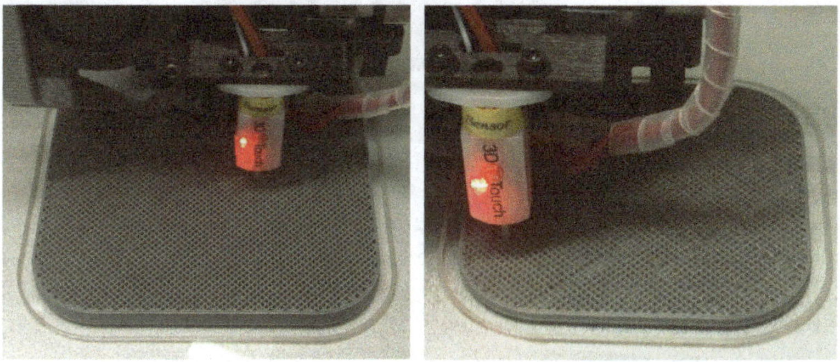

**(a)** 210 °C **(b)** 200 °C

**Fig. 2** Examples of specimen internal defectiveness for two values of the production temperature

temperature of 210 °C. Therefore, a temperature of 210 °C has been chosen to manufacture a sound B-1 specimen. Instead, for the other two specimens the temperature of 210 °C was used only for the manufacturing of both the external upper and lower surfaces, while a temperature of 200 °C and 195 °C was respectively set to print the internal layers of the two specimens B-2 and B-3 (Table 1). This should allow for limiting the defectiveness through the specimen thickness without affecting the external surfaces.

## 3 Testing with Lock-In Thermography

All the specimens have been subjected to non-destructive evaluation with lock-in thermography in reflection [1]. The specimens are thermally stimulated by means of two halogen lamps of 1000 W each. The used infrared camera is the Flir SC6000, which is equipped with long wave QWIP detector and with the Lock-in module

**Fig. 3** LT test setup

that drives the halogen lamps to generate a sinusoidal thermal wave of selectable frequency $f$ and the IRLock-In© software for data analysis. The test setup is shown in Fig. 3.

The basic LT relationship is given by:

$$p = 1.8\sqrt{\frac{\alpha}{\pi f}} \tag{1}$$

where $p$ is the depth at which a defect is located, $\alpha$ is the material thermal diffusivity and $f$ is the heating frequency. Lower is the heating frequency higher is the inspection depth through the thickness. Tests are carried out for several values of the heating frequency in order to inspect the specimens at different depths starting from the surface viewed by the infrared camera.

First of all, the samples thermal diffusivity has been evaluated with the same lock-in thermography according to Ref. [7]. To this end a metallic target has been placed in perfect contact with the specimen rear surface. The heating frequency value at which the metallic target becomes visible in the phase image for the first time, represents the frequency value $f^*$ at which the specimen thickness $p$ has been entirely crossed. This value is introduced in Eq. 1 to evaluate the average through the thickness material thermal diffusivity $\alpha_m$.

## 4 Results

First of all, from tests performed with the metallic target a mean thermal diffusivity $\alpha_m = 0.0038\ \text{cm}^2/\text{s}$ has been obtained and used to set the heating frequency to inspect the material at different $p$ values.

As first observation in all the obtained phase images for both types of specimens it is possible to clearly recognize the sample internal honeycomb structure. Some phase images relative to the A-type specimens are shown in Figs. 4 and 5 with indication of the heating frequency and the approximate depth. Looking at Fig. 4a it is possible to notice that the artificial defect becomes visible already for $f = 1.14$ Hz which corresponds to $p \approx 0.6$ mm that is lower than the nominal defect depth (Table 1). This is probably due to unexpected manufacturing problems. More specifically, some fused PLA filaments collapsed on the underlying layers; this has caused loss of the artificial defect planarity with reduction of the real defect distance from the surface with respect to its nominal depth. In fact, the two white spots at the center of the defect are indicative of this collapse.

From Fig. 5 it is possible to see that the defect appears visible at the $f$ value corresponding practically to its nominal depth. It is worth noting that a data deviation is admissible and due to having assumed a unique average thermal diffusivity value

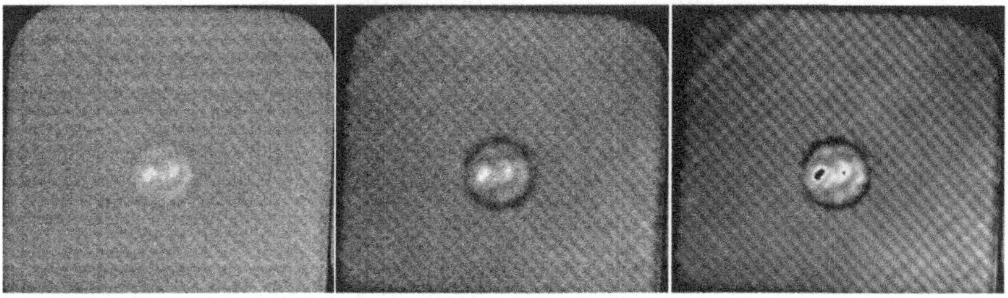

(a) $f = 1.14$ Hz, $p \approx 0.6$mm (b) $f = 0.26$ Hz, $p \approx 1.2$mm (c) $f = 0.10$ Hz, $p \approx 2.0$mm

**Fig. 4** Some phase images of sample A-1

(a) $f = 0.15$ Hz, $p \approx 1.6$mm (b) $f = 0.05$ Hz, $p \approx 3.0$mm (c) $f = 0.02$ Hz, $p \approx 4.4$mm

**Fig. 5** Some phase images of sample A-2

for all specimens. Comparing Figs. 4 and 5 it is also possible to underline some blurred regions that account for non uniform material distribution due to production issues.

Phase images of specimens type B taken at $f = 0.1$ and 0.02 Hz are reported in Figs. 6 and 7 respectively. The phase images relative to the specimens B-1 (Figs. 6a and 7a) indicate a more homogeneous material distribution as expected since the correct production temperature $T = 210$ °C value was used.

In the phase images relative to the B-3 sample, which are shown in Figs. 6c and 7c, it is possible to notice the highest value of contrast level and also some dark stains that are located mainly on the top of the image. These dark stains are likely to represent regions with severe anomalies in the material internal structure where the lower temperature value, chosen for the production, has prevented a correct deposition of the molten filament with the consequent formation of wide void regions and interruption of the internal regular structure (see also Fig. 2b).

(a) B -1 (b) B-2 (c) B-3

**Fig. 6** Phase images of type B specimens taken at $f = 0.1$ Hz

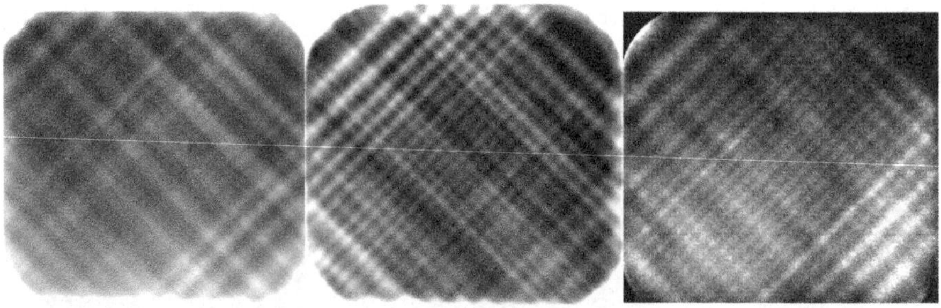

**(a) B-1** **(b) B-2** **(c) B-3**

**Fig. 7** Phase images of type B specimens taken at $f = 0.02$ Hz

## 5 Conclusions

In this work some 3D printed PLA items have been considered to assess the possibility to detect material internal defects, which are caused by manufacturing errors, non-destructively by means of lock-in thermography. In particular, the attention was driven towards two types of problems. One is the formation of local agglomerates buried trough the thickness as those simulated by type A specimens. Another is the loss of regular structure and random spot clouding, which are caused by the incorrect temperature setting during the manufacturing process.

The obtained results are promising about the LT capacity to evaluate the 3D printed items internal structure. Of course, this is only a preliminary investigation restricted to specific geometry and production method. Further tests are necessary by considering a wide range of polymers, production methods, internal structures and geometries as well for a complete assessment of the LT effectiveness for NDT of 3D printed items.

## References

1. Meola C, Boccardi S, Carlomagno GM (2016) Infrared thermography in the evaluation of aerospace composite materials, 180 p. ISBN: 9781782421719
2. Meola C, Carlomagno GM (2010) Impact damage in GFRP: new insights with infrared thermography. Compos Part A: Appl Sci Manuf 41:1839–1847
3. Meola C, Boccardi S, Carlomagno GM (2018) Composite material overview and its testing for aerospace components (Chap. 5). In: Jawaid M, Thariq M (eds) Sustainable composites for aerospace applications. ISBN: 978-0-08-102131-6 (print)
4. Mitchell A, Lafont U, Hołyńska M, Semprimoschnig C (2018) Additive manufacturing—a review of 4D printing and future applications. Addit Manuf 24:606–626
5. Wohlers T (2016) Wohlers report 2016. 3D printing and additive manufacturing state of the industry
6. Kio OJ, Yuan J, Brooks AJ, Knapp GL, Ham K, Ge J, Van Loo D, Butler LG (2018) Non-destructive evaluation of additively manufactured polymer objects using X-ray interferometry. Addit Manuf 24:364–372
7. Meola C, Carlomagno GM, Squillace A, Giorleo G (2002) Non-destructive control of industrial materials by means of lock-in thermography. Meas Sci Technol 13:1583–1590

# In Vitro Discrimination of Bacterial Volatile Compound Patterns Using a Gas Sensor Array

**R. Capuano, Elisabetta Delibato, Ana Carolina Domakoski, Alexandro Catini, Eugenio Martinelli, Roberto Paolesse and C. Di Natale**

**Abstract** This paper illustrates that an electronic nose applied to the measure of the headspace cultures of *Salmonella* serotypes can discriminate among different strains. S*almonella* is one of the main bacteria causing food borne illnesses, this there is a great interest for fast identification methods. Standard detection methods for *Salmonella* food contamination have limited performance in terms of costs and rapidity. Electronic noses are among the technologies that supposedly can improve the management of *Salmonella* infections. Preliminary experiments are discussed in this paper; results confirm that electronic noses offer a viable technology for rapid identification of bacteria.

**Keywords** Salmonella · Volatile organic compounds · Gas sensor array

## 1 Introduction

*Salmonella* is a major pathogen infecting foods and causing acute onset of fever, and gastroenteric symptoms such as abdominal pain, diarrhea, nausea and sometimes vomiting [1]. The currently used procedure for *Salmonella* detection is based on standard cultural method (EN/ISO 6579) [2]. This protocol involves a non-selective pre-enrichment, a selective enrichment and a plating on selective agars. Suspect colonies are then confirmed by biochemical and serological tests. Although this method is very sensitive and inexpensive, it is labor-intensive and extremely time consuming (up to 5 days to obtain results).

R. Capuano (✉) · A. C. Domakoski · A. Catini · E. Martinelli · C. Di Natale
Department of Electronic Engineering, University of Rome Tor Vergata, via del Politecnico 1, 00133 Rome, Italy
e-mail: capuano@ing.uniroma2.it

E. Delibato
Department of Food Safety, Nutrition and Veterinary Public Health, Istituto Superiore di Sanità, Viale Regina Elena 299, 00161 Rome, Italy

R. Paolesse
Department of Chemical Science and Technology, University of Rome Tor Vergata, via della Ricerca Scientifica, 00133 Rome, Italy

 
G. Di Francia et al. (eds.), *Sensors and Microsystems*, Lecture Notes in Electrical Engineering 629, https://doi.org/10.1007/978-3-030-37558-4_23

As a consequence, there is an urgent need for early and readily applicable methods to detect *Salmonella*-infected food and to distinguish between pathogenic and harmless species.

Bacteria are known to release volatile organic compounds (VOCs) as products of their metabolism during growth, as signaling molecules for intercellular communications, or as protection against antagonists [3, 4]. Analytical studies, essentially based on gas chromatographic (GC) techniques, demonstrate the existence of characteristic VOCs patterns for each different kind of microorganisms [5]. However, many volatile compounds are common to several bacterial species and the differences between samples often relate to total VOC patterns rather than to a major change in one or two constituents [6, 7]. In particular, the pattern of VOCs released by *Salmonella* contains, among the others, alcohols, ketones, and sulphur containing compounds [8, 9].

However, also GC analysis is time consuming and it is operated by specialized personnel.

Electronic noses might, in several cases, provide analysis and discrimination between different complex mixture of volatile compounds in several application fields [10], therefore they are considered a viable alternative to GC for fast classification of samples according to their VOCs pattern composition.

In this paper an electronic nose based on an array of porphyrinoid coated quartz microbalances [11] has been applied to the identification of microorganisms. In particular, experiments were aimed at discriminating different *Salmonella* serotypes grown in trypticase soy broth (TSB) respect to other bacteria.

## 2 Materials and Methods

Bacteria cultures of various *Salmonella* serovars (*S. Enteriditis*, *S. Typhimurium*, *S. Napoli* and *S. Thompson*) and other bacteria, such as *E. coli*, and *Enterobacter cloacae* have been inoculated in 10 mL of trypticase soy broth (TSB) at 37 °C for 24 h. The final bacterial concentration is around $10^9$ CFU/ml. Blank TSB has been used as control sample.

The broth cultures have been diluted in order to obtain $10^7$, $10^5$ and $10^3$ CFU/ml.

For electronic nose measurements, 5 ml of culture liquids have been transferred into previously autoclaved 20 ml vial (Flat bottom headspace vial, SUPELCO, Bellefonte, PA, USA) and sealed with a PTFE/Silicone crimp seal (SUPELCO, Bellefonte, PA, USA). Headspace composition has been equilibrated maintaining the samples at constant temperature of 37 °C in a thermostatic water bath.

The electronic nose was made of an array of twelve quartz microbalances (QMBs) each with a fundamental frequency of 20 MHz and functionalized using different metalloporphyrins and corroles [11].

For this kind of sensors, the change of the frequency ($\Delta f$) of the electrical output signal of the oscillator circuit is the consequence of a change of the mass load ($\Delta m$)

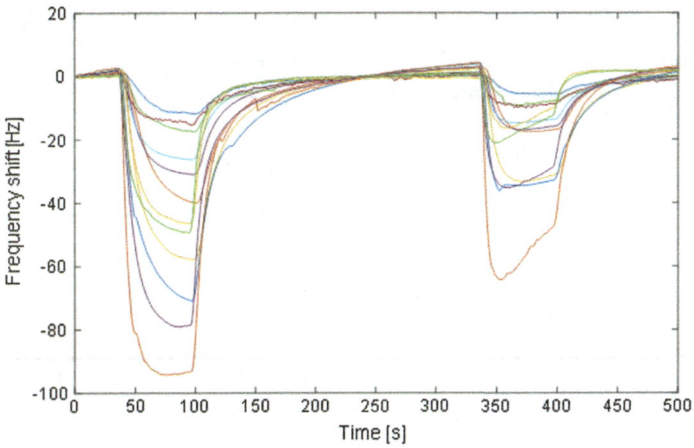


**Fig. 1** Pattern of frequency shift registered by electronic nose during bacterial culture headspace measurements

on the quartz surface. In low perturbation regime, $\Delta m$ and $\Delta f$ are linearly related each other and the relationship between them is described by Sauerbrey equation [12].

Electronic nose has been operated with a constant flow of 50 sccm (standard cubic centimeters per minute). Sensor baseline signal has been achieved using ambient air filtered with a $CaCl_2$ humidity trap. Bacterial culture headspace has been measured by the aspiration of the air contained in the vials for one minute. After each measurement, sensor baseline condition has been restored in about four minutes.

Figure 1 shows examples of electronic nose sensors signals related to two different samples. The difference between the frequency of the signals recorded during the exposure to filtered reference air and those obtained during the exposure to the bacterial sample, has been considered as features of the sensors.

## 3 Results and Discussion

The ability of the electronic nose to distinguish *Salmonella* serotypes from other bacteria (*E. coli* and *Enterobacter cloacae*) and respect to the TSB, has been evaluated separately for each bacteria concentration. In practice, the identification of *Salmonella* samples respect to the other bacteria and the TSB background has been evaluated keeping constant the number of colonies forming units.

K-nearest neighbours (k-NN), with $k = 1$, has been used as classifier. k-NN models have been cross-validated using a 5-fold cross-validation.

Figure 2 shows the confusion matrices for the four analyzed cases.

(a)

| Real Class | NO SAL | SAL | TSB |
|---|---|---|---|
| NO SAL | 27 | 0 | 0 |
| SAL | 0 | 48 | 0 |
| TSB | 1 | 1 | 17 |

(b)

| Real Class | NO SAL | SAL | TSB |
|---|---|---|---|
| NO SAL | 27 | 2 | 0 |
| SAL | 1 | 47 | 2 |
| TSB | 0 | 1 | 18 |

(c)

| Real Class | NO SAL | SAL | TSB |
|---|---|---|---|
| NO SAL | 18 | 0 | 1 |
| SAL | 1 | 30 | 0 |
| TSB | 1 | 0 | 18 |

(d)

| Real Class | NO SAL | SAL | TSB |
|---|---|---|---|
| NO SAL | 25 | 1 | 0 |
| SAL | 4 | 42 | 2 |
| TSB | 1 | 3 | 15 |

(Column headers: Predicted class)

**Fig. 2** Confusion matrices obtained by k-NN classifier for culture at different bacterial loads conditions: **a** $10^9$ CFU/ml; **b** $10^7$ CFU/ml; **c** $10^5$ CFU/ml; **d** $10^3$ CFU/ml. NO SAL: *E. coli* and *Enterobacter cloacae*; SAL: *Salmonella*; TSB: Trypticase soy broth

The accuracy is barely larger than 90% in all the different considered conditions. The same performance is obtained in terms of sensitivity of *Salmonella* identification (ratio of *Salmonella* samples correctly classified respect to the total *Salmonella* broth cultures).

## 4 Conclusion

Preliminary results show that *Salmonella* serotypes could be discriminated by other bacterial culture by the analysis of their released volatile compounds. In this paper we show that an electronic nose based on porphyrinoids coated QMBs provides a good accuracy in the identification of bacteria growing in TSB. However, the detection is dependent on the amount of CFU of each sample, and a proper identification is only achieved where the CFU of *Salmonella* and other bacteria are kept constant.

Compared to colony counting methods used as conventional *Salmonella* detecting method, or to analytical technique used for VOC analysis, the electronic nose might meet the analytical requirements needed in terms of portability, rapidity and low costs.

This system should be potentially applied to direct food contamination analysis.

## References

1. World Health Organisation (2013) Salmonella (non-typhoidal) fact sheet. http://www.who.int/mediacentre/factsheets/fs139/en/
2. Commission Regulation (EC) No 2073/2005 on microbiological criteria for foodstuffs. (2005). Available: http://eur-lex.europa.eu/legal-content/IT/TXT/?uri=celex:32005R2073
3. Schulz S, Dickschat JS (2007) Bacterial volatiles: the smell of small organisms. Nat Prod Rep 24:814–842
4. Kai M, Haustein M, Molina F, Petri A (2009) Bacterial volatiles and their action potential. Appl Microbiol Biotechnol 81:1001–1012
5. Chen J, Tang J, Shi H, Tang C, Zhang R (2017) Characteristics of volatile organic compounds produced from five pathogenic bacteria by headspace-solid phase micro-extraction/gas chromatography-mass spectrometry. J Basic Microbiol 57(3):228–237
6. Borjesson T (1996) Electronic nose for odor classification of grains. Cereal Chem 73:457–461
7. Sapsford KE, Ngundi MM, Moore MH, Lassman ME, Shriver-Lake LC, Taitt CR, Ligler FS (2006) Rapid detection of foodborne contaminants using an array biosensor. Sens Actuators B: Chem 113(2):599–607
8. Siripatrawan U (2008) Rapid differentiation between *E. coli* and *Salmonella* typhimurium using metal oxide sensors integrated with pattern recognition. Sens Actuators B: Chem 133:414–419
9. Tait E, Perry JD, Stanforth SP, Dean JR (2014) Identification of volatile organic compounds produced by bacteria using HS-SPME-GC–MS. J Chromatogr Sci 52(4):363–373
10. Röck F, Barsan N, Weimar U (2008) Electronic nose: current status and future trends. Chem Rev 108:705–713
11. Paolesse R, Nardis S, Monti D, Stefanelli M, Di Natale C (2017) Porphyrinoids for chemical sensor applications. Chem Rev 117:2517–2583
12. Ballantine DS, White RM, Martin SJ, Ricco AJ, Zellers ET, Frye GC, Wohltjen H (1997) Acoustic wave sensors: theory, design and physico-chemical applications. Academic Press, Cambridge

## References

# A Low Cost Resistive Gas Sensor Network Based on Zn-Al Doped and $Co_3O_4$ Nanopowder Composite

**T. Addabbo, A. Fort, M. Mugnaini, E. Panzardi, A. Pozzebon, M. Hjiri and M. S. Aida**

**Abstract** In this paper a network architecture for the monitoring of $NO_2$ emissions in the industrial scenarios is presented. The proposed system is based on chemoresistive gas sensors realized by mixing nano-composite of ZnO, doped by Al (5%) and $Co_3O_4$. The obtained compound presents appreciable sensitivity to $NO_2$ exposure at room temperature which makes it suitable for low-power applications. The sensor is embedded in an IoT framework exploiting the Long Range (LoRa) LPWAN connectivity with the associated LoRaWAN protocol, granting data transmission over a wide area. Thus, the acquired data will be forwarded to a Cloud platform to be collected and remotely managed.

**Keywords** $NO_2$ sensor · Resistive gas sensor · Gas sensors network · Industrial plants monitoring

## 1 Introduction

Toxic gas emission monitoring is a crucial issue involving different research areas, thus attracting a growing interest not only for environmental purposes but also other scenarios like the industrial one. The monitoring of exhaust gases in combustion processes is one of the techniques used in the condition monitoring of industrial plants [1, 2]. The collected information, combined with other measured quantities related to the machine operation, may provide an overview of the plant health. In this work, we propose an IoT framework composed by resistive gas sensors for $NO_2$ detection in industrial plants [3, 4]. Conventional and commercial resistive metal oxide gas sensors operate at high temperature and require a significant power being thus not suitable for low-power wireless sensing applications. Recently, research advancements related to the development of new materials have opened some new

---

T. Addabbo · A. Fort · M. Mugnaini · E. Panzardi (✉) · A. Pozzebon
Dipartimento di Ingegneria dell'Informazione e Scienze Matematiche, University of Siena, Via Roma 56, 53100 Siena, Italy
e-mail: panzardi@diism.unisi.it

M. Hjiri · M. S. Aida
Department of Physics, Faculty of Sciences, King Abdulaziz University, Jeddah, Saudi Arabia



G. Di Francia et al. (eds.), *Sensors and Microsystems*, Lecture Notes in Electrical Engineering 629, https://doi.org/10.1007/978-3-030-37558-4_24

perspectives. In particular, nano-structured materials and the combination of different nano-structured materials allow to modify the gas sensing characteristics according to the considered application [5]. In this context, using mixed nano-composites can lead to the development of low-power resistive gas sensors, with satisfactory gas sensitivity, selectivity and stability. In particular, in this work a resistive $NO_2$ gas sensor, based on a nano-composite of ZnO, doped by Al (5%) (ZnO-Al5), and $Co_3O_4$ useful for room temperature measurement applications is proposed. An architecture integrating this sensor in a low power wireless data acquisition and transmission platform is also described. This solution is based on the Long Range (LoRa) data transmission technology, whose use has already been proposed for the realization of industrial monitoring infrastructures [3, 6, 7].

The characteristics of the used materials have been deeply investigated by the authors in [8] where the gas sensing features of the nanomaterial have been studied for different environmental conditions such as temperature, humidity, concentration and nature of gas exposure. The obtained results showed satisfactory response in speed and stability also at room temperature. These characteristics, which are peculiar of the composite material, allow for the use of resistive gas sensors in low power applications.

## 2 Materials and Methods

The ZnO-Al5 and $Co_3O_4$ nano-powders were prepared by means of sol-gel and hydrothermal method, respectively. The results of the material analysis by SEM method and the XRD pattern are reported in Figs. 1 and 2.

The powders were mixed with different proportions such as 30% of ZnO-Al5 and 70% of $Co_3O_4$ (w/w) with the addition of deionized water. The obtained solution was deposited on alumina substrates across two electrodes by drop casting method (2 μL) using a micropipette. The deposited film was dried in free air at RT and then heated at 400 °C. The sensors were characterized by means of a system which allows for accurately controlling the temperature, the gas flow, the concentration and

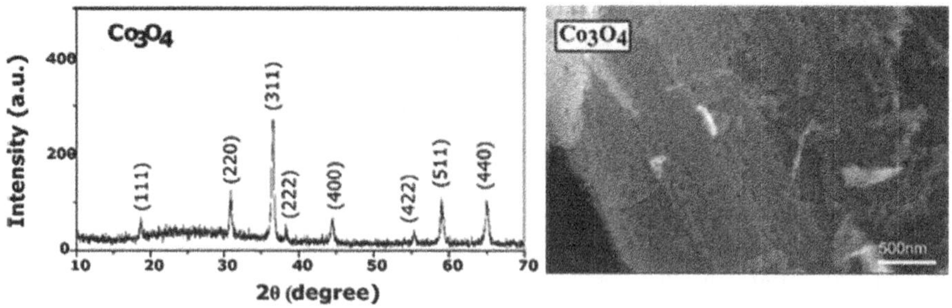


**Fig. 1** SEM image and XRD pattern of the $Co_3O_4$ nano-powder

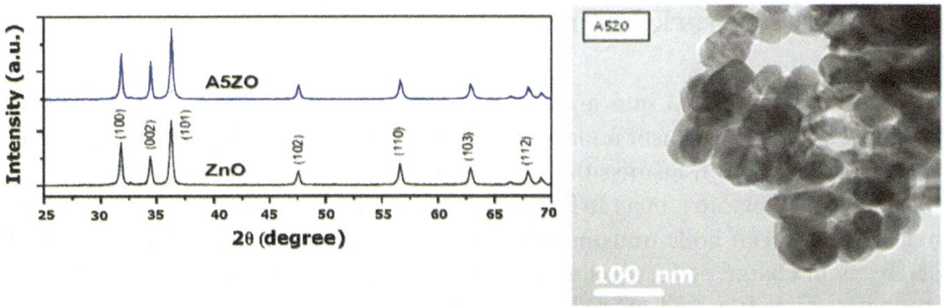


**Fig. 2** SEM image and XRD pattern Zn-Al5% doped nano-powder

the humidity, under a constant flow (200 mL/min) of mixtures of $NO_2$ in different concentrations and air as carrier gas, humid and dry. Figure 3 shows the sensor responses to $NO_2$ as a function of time at room temperature.

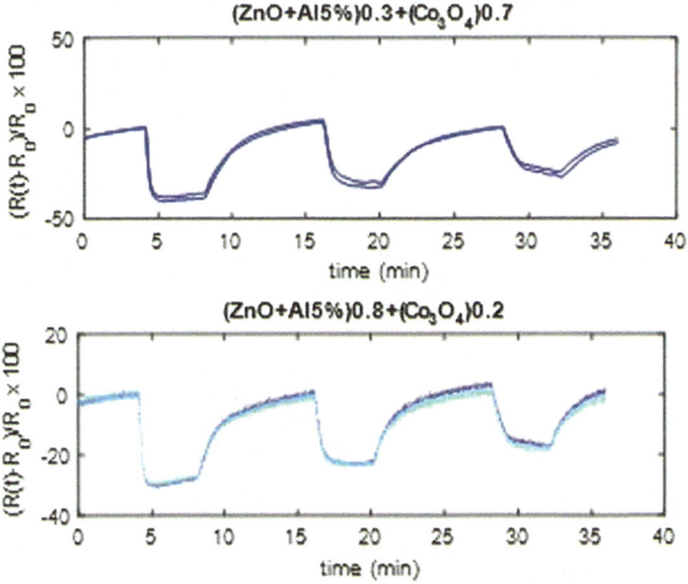


**Fig. 3** Transient sensor responses as a function of time at room temperature. $R_0$ is the resistance baseline value in air at room temperature. The measurement protocol provides 3 alternating phases of 4 min to $NO_2$ exposure with the concentration of 50 ppm, 25 ppm and 12 ppm, respectively

## 3 Sensors Network Framework

The sensor is integrated into a node in charge of collecting and transmitting the measured values. The sensor node is provided with LoRa LPWAN (Low Power Wide Area Network) data transmission module, implementing the LoRaWAN protocol: this allows to integrate it in an IoT framework composed of a large quantity of sensor nodes. Each sensor node transmits the measured values to one or more LoRaWAN Gateways in charge of forwarding the data to a LoRaWAN Network server acting as a data acquisition and management Cloud platform, as shown in the summary diagram of Fig. 4. The sensor node, in addition to the sensor itself, is composed of a microcontroller that manages the gas sensing biasing circuit by a DAC (12-bit) and reads the measured values by means of an ADC (12-bit). The node is also provided with a temperature and a humidity sensor, useful in case reference data are needed. A schematic diagram of the sensor node architecture is reported in Fig. 5. A Semtech SX1276 LoRa module is connected to the microcontroller, providing the LoRaWAN data transmission features. Finally, the node features a power management system in charge of disconnecting its sub-systems when not used. Indeed, by adopting a strict duty-cycling policy, sampling for example the sensor each hour and powering the node with off-the-shelf 1.5 V AA batteries, a lifetime node up to 2 years can be achieved.

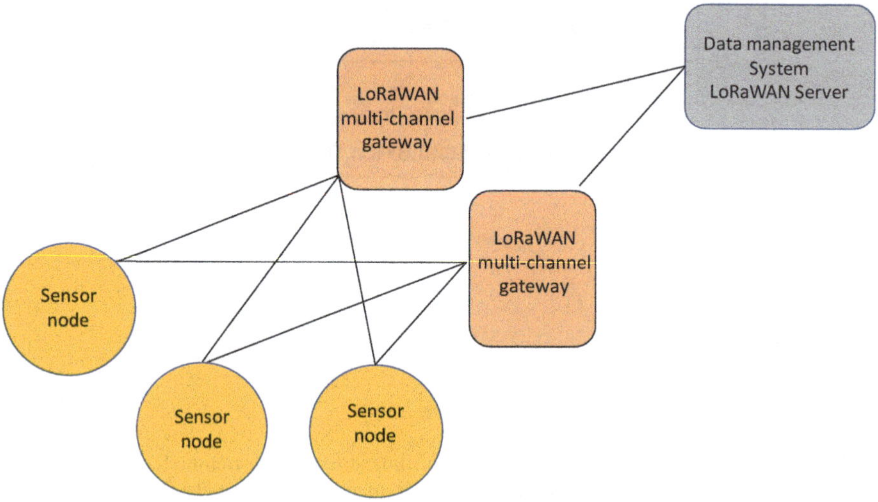


**Fig. 4** IoT network architecture

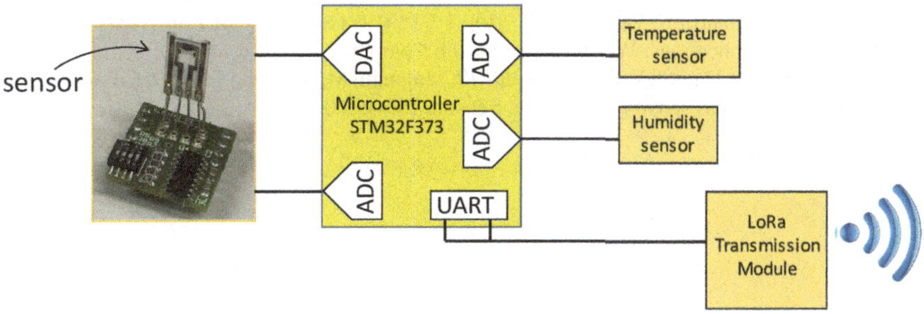


**Fig. 5** Sensor node structure

## 4 Conclusion

In this paper the authors have discussed the architecture of a gas sensor based on chemoresistive gas sensors realized by mixture of nanopowder of ZnO, doped by Al (5%) and $Co_3O_4$ for the monitoring of $NO_2$ in industrial emissions. The gas sensing features of the realized sensor have been investigated, while its integration into a larger IoT framework, exploiting a LoRaWAN LPWAN transmission infrastructure, has been developed. The network architecture provides the possibility to transmit the acquired data to a Cloud platform to be remotely managed. The developed monitoring system may be employed in any industrial or similar scenarios where the supervision of $NO_2$ concentration in the emitted gas is required and the deployment of a large quantity of low cost and low power sensor nodes is needed.

## References

1. Deshmukh S, Bandyopadhyay R, Bhattacharyya N, Pandey RA, Jana A (2015) Application of electronic nose for industrial odors and gaseous emissions measurement and monitoring—an overview. Talanta 144:329–340
2. Potyrailo RA (2016) Multivariable sensors for ubiquitous monitoring of gases in the era of internet of things and industrial internet. Chem Rev 116(19):11877–11923
3. Addabbo T, Fort A, Mugnaini M, Parri L, Parrino S, Pozzebon A, Vignoli V (2018) An IoT framework for the pervasive monitoring of chemical emissions in industrial plants. In: 2018 Workshop on metrology for industry 4.0 and IoT, Apr 2018, pp 269–273
4. Addabbo T, Fort A, Mugnaini M, Panzardi E, Rocchi S, Vignoli V (2016) Automated testing and characterization of electrostatic measurement systems for the condition monitoring of turbo machinery. In: Proceedings of 3rd IEEE international workshop on metrology for aerospace, MetroAeroSpace 2016, Article no 7573225, pp 271–275
5. Korotcenkov G, Cho BK (2017) Metal oxide composites in conductometric gas sensors: achievements and challenges. Sens Actuators B: Chem 244:182–210
6. Rizzi M, Ferrari P, Flammini A, Sisinni E, Gidlund M (2017) Using LoRa for industrial wireless networks. In: 2017 IEEE 13th international workshop on factory communication systems (WFCS), May 2017, pp 1–4

7. Luvisotto M, Tramarin F, Vangelista L, Vitturi S (2018) On the use of LoRaWAN for indoor industrial IoT applications. Wirel Commun Mob Comput
8. Fort A, Panzardi E, Vignoli V, Hjiri M, Aida MS, Mugnaini M, Addabbo T (2019) $Co_3O_4$/Al-ZnO nano-composites: gas sensing properties. Sensors 19(4):760

# Car-Driver Drowsiness Monitoring by Multi-layers Deep Learning Framework and Motion Analysis

**Francesco Rundo, Sabrina Conoci, Francesca Trenta and Sebastiano Battiato**

**Abstract** Recent developments in the automotive industry have led to an interest in monitoring car driver drowsiness. The purpose is to develop an efficient system for the detection of bad psychophysical states in order to reduce the number of fatigue-related car accidents. Much of the current literature pays particular attention to the study of physiological signals to obtain information about cardiac activity by measuring the Heart Rate Variability (HRV). In fact, the HRV represents a useful indicator for evaluating physiological stress because it provides information about the cardiovascular system activity controlled by the Autonomic Nervous system. The present study is designed to analyze the skin micro-movements caused by blood pressure by extracting facial landmarks in order to reconstruct the photoplethysmogram (PPG) signal in a robust way. To conclude, we obtained evidence from the validation results to support the idea that the PPG signal detected by sensors and the reconstructed PPG by using facial landmarks are strongly correlated.

**Keywords** Machine learning · LSTM · Driver-drowsiness

F. Rundo (✉) · S. Conoci
STMicroelectronics-ADG Central R&D, Catania, Italy
e-mail: francesco.rundo@st.com

S. Conoci
e-mail: sabrina.conoci-ext@st.com; sabrina.conoci@unime.it

F. Trenta · S. Battiato
IPLAB, Department of Mathematics and Computer Science, University of Catania, Catania, Italy
e-mail: francesca.trenta@unict.it

S. Battiato
e-mail: battiato@dmi.unict.it

 
G. Di Francia et al. (eds.), *Sensors and Microsystems*, Lecture Notes in Electrical Engineering 629, https://doi.org/10.1007/978-3-030-37558-4_25

# 1 Introduction

Drowsiness is a physiological state characterized by the reduction of the level of consciousness and difficulty in maintaining the wakeful state. According to National Safety Council, the percentage of fatal accidents due to driver drowsiness is dramatically increasing in the United States [1]. As a consequence, the development of an effective warning system that can deliver early detection of the inadequate physiological condition of a car driver would be significant. As reported, research suggests that the Heart Rate Variability is correlated with drivers' level of attention [2]. In fact, cardiac variability represents an essential indicator of the degree of the individual's physiological resilience, and behavioral flexibility. The cardiac activity is evaluated by using the PPG signal in order to detect the blood pressure and, consequently, the heart rate variability. In fact, the PPG signal consists of peaks representing a cardiac cycle which can be evaluated by using a source LED which emits light at different parts of the skin and a photodiode that evaluates the re-transmitted amount of light [3]. Despite physiological signals allowing us to monitor the drowsiness level, recent researches focus on Computer Vision technologies to assess car drivers' fatigue [4]. Although the development of face detection system in a car environment is certainly challenging, many of these approaches are based on using video camera to determine the eye blink rate [5]. Different from other works, our method focuses on Computer vision techniques to detect and extract facial landmarks in order to define landmarks time series by considering the variation of pixel intensities during the video sequence which we previously recorded. More specifically, our approach is based also on "Video Magnification" to reveal facial micromovements caused by blood pressure. Regarding these motivations, the objective of this study is to construct the PPG signal by defining landmarks time-series and without using sensors.

The remainder of the paper is structured as follows. In Sect. 2 we present the related works. In Sect. 3 we proposed an overview about PPG signal and present our pipeline based on Long Short-Term Memory and Convolutional neural networks. Section 4 explains how the experiments are performed. Finally, in Sect. 5 we discuss about advantages of our method and future works.

# 2 Related Works

A considerable amount of literature has been published on driver drowsiness detection through physiological signals to achieve major accuracy. In fact, several studies have shown that methods based on only Computer Vision techniques may not necessarily represent a valid solution. Especially the approaches focused on analyzing road signs fail in the case of non-optimal conditions of road surfaces.

One of the PPG-detection methods is based on the work of Xu et al. [6] where the authors reached effective results by using a low power wireless PPG sensors. Another approach is [7] in which the authors utilized the Low and High Frequency of PPG detected at fingers and earlobes. The works reported in this document are mainly based on the study of ECG and PPG signals in order to evaluate the HRV signal. However, the cited methods involve high computational requirements and expensive devices to be integrated into the vehicle. Although the integrated sensors are not necessarily an invasive tool, it is required that the driver put his/her hands or another part of his/her body, such as the earlobes or the fingers over the sensors in order to acquire the physiological signals accurately, which represents a limitation. Different from previous methods, the proposed framework is based on reconstructing the PPG signal by using facial landmarks from an image of a face of a driver in order to assess the HRV signal and the drowsiness level.

## 3 Background and the Proposed Pipeline

As previously mentioned, we propose an innovative method to determine the drowsiness state of a driver without using sensors for the acquisition of the PPG signal. The work of Wu et al. [8] illustrated how Video Magnification can reveal facial motion changes in ordinary videos by amplifying them whereas the color changes in different spots of the skin are caused by blood pressure during a cardiac cycle. Studies have shown Autonomous Nervous System activity, which regulates certain body processes, such as blood pressure and the rate of breathing, can be measured non-invasively from the Heart Rate Variability signal that presents alterations during stress, extreme fatigue and drowsiness episodes. The HRV can be evaluated by using any biofeedback tool or software, with good-quality sensors for detecting the cardiac signal and a sufficiently powerful processor to manage a large amount of data. Traditionally, the ECG signal is used to assess the Heart Rate Variability signal. However, there are certain drawbacks associated with its use. Despite the effectiveness of the ECG signal, it shows some noise and artifacts due to micro-movements of the person during the acquisition data (data sampling). In order to overcome the problems related to ECG, the PPG signal has been proposed as a reliable solution. The PPG's ability to detect volumetric changes of the heart makes it a very powerful method also for detecting cutaneous micro-movements difficult to observe with the naked eye. In particular, the analysis of the PPG signal allows us to define the variation of the heartbeats over a specific time frame in order to indicate that both branches (parasympathetic and sympathetic) of the Autonomic Nervous system are functioning properly. Generally, a low HRV value refers to a constant interval between heartbeats, otherwise, a high HRV value is provided. A loss of attention due to a condition of chronic stress is determined by a very regular heart rhythm and few variations between heartbeats. However, there is no a single normal HRV value because it differs between individuals.

Taking this into account, the implemented system is based on the combination of Long-Term Short Memory (LSTM) neural networks [9] and Convolutional Neural Network (CNN) [10] for evaluating driver drowsiness. The proposed pipeline represents an improvement in assessing cardiac activity because it requires the use of a video camera with a low frame rate (25fps) for detecting and extracting fundamental points of interest in an image of a face and analyzing their pixel variation for each video frame. In fact, LSTM represents a powerful solution to evaluate the hidden non-linear correlation between data.

In particular, the output of LSTM pipeline represents predicted landmarks time series considering the original PPG, detected by sensors, as target data.

Furthermore, the accurate classification achieved by the CNN model represents the validation for LSTM prediction in order to confirm the level of attention of a car driver.

## 4 Experiments

A total of 71 subjects have been involved to perform our LSTM-CNN pipeline. More specifically, the dataset provides PPG samples from patients/drivers with different genders, ages (between 20 and 70 years old) and pathologies. In this instance, we collected not only healthy subjects but also sick ones who had different issues such as hypertension, diabetes, etc. Their respective PPG signal samples were measured considering two different drowsiness states. Specifically, we emulate a full wakeful and drowsy scenario confirmed by simultaneously ECG signal sampling which showed Beta and Alpha waves respectively confirming high wakeful and drowsy brain activity. Time intervals of 5 min for each scenario were set up to ensure the preliminary system calibration and for real time continuous learning. Simultaneously, we recorded a video sequence of a face of a car driver by using a low frame rate (25fps) Full-HD video camera. As already reported, we detected and extracted facial landmarks from frames of video sequence by using dlib library which represents implementation of algorithm proposed by Kazemi and Sullivan [11]. Next, we calculated the intensity of the pixels associated with each landmark and the variation of its values for each frame with the aim to determine landmarks time series as input of the LSTM network.

### *4.1 CNN Pipeline*

This section describes in more detail the architecture of the CNN model used in the experiments. The proposed CNN architecture represents a reliable support to validate LSTM prediction. In particular, our CNN model is able to track and learn facial expressions of the car driver for improving the drowsiness detection level. We used a batch of size 32 and an initial learning rate of 0.0001 in order to perform the training of the model. Also, we used 32 neurons in the hidden layers and 2 output neurons for binary classification.

The experimental results are very promising as we reached an accuracy of 80%.

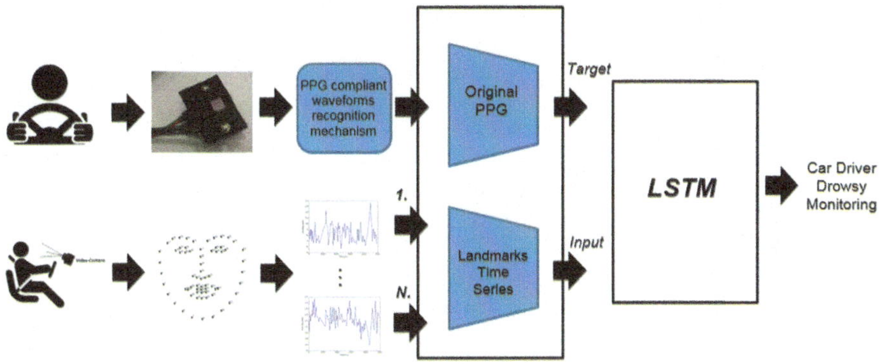


**Fig. 1** LSTM pipeline

### 4.2 *LSTM Pipeline*

Regarding the ability of LSTM to detect dependencies in sequential data (time-series), we implemented LSTM model in order to reconstruct PPG signal by using facial landmarks time-series as input data and original PPG signal as target data (Fig. 1). After scaling all time-series values in the range (0.2, 0.8), with the use of MinMaxScaler algorithm, we performed the training of the model by considering the following parameters. The training is performed using 256 neurons and a batch of size 128. The initial learning rate and the dropout ratio were set up to 0.001 and 0.2, respectively. In order to evaluate the robustness of the reconstructed PPG signal, we calculated the frequency (Fourier Spectrum) of the PPG minimum points. In particular, we considered the distance of these points by comparing the distance of the original PPG minimum points to the distance of the reconstructed PPG minimum points.

## 5 Conclusion

To conclude, we provided an effective LSTM-CNN based system to determine driver drowsiness by evaluating cardiac activity through the PPG signal. Different from other approaches, our method is based on reconstructing the PPG signal from facial landmarks without involving a sensor system. As reported, we implemented the LSTM pipeline in order to prove the robustness of the reconstructed PPG by using facial landmarks time-series as input data and the PPG detected by sensors as target data. In addition, we implemented a CNN model not only to classify the physiological state of a driver but also to validate the LSTM prediction. Finally, we calculated Fast Fourier Transform (FFT) spectrum of the original PPG minimum points and FFT spectrum of the reconstructed PPG minimum points (Fig. 2). The results showed very promising performances of the proposed approach as we were able to discriminate

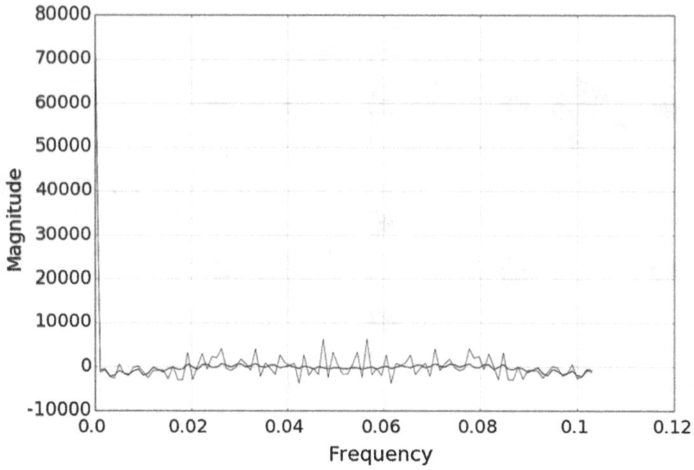


**Fig. 2** Fast Fourier Transform (FFT) spectrum of the original PPG minimum points (blue) and FFT spectrum of the reconstructed PPG minimum points (green)

drowsy subject from wakeful ones with accuracy near to 100% which coincides with the average performance obtained by similar pipeline reported in scientific literature. The authors are investigating such improvements of the proposed pipeline based on the use of modified PPG sensor [12] as well as further processing of PPG signal by means of ad hoc features learned by Stacked-AutoEconder architecture [13–15].

## References

1. National Safety Council. https://www.nsc.org/road-safety/safety-topics/fatality-estimates
2. Melo HM, Nascimento LM, Takase E (2017) Mental fatigue and heart rate variability (HRV): the time-on-task effect. Psychol Neurosci 10:428–436
3. Rundo F, Ortis A, Battiato S, Conoci S (2018) Advanced bio-inspired system for noninvasive cuff-less blood pressure estimation from physiological signal analysis. Computation 6:46
4. Zhang W, Cheng B, Lin Y (2012) Driver drowsiness recognition based on computer vision technology. Tsinghua Sci Technol 17:354–362
5. Haq ZA, Hasan Z (2016) Eye-blink rate detection for fatigue determination. In: 1st India international conference on information processing (IICIP), Delhi, India, pp 1–5
6. Xu YJ, Ding F, Wu Z, Wang J, Ma Q, Chon K, Clancy E, Qin M, Mendelson Y, Fu N, Assad S, Jarvis S, Huang X (2012) Drowsiness control center by photoplythesmogram. In: Proceedings of 38th annual northeast bioengineering conference (NEBEC), Philadelphia, USA, pp 430–431
7. Koh S, Cho BR, Lee J-I, Kwon S-O, Lee S, Lim JB, Lee SB, Kweon H-D (2017) Driver drowsiness detection via PPG biosignals by using multimodal head support. In: Proceedings of the 4th international conference on control, decision and information technologies (CoDIT), Barcelona, Spain, pp 383–388
8. Wu H-Y, Rubinstein M, Shih E, Guttag J, Durand F, Freeman WT (2012) Eulerian video magnification for revealing subtle changes in the world. ACM Trans Graph 31:1–8
9. Hochreiter S, Schmidhuber J (1997) Long short-term memory. Neural Comput 9:1735–1780

10. Lecun Y, Bengio Y (1995) Convolutional networks for images, speech, and time-series. In: The handbook of brain theory and neural networks
11. Kazemi V, Sullivan J (2014) One millisecond face alignment with an ensemble of regression trees. In: 2014 IEEE conference on computer vision and pattern recognition, Columbus, USA, pp 1867–1874
12. Mazzillo M, Maddiona L, Rundo F, Sciuto A, Libertino S, Lombardo S (2018) Characterization of SiPMs With NIR long-pass interferential and plastic filters. IEEE Photonics J 10:1–12
13. Rundo F, Conoci S, Banna GL, Ortis A, Stanco F, Battiato S (2018) Evaluation of Levenberg-Marquardt neural networks and stacked autoencoders clustering for skin lesion analysis, screening and follow-up. IET Comput Vis 12:957–962
14. Rundo F, Ortis A, Battiato S, Conoci S (2018) Advanced bio-inspired system for noninvasive cuff-less blood pressure estimation from physiological signal analysis. Computation 6(3):46
15. Rundo F, Spampinato C, Conoci S (2019) Ad-hoc shallow neural network to learn hyper filtered photoplethysmographic (PPG) signal for efficient car-driver drowsiness monitoring. Electronics (Switzerland) 8:890

# Integration of Two-Dimensional MEMS Ultrasonic Transducer Arrays with Front-End Electronics for Medical Ultrasound Imaging

**Alessandro S. Savoia, Barbara Mauti and Giosuè Caliano**

**Abstract** Electrostatically actuated MEMS ultrasonic transducers, also known as Capacitive Micromachined Ultrasonic Transducers (CMUTs), offer better performance with respect to conventional piezoelectric transducers in terms of bandwidth and efficiency. The enhanced compatibility between MEMS and standard integrated circuit technologies enables the integration of the transducer and the front-end electronics, and is currently leveraged in the development of new-generation diagnostic systems for clinical, emergency and point-of-care uses. This paper reports on the development of an integrated 2-D CMUT array designed for volumetric ultrasound imaging. An acoustically optimized 3-D packaging technology has been applied to the design, fabrication and characterization of a Multi-Chip Module (MCM) obtained by 3-D interconnection of a 256-element CMUT spiral array and a 256-channel analog front-end Application Specific Integrated Circuit (ASIC) with integrated pulsers, low-noise receivers and a programmable TX beamformer.

**Keywords** CMUT · ASIC · Medical ultrasound imaging · Analog front-end · Spiral array · Integrated beamformer · Low-noise amplifier · 3-D packaging

## 1 Introduction

Electrostatically actuated MEMS-based ultrasonic transducers for medical imaging applications, also known as Capacitive Micromachined Ultrasonic Transducers (CMUTs), offer better performance with respect to conventional piezoelectric transducers in terms of bandwidth and efficiency. In vivo ultrasound imaging using one-dimensional (1-D) CMUT arrays was demonstrated more than one decade ago [1]. Two-dimensional (2-D) CMUT arrays with front-end electronics have been demonstrated earlier [2] using conventional flip-chip bonding methods, which, however, are not suitable for the interconnection of large-area dice and for wafer-level packaging. Moreover, conventional flip-chip bonding approaches do not represent a preferable

---

A. S. Savoia (✉) · B. Mauti · G. Caliano
Dipartimento di Ingegneria, Università degli Studi Roma Tre, Rome, Italy
e-mail: alessandro.savoia@uniroma3.it



G. Di Francia et al. (eds.), *Sensors and Microsystems*, Lecture Notes in Electrical Engineering 629, https://doi.org/10.1007/978-3-030-37558-4_26

solution for the integration of medical ultrasound imaging transducers due to the negative impact of the package characteristics on the acoustic performance. In this paper, the development of an integrated 2-D CMUT array for volumetric ultrasound imaging is reported. A recently proposed acoustically optimized 3-D packaging technology [3] has been applied to the design and fabrication of a Multi-Chip Module (MCM) obtained by interconnecting a 256-element CMUT spiral array and a 256-channel analog front-end Application Specific Integrated Circuit (ASIC) with integrated pulsers, low-noise receivers and a programmable TX beamformer. The fabricated MCM is packed in a prototype ultrasound probe head and acoustic characterization results are shown.

## 2 Design and Fabrication of the Integrated Array

### *2.1 MEMS Ultrasonic Transducer Array*

The array configuration was designed following a deterministic approach, in which the array layout is achieved by positioning 256 one-wavelength-wide transducer elements according to a Fermat's spiral. The aperiodicity characteristics of this layout, together with a decreasing spatial density from the center to the edges of the aperture, allows a reduction of the number of elements by a factor of 10 compared to a dense array of equivalent dimensions [4]. The array was designed for a total aperture of 45 wavelengths, allowing volumetric beam steering over a 64° angular sector with a maximum one-way side lobe level of −20 dB. The resulting 256-element spiral array size was 10 mm and the single element size was 220 μm. The CMUT microstructure was designed for a two-way frequency response centered around 7 MHz with a −6dB fractional bandwidth of 100%. Array microfabrication was carried out using CMUT Reverse Fabrication Process [5]. Figure 1 shows show a detail of the single array element composed by 19 CMUT cells arranged in hexagonal tiling and electrically routed to a bond pad for ASIC interconnection, and of the 256-element spiral array.

### *2.2 Front-End Electronics*

The developed integrated analog front-end, based on the transceiver architecture shown in Fig. 2, comprises a high-voltage unipolar pulser, a T/R switch, and a capacitive feedback low noise amplifier with state-of-the-art noise and power performance. The specific circuit topologies used in the co-design of the single CMUT array element and the transceiver integrated in the ASIC are detailed in [6–9].

The new ASIC features 256 transceivers provided with programmable transmit (TX) pulse shaping parameters [9], including pulse frequency, number of cycles, duty cycle, and rising and falling edge slope for CMUT harmonic distortion compensation,

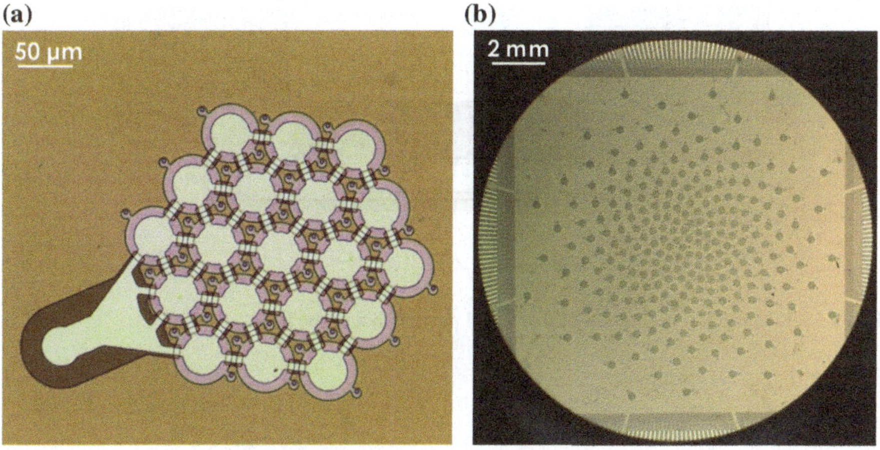


**Fig. 1** Optical microscopy of one array element composed of 19 circular CMUT cells (**a**) and of the 256-element spiral array (**b**)

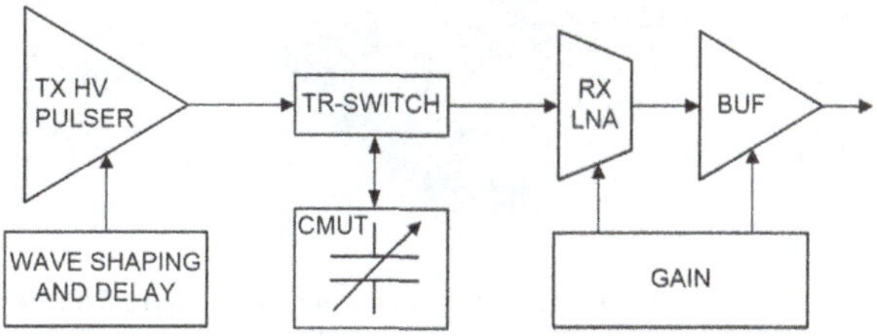


**Fig. 2** Architecture of the transceiver integrated in the 256-channel front-end ASIC

and with programmable RX gain parameters. The ASIC integrates a TX beamformer with a time-delay resolution of 10 ns. The 256 transceivers and the beamformer are programmed and controlled by a digital interface based on four two-channel SPIs and a daisy-chain clocked at 100 MHz. The ASIC was fabricated using a BCD-SOI technology from STMicroelectronics (Agrate, Italy), modified to achieve a surface topography optimized for the proposed 3-D packaging process.

## 2.3 *Integration and Packaging*

A MEMS-ASIC multi-chip module (MCM) was obtained by processing the fabricated CMUT and ASIC wafers following the packaging procedure described in [3], in which Cu pillars and SnAg solder reflow are used for electrical interconnection, and patterned benzocyclobutene (BCB) for mechanical bonding. The resulting MCM [10] is accessible through 320 wire bonding pads located on the four sides of the MEMS die.

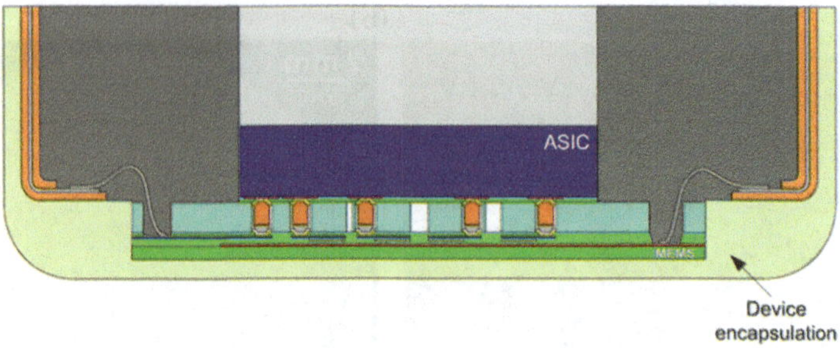


**Fig. 3** Schematic representation of the MCM packed in a probe head

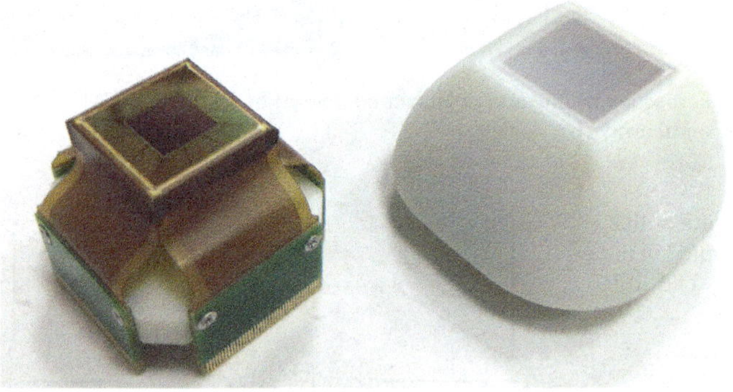

**Fig. 4** The MCM is electrically interconnected to a rigid-flex PCB and the resulting assembly (left) is packed in a prototype probe head (right)

The fabricated MCM was packed according to the Reverse Fabrication Packaging procedure [5], as schematically shown in Fig. 3: wire bonding was used to electrically interconnect the MCM to a rigid-flex PCB; an acoustic backing with thermal dissipation features was applied using bonding encapsulation material; the CMUT array was finally released by etching the silicon fabrication substrate. The packed MCM was finally integrated in the prototype probe head shown in Fig. 4.

## 3 Characterization

Acoustic characterization was carried out by connecting the prototype probe head to a test board for ASIC power supply and programming. The test board was interfaced to a computer-controlled digital pattern generator, and ASIC programming sequences were implemented and used to configure the probe head for TX pressure and pulse-echo testing of a single array element using a 7 MHz single-cycle pulse. The TX pressure signal generated from a single array element was acquired at 3 mm using a

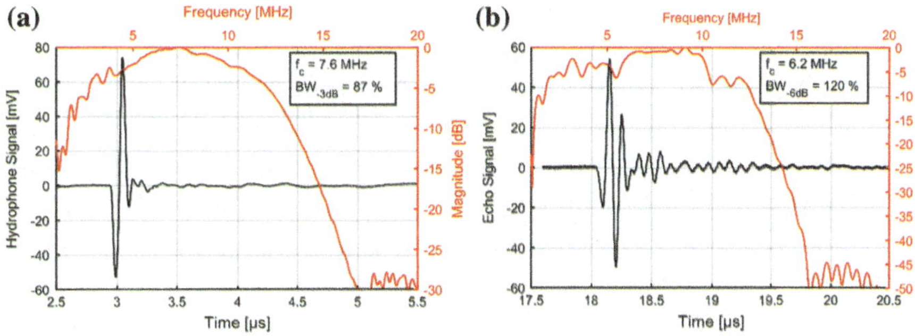


**Fig. 5** TX response **a** measured using a hydrophone and **b** pulse-echo signal received by one array element

needle-type MHA9-150 hydrophone (Force Technology, Brøndby, Denmark). Pulse-echo testing was carried out by using a planar reflector. Figure 5a, b show the hydrophone and echo signals and the relative spectra, respectively, demonstrating the functionality of the MCM.

## 4 Conclusion

A 2-D MEMS ultrasonic transducer array with front-end electronics for medical ultrasound imaging has been developed. A MCM including a 256-element CMUT 2-D array and a 256-channel analog front-end ASIC was obtained using a wafer-level compatible and acoustically optimized 3-D packaging approach. The CMUT array layout was designed for volumetric ultrasound imaging. The front-end electronics integrates high-voltage pulsers, low-noise receivers and a programmable TX beamformer. Acoustic characterization, including TX pressure and pulse-echo measurements, was successfully carried out and demonstrated the functionality of the CMUT array and the ASIC. The integration of the probe head in a 3-D imaging system [11] is ongoing.

## References

1. Savoia A, Caliano G, Carotenuto R, Longo C, Gatta P, Caronti A, Cianci E, Foglietti V, Pappalardo M (2005) Enhanced echographic images obtained improving the membrane structural layer of the cMUT probe. In: Proceedings of IEEE ultrasonic symposium, vol 4, pp 1960–1963
2. Wygant IO et al (2009) An integrated circuit with transmit beamforming flip-chip bonded to a 2-D CMUT array for 3-D ultrasound imaging. IEEE Trans Ultrason Ferroelectr Freq Control 56:2145–2156
3. Savoia AS et al (2017) A 3D packaging technology for acoustically optimized integration of 2D CMUT arrays and front end circuits. In: Proceedings of IEEE ultrasonic symposium

4. Ramalli A, Boni E, Savoia AS, Tortoli P (2015) Density-tapered spiral arrays for ultrasound 3-D imaging. IEEE Trans Ultrason Ferroelectr Freq Control 62:1580–1588
5. Savoia AS, Caliano G, Pappalardo M (2012) A CMUT probe for medical ultrasonography: from microfabrication to system integration. IEEE Trans Ultrason Ferroelectr Freq Control 59:1127–1138
6. Sautto M et al (2014) A CMUT transceiver front-end with 100-V TX driver and 1-mW low-noise capacitive feedback RX amplifier in BCD-SOI technology. In: Proceedings of European solid state circuits conference (ESSCIRC), pp 407–410
7. Savoia AS et al (2014) An ultra-low-power fully integrated ultrasound imaging CMUT transceiver featuring a high-voltage unipolar pulser and a low-noise charge amplifier. In: Proceedings of IEEE ultrasononic symposium, pp 2568–2571
8. Sautto M et al (2017) A comparative analysis of CMUT receiving architectures for the design optimization of integrated transceiver front ends. IEEE Trans Ultrason Ferroelectr Freq Control 64:826–838
9. Savoia AS et al (2015) Second-harmonic reduction in CMUTs using unipolar pulsers. In: Proceedings of IEEE ultrasonic symposium
10. Savoia AS et al (2018) A 256-element spiral CMUT array with integrated analog front end and transmit beamforming circuits. In: Proceedings of IEEE ultrasonic symposium, pp 206–212
11. Boni E et al (2016) ULA-OP 256: a 256-channel open scanner for development and real-time implementation of new ultrasound methods. IEEE Trans Ultrason Ferroelectr Freq Control 63:1488–1495

# Enhancement of PCR Reaction Efficiency by Gold-Nanoparticles Immobilized at Microreactor Surface

**Salvatore Petralia, Giuseppe Nocito, Sabrina Conoci and Salvatore Sortino**

**Abstract** Metal nanoparticles are widely employed in molecular biology as PCR-facilitators for the improvement of the polymerase chain reaction efficiency. Various metal nanoparticles were investigated such as Platinum, Silver and Gold, with various capping approach. The PCR enhancement mechanism could be related to the excellent heat dispersion of metal nanomaterial, to the surface interaction with the DNA template and with the enzyme polymerase. Herein we report the Real Time PCR facilitator effect of Gold nanoparticles immobilized at microreator surface. The Gold nanoparticles prepared with the standard chemical reduction method have been immobilized on epoxy-silanized plastic microreactor surface. The PCR efficiency was evaluated in presence of different amounts of nanoparticles. The results demonstrated that the Gold nanoparticles immobilized at surface were able to increase the polymerase chain reaction efficiency mainly by metal thermal efficiency mechanism, determining a sensitivity improvement of about 3 CT (corresponding to about one magnitude order). The proposed approach is very promising for the application on molecular diagnostics applications.

**Keywords** Gold nanoparticles · Polymerase chain reaction · Nucleic acids detection

## 1 Introduction

The Polymerase Chain Reaction (PCR) discovered by Mullis (Nobel Prize in Chemistry in 1993) [1] is the most famous in vitro molecular biology method for the amplification of DNA regions. It is used in many applications such as genotyping, sequencing and molecular diagnostics. The Real Time PCR is the most innovative and useful amplification method for the direct Nucleic Acids detection, largely used on Point-of-care system [2].

S. Petralia (✉) · S. Conoci
STMicroelectronics, Stradale Primosole, 50, 95121 Catania, Italy
e-mail: salvatore.petralia@st.com

G. Nocito · S. Sortino
Laboratory of Photochemistry, Department of Drug Sciences, University of Catania, Catania, Italy

 
G. Di Francia et al. (eds.), *Sensors and Microsystems*, Lecture Notes in Electrical Engineering 629, https://doi.org/10.1007/978-3-030-37558-4_27

In 2005 Li and coworkers provided the first example of noble metal nanoparticle-assisted-PCR to increase the amplification efficiency and to avoid the unspecific amplification [3]. The main effects of metal nanoparticles as PCR-facilitators could be explained by their excellent thermal efficiency [4], by metal surface interaction with PCR primers and with enzyme DNA polymerase [5]. The main limitation of NPs on biological application are addressable to their low stability and solubility in aqueous media. To prevent particle aggregation some stabilizing agents such as Thioglycolic acid, Polyvinylpyrrolidone, and Citrate are usually added during the NPs preparation, to create a capping layer.

Although the positive effects of colloidal metal nanoparticles are widely reported in literature [6], few examples of PCR facilitators effect for metal nanoparticles immobilized on microreactor surface are disclosed. Herein we report the first example of Real Time PCR facilitators effect of Gold nanoparticles (Au-NPs) immobilized at microreator surface. The Au-NPs were prepared with the standard chemical reduction method and have been immobilized on epoxy-silanized plastic microreactor surface. In this case the capping layer is involved on interaction with the epoxy-layer surface.

The PCR efficiency was evaluated in presence of different amounts of both nanoparticles. The results demonstrated that the Gold nanoparticles immobilized at surface were able to increase the polymerase chain reaction efficiency mainly by metal thermal efficiency mechanism, determining a sensitivity improvement of about 3 CT (corresponding to about one magnitude order), the proposed approach is very promising for the application on molecular diagnostics applications.

## 2 Materials and Methods

### *2.1 Chemicals and Instruments*

The glycidoxypropyltrimethoxysilane (GPTS) for silanization process and all reagent for AuNPs preparation, Gold(III) chloride trihydrate and sodium citrate trihydrate, were purchased by Sigma-Aldrich. Deionized water was used for cleaning steps. All reagents for the Real time PCR experiments were purchased from Clonit and used according to the instruction for use. The target used was the Hepatitis B virus (HBV) clone (ref. product CLO-05960116 HBV Complete Genome). Real time PCR experiments were performed using standard Applied Biosystem thermocycler. The 384-PCR-microplates were purchased by Corning. The TEM analysis was performed by JEOL JEM-2010 instrument equipped with a thermo-ionic source of Lanthanum exaboride (LaB6) and using bright field in conventional parallel beam mode. The samples were prepared dispensing large and low amounts of NPs on Copper grid surface. The Dynamic Light Scattering (DLS) measurements were carried out by the Horiba LB-550 instrument with a 1 cm quartz cell. Each measurement was performed three times. The UV/Vis spectra were recorded using a double-beam Perkin Elmer Lambda 365 spectrophotometer in a 1 cm quartz cell.

## 2.2 Gold Nanoparticles Preparation

The Au-NPs were prepared suing the standard Turkevich's method, resumed and modified by Polte et al. [7], when small and spherical AuNPs were formed by the reduction of the Sodium citrate dihydrate to the gold precursor ($HAuCl_4$ - Gold(III) chloride trihydrate) that plays at the same time the role of the capping agent. In detail a 35 mL aliquot of $HAuCl_4$ $5 \times 10^{-4}$ M, prepared from dilution by the stock solution, was heated, using a temperature-controlled oil bath, in a flask at 75 °C and under stirring was added the same volume of a sodium citrate solution $5 \times 10^{-3}$ M (51.45 mg) pre-heated under the previous conditions. The reaction was conducted at 75–80 °C for 20 min observing rapidly the discoloration of the initial yellow for changing, to the end, in a light purple that becomes a purple-red when the solution was cooled at room temperature.

## 2.3 Real Time PCR Experiments

Real time PCR experiments were performed using a Master Mix solution composed by Clonit buffer 1× and Taq DNA polymerase, 0.5 μM of forward and reverse primers, 2 μL of HBV clone ($10^3$ copies/μL). The PCR cycling was performed in standard AB thermocycler by using the following thermal program: preheating period of 10 min at 95 °C followed by a maximum of 45 cycles of 95 °C for 15 s and 60 °C for 60 s. Different temperatures of annealing were used for testing (60–63 °C). Negative samples were prepared substituting the HBV clone with the same amount of water. Each experiment was repeated 5 times.

# 3 Discussion

The 384-PCR-microplates were cleaned by Plasma-$O_2$ method at 100 W for 10 min and silanized with GPTS in vapor phase mode at 120 °C, 0.1 Atm for 4 h. After that various amount of pre-formed Au-NPs (0, 6, 12, 18 and 24 μg) were immobilized on epoxy-silanized microreactors by overnight incubation at 35 °C and dried at 30 °C.

The effectiveness of the silanization chemical process was demonstrated by the water contact angle measurements on test chips. In details, after the plasma-$O_2$ cleaning step a very low water contact angle value of about 5° was measured to indicate a hydrophilic surface; this value increase to about $80 \pm 5°$ upon the silanization to indicate the formation of the hydrophobic epoxy-silane layer.

The Gold-NPs diameter was investigated by DLS and TEM techniques. The TEM analysis indicates an Au-NPs size of $7 \pm 3$ nm (Fig. 1 left) while the DLS measurement reports a medium hydrodynamic diameter of $27 \pm 0.7$ nm. The UV-Vis optical

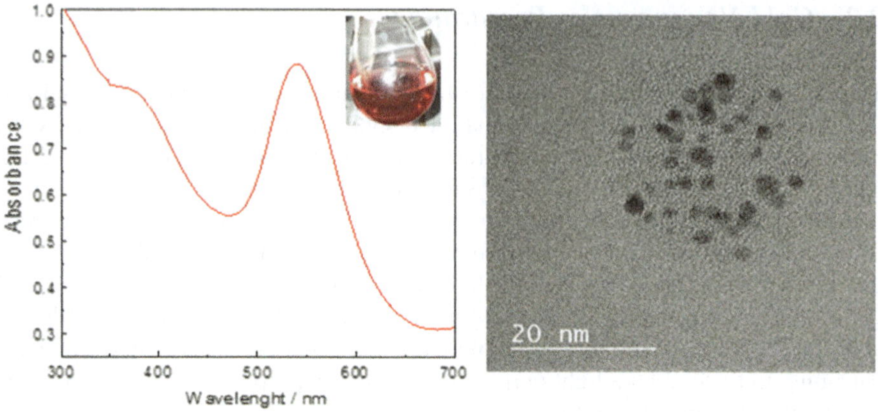


**Fig. 1** From left: the UV-Vis absorption spectrum shows the typical plasmon absorption band centered at 530 nm and the representative TEM image of Au-NPs

absorption spectrum shows the typical plasmon absorption band centered at 530 nm (Fig. 1 right).

The Real Time PCR experiments, performed on 384-PCR-microplates containing various amounts of Au-NPs (Fig. 2 left), indicate a noticeable increasing of DNA amplification with the increasing of Au-NPs amounts (Fig. 2 right). In particular, straightforward sensitivity improvement of 10, 1.5 and 3 CT were recorded for the AuNPs amount of 12 μg, 18 μg and 24 μg respectively.

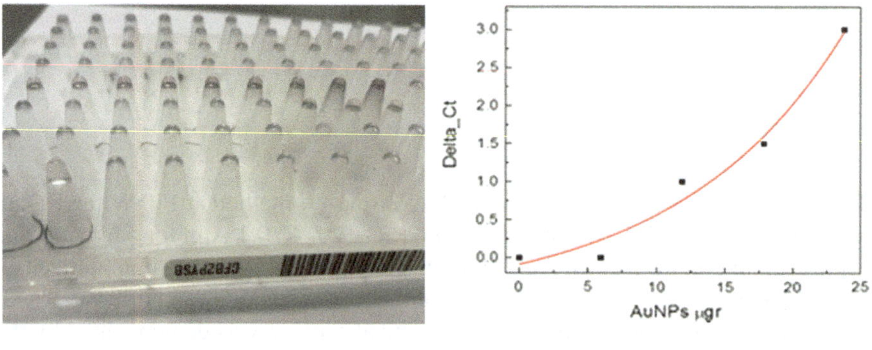


**Fig. 2** From left: the 384-PCR-microplates treated with Au-NPs and the decreasing of CT versus the Au-NPs amount

## 4 Conclusion

In this contribution, we report the PCR facilitators effect of Au-NPs immobilized at 384-microreactors surface chemical finished with epoxy-silanized layer. The data demonstrated that AuNPs immobilized at surface were able to increase the PCR efficiency mainly by metal thermal efficiency mechanism, determining a sensitivity improvement of about 3 CT (corresponding to about one magnitude order of sensitivity). The proposed approach is very promising for the application on molecular diagnostics applications.

## References

1. Saiki RK, Scharf S, Faloona F, Mullis KB, Horn GT, Erlich HA, Arnheim N (1985) Science 230:1350–1354
2. Petralia S, Conoci S (2017) ACS Sens 2:876
3. Li HK, Huang JH, Lv JH, An HJ, Zhang XD, Zhang ZZ, Fan CH, Hu J (2005) Angew Chem Int Ed 44:5100–5103
4. Keblinski P, Phillpot SR, Choi SUS et al (2002) Int J Heat Mass Transf 45(4):855–863
5. Mi LJ, Zhu HP, Zhang XD, Hu J, Fan CH (2007) Chin Sci Bull 52(17):2345–2349
6. Petralia S, Barbuzzi T, Ventimiglia G (2012) Mater Sci Eng C 32:848–850
7. Polte J, Tornsten A, Kraehnert R (2010) JACS 132:1297

# Analysis of the Potential Barrier on the Behaviour of 4H-SiC JBS Temperature Sensors

**Luigi Di Benedetto, Gian Domenico Licciardo and Alfredo Rubino**

**Abstract** In this paper we report the dependency of the potential barrier height in 4H-SiC Junction Barrier Schottky diodes on the channel width and on the junction temperature. The model takes onto account geometrical parameters of the device and physical parameters, like the work function of the Schottky metal and the doping concentration of the epilayer. The obtained expressions are fully analytical so that the dependency is well described. Moreover, the model results are compared with numerical simulations in order to understand the goodness of the proposed model. It is also shown the different effect of the temperature on the potential barrier height and, in particular, the barrier reduces its values with the temperature when the channel width is lower than 1 μm. That expects a different behavior of the devices as temperature sensors, like sensitivity, in terms of the only channel width making the design of the sensor easier.

**Keywords** 4H-SiC semiconductor device · Modeling · Potential barrier · Temperature sensor

## 1 Introduction

4H-polytype Silicon Carbide semiconductor is a wide-bandgap material with superior physical properties respect to Silicon, like high critical electric field, saturation velocity and radiation hardness [1]. Its relevant interest for the power electronics permitted to improve its manufacturing processes developing more robust and stable electronic devices [2–5]. Such aspects i.e. mature technology and superior properties, allowed the development of new devices applied to several other applications, like

L. Di Benedetto (✉) · G. D. Licciardo · A. Rubino
Department of Industrial Engineering, University of Salerno, Via Giovanni Paolo II, 132, 84084 Fisciano, SA, Italy
e-mail: ldibenedetto@unisa.it

G. D. Licciardo
e-mail: gdlicciardo@unisa.it

A. Rubino
e-mail: arubino@unisa.it

 
G. Di Francia et al. (eds.), *Sensors and Microsystems*, Lecture Notes in Electrical Engineering 629, https://doi.org/10.1007/978-3-030-37558-4_28

optoelectronics [6] and sensing [7]. In particular, 4H-SiC Schottky and pin diodes have been developed as temperature sensors working, respectively, up to 440 K [8] and 873 K [9]. In this context, 4H-SiC Junction Schottky Barrier diodes can be used to obtain temperature sensors with different characteristics, i.e. sensitivity and linearity, respect to 4H-SiC Schottky and to 4H-SiC pin diodes thanks to the potential barrier developed into the channel [10, 11]. Indeed, observing Fig. 1 a JBS diode structure consists in a Schottky junction surrounded by two $p^{+}/n$ junctions, fabricated by ion implantations [12], in which we define a channel region. Inside this area, due to the interaction of the electric fields at two bipolar junctions, the potential barrier, $V_{SB/CH}$, is formed between the Schottky anode contact and the saddle point in the middle of the channel region, similarly to JFET transistor [13–15] or to the JFET-region of DMOSFET [16, 17], and can have a height higher than the built-in voltage of the Schottky junction, $V_{bi,SCH}$. That causes that the electron flow from the cathode to the anode is controlled by $V_{SB/CH}$ and not by the Schottky junction, whose result is a rectifying characteristic with a turn-on voltage higher than that of a Schottky diode and lower than that of pin diode. The advantages of such device are to design a temperature sensor that it can operate at temperatures higher than that of a Schottky diode and with interesting properties like bipolar diode [18], and that it is not affected by the drawbacks of bipolar diodes, mainly the highest voltage drops in ON-state and the bipolar degradation [19].

In this paper, we report the analysis on $V_{SB/CH}$ as function of the temperature and of $X_{CH}$. Indeed, the value of $V_{SB/CH}$ depends on the design of the channel region and, in particular, on the channel width, $X_{CH}$, and depth, $Y_G$, on the channel doping concentration, $N_{EPI}$, and on the metal work function, *WF*. It is worth to note that, once the device process is fixed, $X_{CH}$ is the only structure parameter that can be changed by designing device layout.

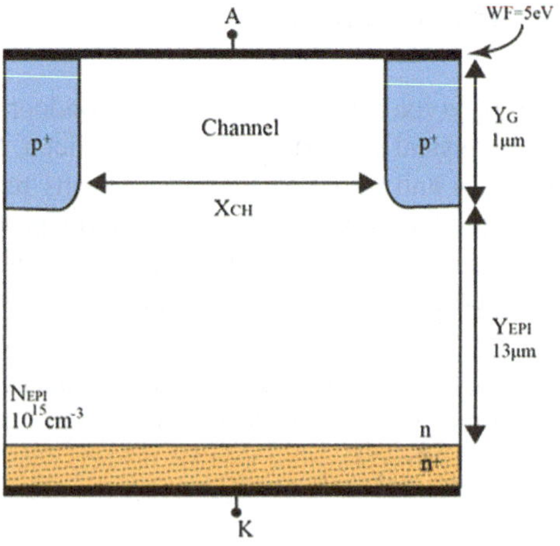


**Fig. 1** Device cell structure and relevant geometrical and physical parameters

## 2 Analytical Model

The device structure of Fig. 1 is a schematized JBS structure where the Schottky contact has a $WF = 5$ eV with a $V_{bi,SCH}$ of 1.03 eV, 0.83 eV and 0.62 eV at a temperature of 300 K, 500 K and 700 K, respectively. The values of the physical and geometrical quantities are reported in Fig. 1.

The model defines the Gate/Channel potential barrier, $V_{G/CH}$, as the lateral voltage drop from the p-region to the saddle point in the centre of the channel [20] and analytically calculates it in terms of all physical parameters [11] as follows:

$$V_{G/CH} = V_A + V_B \tag{1}$$

$V_A$ is the voltage drop in the region at the $p^+/n$ junction where the hole carrier concentration reduces from the effective acceptor doping concentration of p-region down to a concentration of a tens of $N_{EPI}$ and equal to $V_A = V_T \ln\left(N_{A,eff}/20N_{EPI}\right)$, $V_T$ the thermal voltage, $N_{A,eff}$ the effective doping concentration of the p-type regions, $V_B$ mainly drops in the channel region and depends on the horizontal electric field as follows:

$$V_B = \frac{1}{2}\left(E_M \frac{X_{CH}}{2}\right) = \frac{X_{CH}}{4}(E_{LIN} + E_{CYL}) \tag{2}$$

where $E_{LIN}$ is the maximum electric field obtained from 1-D Poisson equation [20] and $E_{CYL}$ takes into account the contribution of the corner effects at the $p^+$-n junctions [13]. Once the $V_{G/CH}$ is evaluated, $V_{SB/CH}$ can be calculated as:

$$V_{SB/CH} = V_{bi,pn} - V_{bi,SCH} - V_{G/CH} \tag{3}$$

where $V_{bi,pn}$ is the built-in voltage of $p^+$-n junction.

## 3 Results

In this session our analytical model is compared with numerical simulation results, obtained with the simulation framework ATLAS Silvaco [21] with the enabling models: incomplete ionization of dopants, band–gap narrowing, field and doping dependent carrier mobilities, SRH phenomenon and thermionic and field emission models for the Schottky contacts [4, 10, 22].

In Fig. 2 the comparison between numerical simulation and analytical model of the potential barrier heights at different $X_{CH}$ is reported. When the channel is narrow, the channel region is completely depleted and $V_{SB/CH}$ is higher than that at the Schottky junction affecting the electron flow from the cathode to the anode; instead, when the channel is enough wider, i.e. $X_{CH} > 2$ μm, $V_{SB/CH}$ disappears and the device behaves as Schottky diode because the electron flow is controlled by the Schottky barrier. It is worth to note that the reference potential is at the anode contact and for $X_{CH} > 2$ µm the potential in the channel is lower than that at the Schottky junction.

In Fig. 3 the effects of the temperature from 300 to 700 K is reported. It is interesting to note that the increasing of the temperature strongly reduces $V_{SB/CH}$ at narrow channel, whereas it weakly affected at wider channel, causing a different temperature behavior of the diode current. That suggests the possibility to engineer the sensor by only design the channel width.

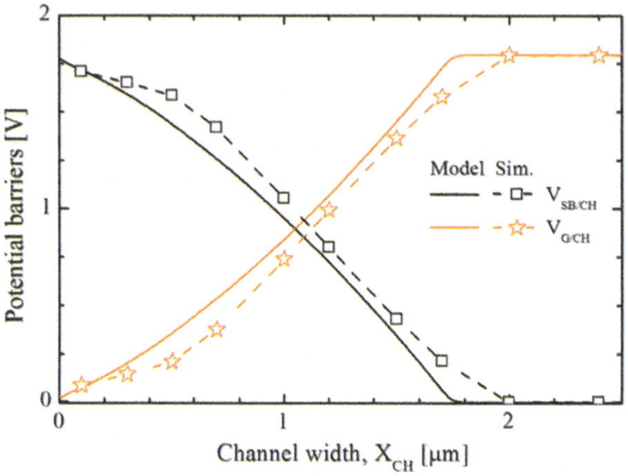


**Fig. 2** Comparison between model and numerical simulation results of $V_{SB/CH}$ and of $V_{G/CH}$ as function of $X_{CH}$ at r.t. and for the device structure of Fig. 1

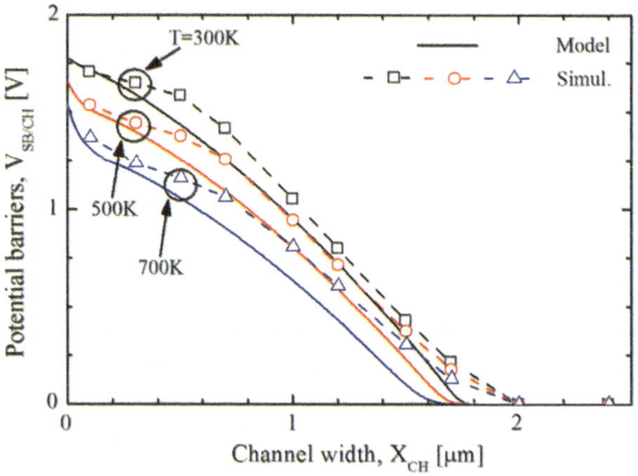


**Fig. 3** Comparison between model and numerical simulation results of $V_{SB/CH}$ as function of $X_{CH}$ at different device temperatures and for the device structure of Fig. 1

## 4 Conclusion

The proposed model permits to calculate the potential barrier height in 4H-SiC JBS diodes in terms of physical and geometrical parameters as well as the temperature. The model is fully analytical and in closed form that can be useful to make explicit the various dependencies. Indeed, it can be used to design the channel region in order to achieve a particular value of the barrier height. Moreover, it is the first step to model the electric characteristics of the devices because the diode current is defined by the electron flow controlled by the potential barrier.

## References

1. Kimoto T, Cooper JA (2014) Fundamentals of silicon carbide technology: growth, characterization, devices, and applications, 1st edn. Wiley-IEEE Press, Singapore
2. Licciardo GD et al (2015) Analytical model of the forward operation of 4H-SiC vertical DMOSFET in the safe operating temperature range. IEEE Trans Power Electron 30:5800–5809
3. GA20JT12-263, GeneSiC semiconductor (2015)
4. Bellone S et al (2014) Design and performances of 4H-SiC bipolar mode field effect transistor (BMFETs). IEEE Trans Power Electron 29:2174–2179
5. Ryu SH et al (2012) Ultra high voltage (>12 kV), high performance 4H-SiC IGBTs. In: Proceedings of the 24th ISPSD, pp 257–260
6. Di Benedetto L et al (2017) Experimental results on lateral 4H-SiC UV photodiodes. In: 7th international workshop on advances in sensors and interfaces (IWASI). IEEE, pp 252–254
7. Matthus C et al (2019) Feasibility of 4H-SiC p-i-n diode for sensitive temperature measurements between 20.5 K and 802 K. IEEE Sens J 19:2871–2878
8. Rao S et al (2016) 85–440 K temperature sensor based on a 4H-SiC Schottky diode. IEEE Sens J 16:6537–6542
9. Zhang N et al (2014) Temperature sensor based on 4H-silicon carbide pn diode operational from 20 °C to 600 °C. Appl Phys Lett 104:073504
10. Di Benedetto L et al (2016) Analytical model and design of 4H-SiC planar and trenched JBS diodes. IEEE Trans Electron Dev 63:2474–2481
11. Bellone S et al (2016) On the analogy of the potential barrier of trenched JFET and JBS devices. Solid State Electron 120:6–12
12. Nipoti R et al (2012) $Al^+$ implanted anode for 4H-SiC p-i-n diodes. Trans Electrochem Soc 50:391–397
13. Bellone S et al (2014) A model of the $I_D$-$V_{GS}$ characteristics of normally off 4H-SiC bipolar JFETs. IEEE Trans Power Electron 29:514–521
14. Brewer RJ (1975) The 'Barrier mode' behaviour of a junction FET at low drain currents. Solid State Electron 18:1013–1017
15. Bellone S et al (2015) A model of the off-behaviour of 4H-SiC power JFETs. Solid State Electron 109:17–24
16. Di Benedetto L et al (2016) A model of electric field distribution in gate oxide and JFET-region of 4H-SiC DMOSFETs. IEEE Trans Electron Dev 63:3795–3799
17. Di Benedetto L et al (2016) Optimized design for 4H-SiC power DMOSFET. IEEE Electron Dev Lett 37:1454–1457
18. Di Benedetto L et al (2014) On the crossing-point of 4H-SiC power diodes characteristics. IEEE Electron Dev Lett 35:244–246
19. Levinshtein ME et al (2011) Bipolar degradation of high voltage 4H-SiC p-i-n diodes in pulse regime. Mater Sci Forum 679–680:539–542

20. Bellone S et al (2012) A quasi-one-dimensional model of the potential barrier and carrier density in the channel of Si and 4H-SiC BSITs. IEEE Trans Electron Dev 59:2546–2549
21. SILVACO Int., Santa Clara, CA, ATLAS User's Manual, 2005 Ver. 5.10R
22. Bellone S et al (2012) An analytical model of the switching behavior of 4H-SiC $p^+$-n-$n^+$ diodes from arbitrary injection conditions. IEEE Trans Power Electron 27:1641–1652

# Photochemical Activation of Non-enzymatic Sensors Based on $Cu/TiO_2$

C. Espro, S. G. Leonardi, S. Marini, G. Neri and D. M. Tobaldi

**Abstract** In this paper are reported the photoelectrochemical properties of $CuO/TiO_2$ for glucose sensing in alkaline media. First, we synthesized by the sol-gel method Cu-$TiO_2$ samples having different Cu loading (from 0 to 10 wt%). The nanometric size of the materials synthesized is evidenced by the characterization done with different methods, including scanning electron microscopy SEM/EDX, TEM and X-ray powder diffraction XRPD. The influence of Cu loading over the photocatalytic performances of $TiO_2$ was examined in detail carrying out electrochemical sensing tests at different illumination conditions. The obtained results demonstrated that the modified electrodes showed a higher specific response signal, lower detection limit and wider linearity range than the blank CuO electrode. The results of this study offer then guidelines for the design of photo-electrochemical screen-printed electrodes based on nano-sized CuO on titania for efficient detection of glucose.

**Keywords** Non-enzymatic sensors · Glucose · Copper nanoparticles · Titanium dioxide

## 1 Introduction

Currently, the development of new screen printed electrochemical sensors for monitoring biochemical substances is a very competitive and stimulating research area and has high relevance and applications in the pharmaceutical and personal care fields. Today, these devices are generally based on enzymes that are sensitive and very specific. However, the activity of enzymes is affected by instability, high cost of enzymes, complicated immobilization procedures and critical operating conditions. To overcome the above mentioned drawbacks, non-enzymatic electrochemical sensors have become of great interest, particularly through the modification of the

C. Espro · S. G. Leonardi · S. Marini · G. Neri (✉)
Department of Engineering, University of Messina, C.da Di Dio, 98166 Messina, Italy
e-mail: gneri@unime.it

D. M. Tobaldi
Department of Materials and Ceramic Engineering, University of Aveiro, Santiago, 3810-193 Aveiro, Portugal



G. Di Francia et al. (eds.), *Sensors and Microsystems*, Lecture Notes in Electrical Engineering 629, https://doi.org/10.1007/978-3-030-37558-4_29

electrode surface with metal oxide nanoparticles. For example, metals oxides, such as copper oxide (CuO) and nichel oxide (NiO), are often recognized for the oxidation of glucose [1, 2]. Copper nanoparticles have been received considerable interest due to the high activity towards glucose oxidation in alkaline conditions, whereas titanium dioxide nanoparticles are well known to exhibits photo-catalytic properties in the oxidation of organic molecules [3]. Moreover $TiO_2$ NPs possess many advantages, i.e. large surface area, excellent semiconducting properties, low-cost, high chemical stability. CuO-$TiO_2$ have been widely used as composite photocatalyst to decrease the recombination of the photo-generated excitement, as well as increase reaction kinetics, whereby charge carriers are transferred along junctions formed at the interface between the semiconductor particles [4]. Therefore, we synthesised CuO-$TiO_2$ samples with different CuO loading with the objective to study this heterojunction system for glucose sensing. Photo-activated electrochemical (PEC) sensors based on photocatalytic systems have been recently exploited for electroanalytical purposes, attracting great attention for obtaining high sensitivity with low background noise [5]. On the other hand, copper oxides have been already reported for PEC sensing of some biomolecules under visible-light irradiation [6, 7]. Fallowing what was said, $TiO_2$ modified screen-printed carbon electrodes (SPCE) were fabricated and tested in dark condition, laboratory illumination and halogen lamp illumination, with the objective to investigate the performances of PEC CuO-$TiO_2$ heterojunctions for an efficient quantification of glucose.

## 2 Experimental

### 2.1 *Synthesis of CuO/$TiO_2$ Nanoparticles*

Aqueous titanium(IV) hydroxide sols were made via the carefully controlled hydrolysis and peptization of titanium(IV) isopropoxide (Ti-i-pr, $Ti(OCH\text{-}(CH_3)_2)_4$) with distilled water diluted in isopropyl alcohol (IPA, propan-2-ol), following a protocol previously reported in detail [8]. CuO modified $TiO_2$ sols, with a $TiO_2$:CuO molar ratio equal to 1/0.05 and 1/0.10, were prepared adding copper(II) nitrate trihydrate (Aldrich, $\geq$98.5%) to the $TiO_2$ sol. Afterwards, dried gels were thermally treated at 450 °C under a static air flow, at heating/cooling rate of 5 °C $min^{-1}$ [9].

### 2.2 *Characterization*

The morphology of 10% Cu/$TiO_2$ nanoparticles was observed using a ZEISS 1540XB FE SEM/EDX (Zeiss, Germany) instrument operating at 10 kV. Images of higher magnification were obtained using a JEM-2100Plus Transmission Electron Microscope (TEM) using an acceleration voltage of 200 keV. Analysis using X-ray powder diffraction (XRD) was carried out to quantify the weight percentage of crystalline phases in the specimens prepared in addition to microstructural features.

XRD analysis were performed on the NIST SRM 660b standard ($LaB_6$) with data collected under the same conditions as those used for the $TiO_2$ samples. The angular range 20–145° 2θ was investigated, with virtual time per step of 500 s, so as to employ data with high signal-to-noise ratio.

### 2.3 *Sensor Fabrication*

For the preparation of the modified sensors, 5 mg of each CuO-$TiO_2$ nanocomposites were dispersed in double distilled water (1 ml) then sonicated until homogenous suspensions were obtained. The working electrode of commercial screen printed carbon electrodes (SPCEs), purchased from DropSens, consisting of a planar substrate equipped with a 4 mm in diameter carbon working electrode, a carbon counter electrode and a silver pseudoreference electrode, was modified by wet impregnation method, casting 5 ml of above suspension on its surface, then allow to dry at room temperature.

## 3 Results and Discussion

SEM images, taken at different magnification, were acquired to investigate the morphology of the synthesized CuO-$TiO_2$ specimens. As shown in Fig. 1 the surface of 10% CuO-$TiO_2$ is rough and constituted of NPs clusters around 20 nm ± 5 nm. EDX analysis maps (not shown) demonstrate that CuO NPs are very well distributed on the $TiO_2$ surface of larger clusters.

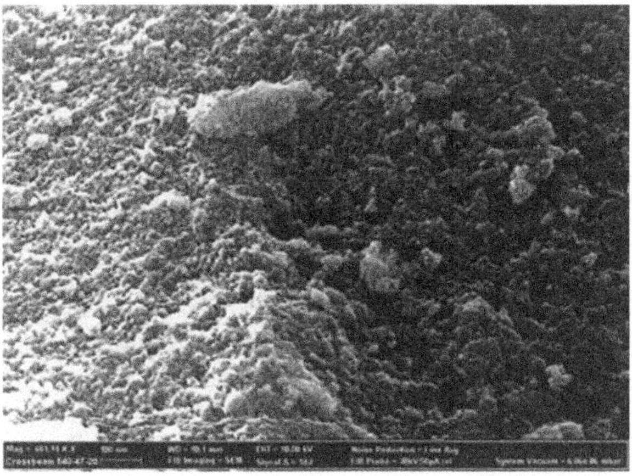

**Fig. 1** SEM image of 10% Cu-$TiO_2$ cluster

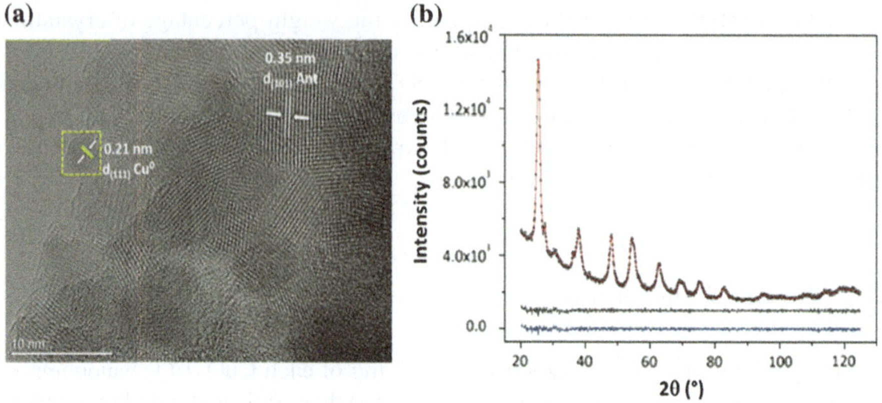


**Fig. 2** **a** High magnification TEM picture of the surface of 10% Cu-$TiO_2$. **b** XRD analysis of 10% Cu-$TiO_2$

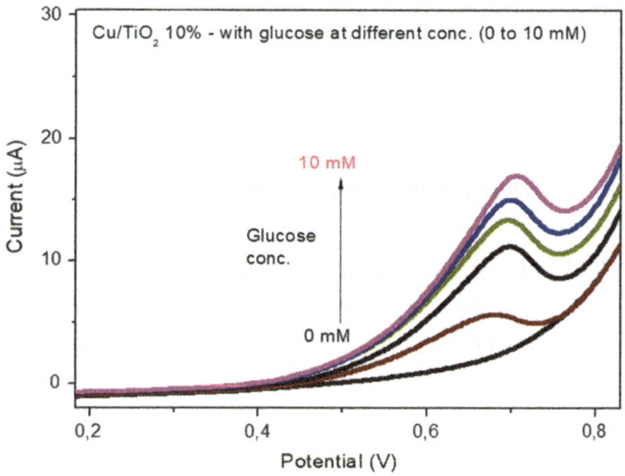


**Fig. 3** Linear sweep voltammetry (LSW) of 10% Cu/$TiO_2$ for different glucose concentration at a scan rate of 50 $mVs^{-1}$

TEM analysis confirmed the nanometer size of CuO/$TiO_2$. In particular Cu (~2 nm) nanocrystals decorate the surface of $TiO_2$ nanoparticles (Fig. 2a). XRD analysis indicates that in 10% Cu/$TiO_2$, titania is composed of 83 wt% of anatase, together a small content of brookite and rutile phase (Fig. 2b). In addition, XRD spectrum confirms that copper does not enter the $TiO_2$ lattice.

Cu/$TiO_2$ nanoparticles were then used to modify the working electrode of commercial screen printed carbon electrodes (SPCEs). The electrochemical behaviour of modified Cu/$TiO_2$/SPCE electrodes was investigated by performing linear sweep voltammetry (LSW) in 0.1 M KOH solution in the range of potential 0–1 V, with different glucose concentration and at a scan rate of 50 $mVs^{-1}$ (Fig. 3). The oxidation potential shifts towards more positive values increasing the concentration of glucose.

Data reported in Fig. 4, collected performing tests under ambient light, shown that the response to glucose is strictly dependent on Cu loading, confirming that this specie acts as the precursor of active sites in the electro-oxidation of glucose in alkaline environment [1].

The 10% $Cu/TiO_2$ sensor has been also tested in dark and in presence of illumination by using a halogen lamp (50 W). Figure 5 shows the current value, subtracted of background current, as function of glucose concentration for the different light conditions.

It can be clearly observed that the 10% $Cu/TiO_2$ electrode irradiated with halogen light exhibits a more pronounced sensitivity to glucose, indicating that the use of halogen light has a promoting effect in the oxidation of glucose.

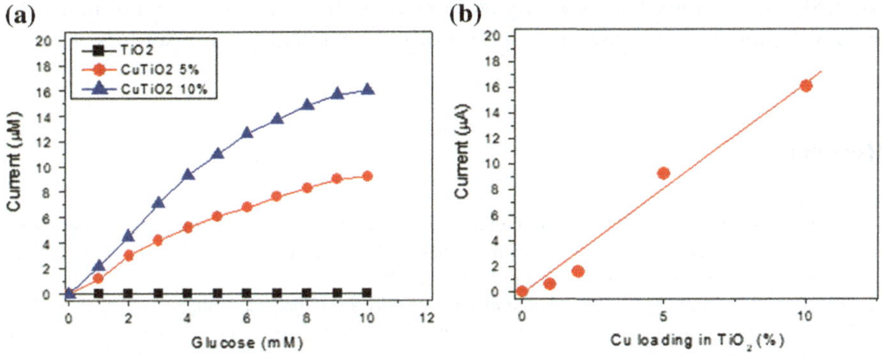


**Fig. 4** **a** Calibration curve for various $Cu/TiO_2$/SPCE electrodes as function of glucose concentration; **b** current versus Cu loading

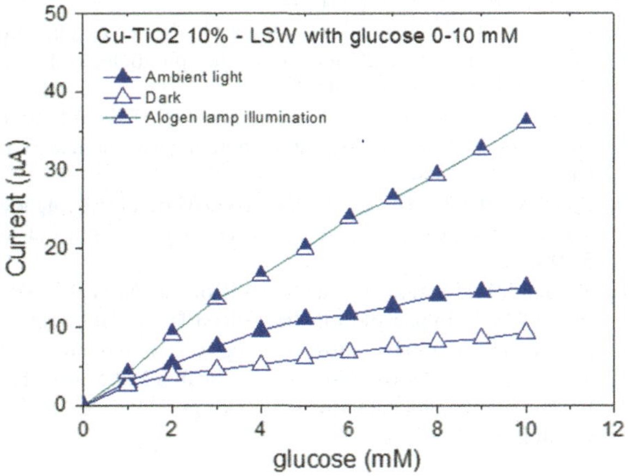


**Fig. 5** Calibration curves for 10% $Cu/TiO_2$/SPCE electrode, operated in different illumination conditions, as function of glucose concentration

# 4 Conclusion

$CuO/TiO_2$ composites with different amount of copper were synthesized by sol-gel method. After thermal treatment, the sample show a rough surface where small Cu nanocrystals are distributed on the surface of titanium dioxide nanoparticles. Electrochemical non-enzymatic glucose sensors were fabricated modifying the working electrode surface of screen printed carbon electrodes. The modified Cu-doped $TiO_2$ electrodes showed promising electrochemical performances in the electro-oxidation of glucose. The glucose oxidation is strictly dependent on Cu loading. It can be also observed that the tested $TiO_2$–CuO electrode, irradiated with halogen light, exhibits a more pronounced sensitivity to glucose, indicating that the use of halogen light has a promoting effect in the oxidation of glucose. The excellent sensing performances coupled with the long-term stability and low cost, make this material promising for the development of non-enzymatic photo-electrochemical glucose sensors.

## References

1. Leonardi SG, Marini S, Espro C, Bonavita A, Galvagno S, Neri G (2017) In-situ grown flower-like nanostructured CuO on screen printed carbon electrodes for non-enzymatic amperometric sensing of glucose. Microchim Acta 184:2375–2385
2. Marini S, Mansour NB, Hjiri M, Dhahri R, El Mir L, Espro C, Bonavita A, Galvagno S, Neri G, Leonardi SG (2018) Non-enzymatic glucose sensor based on nickel/carbon composite. Electroanalysis 30:727–733
3. Fujishima A, Rao TN, Tryk DA (2000) Titanium dioxide photocatalysis. J Photochem Photobiol C 1:1–21
4. de Brito JF, Tavella F, Genovese C, Ampelli C, Zanoni MVB, Centi G, Perathoner S (2018) Role of CuO in the modification of the photocatalytic water splitting behavior of $TiO_2$ nanotube thin films. Appl Catal B 224:136–145
5. Devadoss A, Sudhagar P, Terashima C, Nakata K, Fujishima A (2015) Photoelectrochemical biosensors: new insights into promising photoelectrodes and signal amplification strategies. J Photochem Photobiol C 24:43–63
6. Li H, Li J, Chen D, Qiu Y, Wang W (2015) Dual-functional cubic cuprous oxide for non-enzymatic and oxygen-sensitive photoelectrochemical sensing of glucose. Sens Actuators B Chem 220:441–447
7. Zhan W, Chen Z, Hu J, Chen X (2018) Vertical CuO nanowires array electrodes: Visible light sensitive photoelectrochemical biosensor of ethanol detection. Mater Sci Semicond Process 85:90–97
8. Tobaldi DM, Lajaunie L, Caetano A, Rozman N, Seabra MP, Sever Skapin A, Arenal R, Labrincha JA (2018) Impact of rutile fraction on $TiO_2$ visible-light absorption and visible-light-induced photocatalytic activity. https://doi.org/10.26434/chemrxiv.7461161.v1
9. Tobaldi DM, Pullar RC, Gualtieri AF, Seabra MP, Labrincha A (2013) Sol-gel synthesis, characterisation and photocatalytic activity of pure, W-, Ag- and W/Ag co-doped $TiO_2$ nanopowders. Chem Eng J 214:364–375

# Development of an Efficient Acetone Conductometric Sensor Based on $NdFeO_3$

**Z. Anajafi, M. Naseri, N. Donato, S. G. Leonardi and G. Neri**

**Abstract** In this paper, $NdFeO_3$ nanoparticles were prepared by thermal treatment method and subsequent annealing at 700 °C. The morphological and microstructural features of $NdFeO_3$ nancrystals were investigated by X-ray powder diffraction (XRD) and scanning electron microscopy (SEM). $NdFeO_3$ nanoparticles were used for the fabrication of conductometric sensors for monitoring acetone in air. The effect of $NdFeO_3$ loading and operating temperature on sensing performance was investigated. Results of sensing tests indicate that $NdFeO_3$ nanoparticles exhibit good response to low concentration of acetone (2–40 ppm) in air at 200 °C. Finally, the developed $NdFeO_3$ sensor showed also excellent selectivity to acetone over other interfering gases such as $CH_4$, CO, $CO_2$, $H_2$, $NH_3$, $NO_2$ and NO.

**Keywords** Perovskite · $NdFeO_3$ · Gas sensor · Acetone

## 1 Introduction

In the past years, conductometric gas sensors based on semiconducting metal oxides have been widely used and investigated in the detection of a variety of gas and vapors. Among the gases to be detected, acetone has a prominent importance in the biomedical and environmental field. For example, very low concentration of acetone vapors are found in the breath of patients with diabetes whereas, on the other hand, high concentration of acetone vapors in the environment can lead to vomiting, lethargy, acidosis, and even ephemeral disturbance of consciousness [1]. Today, researchers are therefore looking for new sensing materials to detect this gas by means of simple and cheap conductometric sensors.

$LnBO_3$ perosvskites (Ln: rare earth elements and B: transition metal ions) display interesting properties and useful applications in many fields, including gas sensing [2, 3]. In this paper, we report preliminary efforts made for developing a novel

Z. Anajafi (✉) · M. Naseri
Department of Physics, Faculty of Science, Malayer University, Malayer, Iran

Z. Anajafi · N. Donato · S. G. Leonardi · G. Neri
Department of Engineering, University of Messina, Contrada Di Dio, 98166 Messina, Italy



G. Di Francia et al. (eds.), *Sensors and Microsystems*, Lecture Notes in Electrical Engineering 629, https://doi.org/10.1007/978-3-030-37558-4_30

conductometric sensor for acetone based on $NdFeO_3$ nanoparticles with perovskite structure. Over the past years different methods have been proposed for the synthesis of $NdFeO_3$ nanoparticles materials such as sol-gel [4], solid state reaction [5] and hydrothermal [6].

Previously, we optimized a procedure for the preparation of $NdFeO_3$ nanoparticles by a new thermal treatment method [7]. The sensing properties of these $NdFeO_3$ nanoparticles were investigated with the aim to develop $NdFeO_3$ conductometric sensors for detecting low concentrations of acetone in air and are here reported.

## 2 Experimental

### 2.1 *Synthesis of $NdFeO_3$ Nanoparticles*

An aqueous solution of PVP was prepared by dissolving polymer in 100 ml of deionized water at 90 °C, the metal nitrates with equal molar ratios were added to the PVP solution while the temperature of solution was kept at 90 °C for 2 h. The mixed solution was poured into a glass Petri dish and heated at 100 °C in an oven for 24 h to evaporate the water. The dried solid that remained was crushed and ground in a mortar to form powder. The calcinations of the powders were conducted at 700 °C for 3 h for the decomposition of organic compounds and the crystallization of the nanocrystals.

### 2.2 *Characterization*

X-ray diffraction (XRD) results were collected on powder samples using a Scintag X-II diffractometer with Cu$K\alpha$ radiation ($\lambda = 1.5046$ Å) and a graphite monochromator. Spectra were acquired over the angular range 20°–80°. The morphology of the $LaFeO_3$ nanocrystals was observed using SEM/EDX (Scanning electron microscopy/electron dispersive X-ray analysis, FEI/Philips XL-30).

### 2.3 *Sensor Fabrication*

The devices for the sensing tests were fabricated by drop coating, depositing different loading (0.01, 0.02, 0.04, 0.12 mg) of $NdFeO_3$ on the Pt interdigitated electrode area of the conductometric ceramic platform. Electrical and sensing tests were performed by introducing the sensor in a stainless steel test chamber, controlling the gas concentration by using certified bottles and flow rate by mass flow meters. The sensor response was defined as the ratio $R_g/R_a$ of the sensor resistance in presence of target gases ($R_g$) and that in dry air ($R_a$).

## 3 Results and Discussion

Figure 1a shows the X-ray diffraction pattern of the synthesized $NdFeO_3$ sample by the thermal treatment method and annealed at 700 °C. It can be seen that there are several peaks related to the planes of $NdFeO_3$ nanocrystals which can be indexed to an orthorhombic structure [8]. The average size of the $NdFeO_3$ nanocrystallites, evaluated from XRD line broadening using the Debye-Scherrer equation, was calculated to be about 17 nm.

The, structure, morphology and particle size of the $NdFeO_3$ nanopowder were investigated by SEM. Figure 2 shows SEM micrograph images of the $NdFeO_3$ nanopowder taken at different magnifications. Low magnification image on the left shows that $NdFeO_3$ consists of loosely aggregated grains. High magnification image on the right reveals that these grains are composed of extremely finer particles with average particle size of about 22 nm. Furthermore, SEM images highlighted the high porosity of the $NdFeO_3$ sample.

The results of gas sensing tests are shown in Fig. 3. Figure 3a shows the temperature dependence of baseline resistance, from 200 to 400 °C, for the fabricated $NdFeO_3$ sensors. As expected, the sensor baseline decreases increasing the loading of $NdFeO_3$ sensing layer; from these data it appears clear that more compact the layer is, lower is the resistance of the sensing layer.

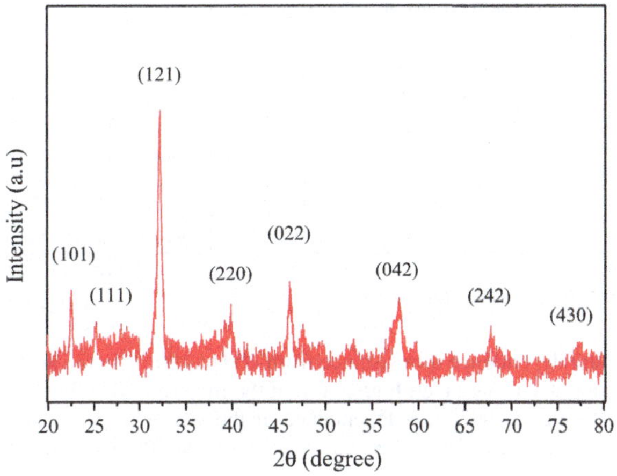


**Fig. 1** XRD pattern of $NdFeO_3$ nanocrystals annealed at 700 °C

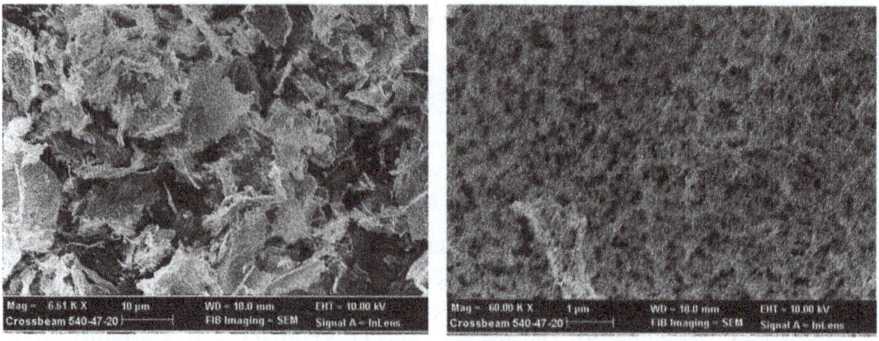


**Fig. 2** SEM images of $NdFeO_3$ powder annealed at 700 °C

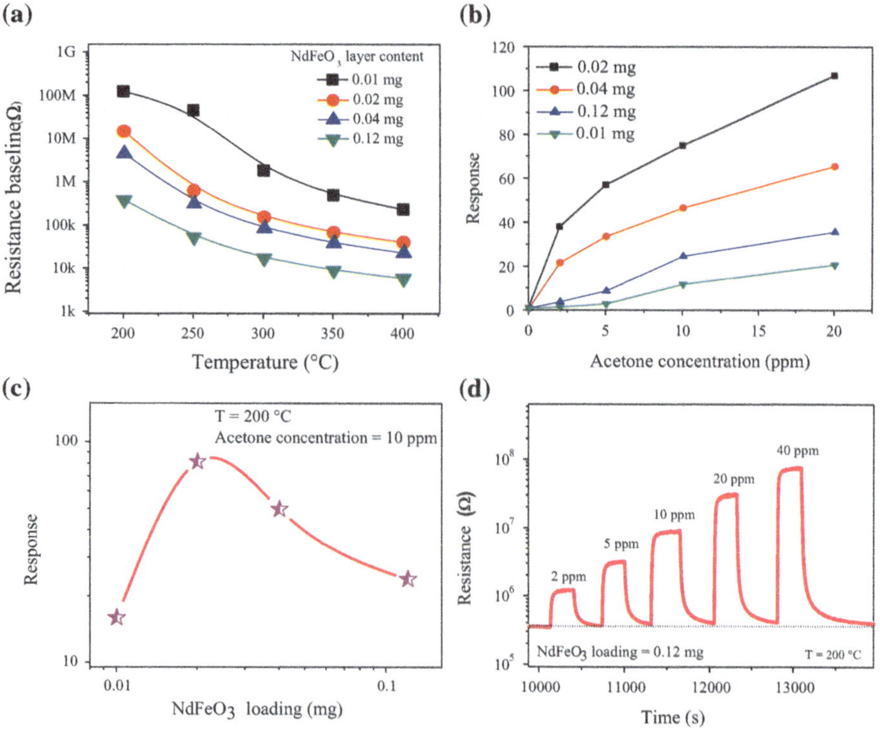


**Fig. 3** **a** Temperature dependence of resistance from 200 to 400 °C in the air for $NdFeO_3$ sensors with different loading of sensing layer; **b** response of the investigated $NdFeO_3$ sensors to different concentration of acetone at 200 °C; **c** dependence of the sensor response versus the loading of $NdFeO_3$; **d** dynamic responses of $NdFeO_3$ sensor, operating at 200 °C, to different concentrations of acetone in air

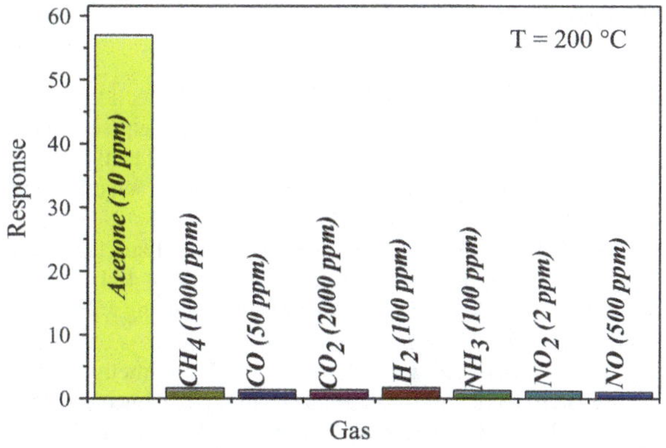


**Fig. 4** Comparison of the response of $NdFeO_3$ sensor to acetone and different interferent gases at 200 °C

Figure 3b displays the response of sensors with different loading of $NdFeO_3$ toward low concentration of acetone in air (from 0 to 20 ppm) at the working temperature of 200 °C. All sensors display good response, with the sensor with 0.02 mg $NdFeO_3$ having the best performance and the sensor with 0.01 mg having the weakest response. That behavior is because thicker films could provide more adsorption sites. However, increasing the loading from 0.04 to 0.12 lead to a decrease of response. The effect of $NdFeO_3$ loading on sensor response is more clearly showed in Fig. 3c. The response increases with increasing $NdFeO_3$ loading from 0.01 to 0.02 mg and decreases for higher loading.

Figure 3d reports the sensing transients of the NdFeO3 sensor to 2–40 ppm acetone recorded at 200 °C. It is evident that the response curves of the sensor increases sharply with increasing concentration of acetone and then returns to the baseline quickly when acetone atmosphere is replaced by air in the closed testing chamber, indicating its quick, reversible and fast response and recovery time. Such a change in the resistance indicates that $NdFeO_3$ behave as p-type semiconductor during acetone sensing.

Finally, the selectivity of $NdFeO_3$ sensor with 0.02 mg $NdFeO_3$ loading was examined and the results are reported in Fig. 4. As it is observed, it exhibits excellent selectivity to acetone with respect to $CH_4$, CO, $CO_2$, $H_2$, $NH_3$, $NO_2$ and NO.

## 4 Conclusion

In summary, $NdFeO_3$ nanoparticles synthesized by the thermal treatment method were used for developing an acetone conductometric sensors. Different loadings of the sensing layer were tested for optimizing the detection of very low concentrations of the target gas (2–40 ppm) in air. At the temperature of 200 °C, the fabricated $NdFeO_3$-based conductometric sensor, demonstrate very good performances (in terms of high response, fast response/recovery and high selectivity toward acetone).

## References

1. Mirzaei A, Leonardi SG, Neri G (2016) Detection of hazardous volatile organic compounds (VOCs) by metal oxide nanostructures-based gas sensors: a review. Ceram Int 42:15119–15141
2. Zhang YQ, Zhou LS, Liu YY, Liu DY, Liu FM et al (2018) Gas sensor based on samarium oxide loaded mulberry-shaped tin oxide for highly selective and sub ppm-level acetone detection. J Colloid Interface Sci 531:74–82
3. Niu X, Du W, Du W (2004) Preparation, characterization and gas-sensing properties of rare earth mixed oxides. Sens Actuators B 99:399–404
4. Toan NN, Saukko S, Lantto V (2003) Gas sensing with semiconducting perovskite oxide $LaFeO_3$. Phys B 327:279–282
5. Chu X, Zhou S, Zhang W, Shui H (2009) Trimethylamine sensing properties of nano-$LaFeO_3$ prepared using solid-state reaction in the presence of PEG400. Mater Sci Eng B 164:65–69
6. Xiao H, Xue C, Song P, Li J, Wang Q (2015) Preparation of porous $LaFeO_3$ microspheres and their gas-sensing property. Appl Surf Sci 337:65–71
7. Naseri MG, Saion EB, Ahanga HA, Shaari AH (2013) Fabrication, characterization, and magnetic properties of copper ferrite nanoparticles prepared by a simple, thermal-treatment method. Mater Res Bull 48:1439–1446
8. Khorasani-Motlagh M, Noroozifar M, Yousefi M, Jahani S (2013) Chemical synthesis and characterization of perovskite $NdFeO_3$ nanocrystals via a co-precipitation method. Int J Nanosci Nanotechnol 9:7–14

# Palmprint Recognition Through a Reliable Ultrasound Acquisition System and a 3D Template

**Antonio Iula and Monica Micucci**

**Abstract** Biometric systems based on ultrasonic images have the merits of being very resistent to spoof attacks. In this work, a 3D recognition procedure based on palmprint ultrasound images acquired through a reliable system, which uses a gel pad as coupling medium between the ultrasound probe and the hand, is proposed and experimentally evaluated. Several 2D palmprint images, at different under-skin depths, are extracted from the acquired volumetric image. For each of them a template is generated through an original procedure. The various templates are then opportunely combined to obtain a 3D template.

**Keywords** 3-D palmprint · Ultrasound imaging · Biometrics · Image processing

## 1 Introduction

In recognition systems that use features of the human hand as biometric characteristics, like fingerprint, palmprint, hand and/or finger geometry, and vein pattern, Ultrasound has several merits over other technologies [1]. It can provide volumetric images of portions of the human body; ink or grease stains on the skin do not affect the quality of images; ultrasound recognition systems are very difficult to counterfeit because they can easily detect liveness by checking blood flow or venous pulse.

Among the other biometric characteristics, ultrasound fingerprint has been the most investigated [2], especially since smartphone producers started to introduce authentication through fingerprint in their devices. Nowadays, several companies produce ultrasound sensors suitable to smartphone specifics.

Biometric systems based on Ultrasound that exploit features like inner hand or finger geometry [3, 4], and vein pattern [5–7] have been proposed as well.

Palmprint is another characteristic that is currently object of intense scientific research [8, 9]. In the last years, one of the authors has widely experimented with ultrasound systems to extract various biometric characteristics, in particular palmprint. Volumetric images of the human palm were acquired by using water as coupling

A. Iula (✉) · M. Micucci
School of Engineering, University of Basilicata, Potenza, Italy
e-mail: antonio.iula@unibas.it

 
G. Di Francia et al. (eds.), *Sensors and Microsystems*, Lecture Notes in Electrical Engineering 629, https://doi.org/10.1007/978-3-030-37558-4_31

medium between human hand and ultrasonic sensor [10, 11]. Several kinds of 3D features were extracted from these images [12] and excellent recognition results were obtained by exploiting the information of under-skin depth of principal lines [13]. Successively, a more reliable system, which uses gel as coupling medium instead of water, was proposed [14]. In the present work, 3D images acquired with this last system are used to extract 2D features at several depths through an original procedure; the 2D templates are then opportunely combined to achieve a 3D template with a similar method as that presented in [13].

## 2 Experimental Setup

Figure 1 shows a photo of the experimental setup used for acquiring 3D ultrasound palmprint images [14]. The user places his hand on in-house developed hand-holder, where there is a window filled with a layer 20 mm thick of commercial ultrasound gel pad. The gel pad works as a coupling medium between the palm of hand and the probe. This system results much more comfortable for the user than the previous wet systems [10–13] and, consequently, the number of invalid acquisitions (mainly caused by hand motion or misplacing) is dramatically reduced. The ultrasound commercial probe LA435 (Esaote S.p.A., Genova, Italy) was driven by an ULtrasound Advanced Platform (ULA-OP) research scanner [15]. 3D ultrasound imaging data of a volume of the palm are acquired by mechanically shifting the probe along the elevation direction. The system is able to scan and acquire a 3D ultrasound image of $38 \times 25 \times 7$ mm$^3$ in about five seconds. The 3D image can be then post-processed in order to extract several 2D images a different under-skin depths. Figure 2 shows an example

**Fig. 1** A photo of the experimental set-up. The ultrasound probe scan the human palm to get a 3D image. A gel pad ensures a good acoustic coupling between probe and hand

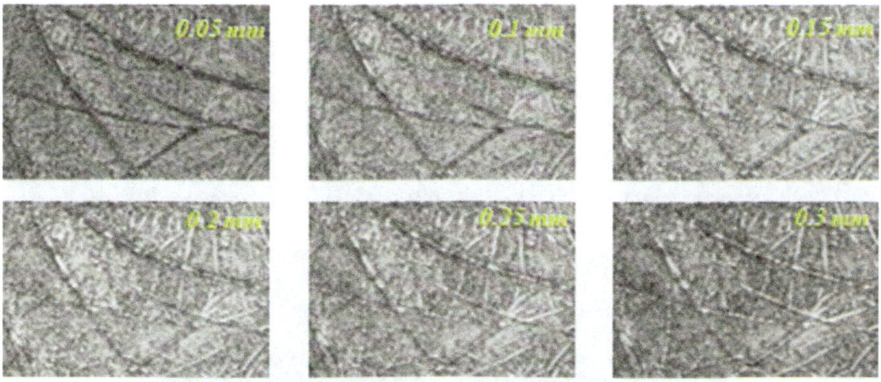


**Fig. 2** 2D under-skin images extracted at several depths

where six 2D images are extracted at growing depths with step of 0.05 mm. As can be seen, palm lines are individuated by dark pixels and the information on wrinkles depth is related to the number of images that present dark pixels in the wrinkles.

## 3 Recognition Procedure

The procedure for generating a 3D template consists first in extracting a 2D template from each 2D image and then in grouping these templates to achieve a 3D one. As a preliminary operation, the first two images, acquired at depths of 0.05 mm and 0.10 mm, respectively, are merged by means of an appropriate filtering operation, which consists in comparing the two images pixel by pixel: all pixels of the deeper image that are darker than the corresponding ones in the other image, are replaced with those ones.

Each of the resulting five 2D images is then filtered with a Speckle Reduction Anisotropic Filter (SRAD) because speckle noise is a main problem in ultrasound images [16, 17]. Subsequently, a bottom-hat operation is performed to extract the darkest object on the background. In the proposed method, four phases of bottom-hat are employed following a pointed direction (0°, 45°, 90°, 135°). The four images are then collected by performing a logical sum. Finally, the image is binarized by using an adaptive threshold and some classical morphological operations (closing, thinning and pruning) are performed. As can be seen from the achieved 2D templates (see Fig. 3), by increasing the under-skin depth secondary traits appears while some principal lines gradually disappear. 2D templates are stored in a matrix $T_k(X,Y)$, where k is between 1 and 5. In order to achieve a 3D template, the first 2D template is dilated and stored in $T_1(X,Y)$; then it is compared via the logical AND operation with the second one. The result is dilated, stored in $T_2(X,Y)$ and then compared with the third layer and so on. The dilate operation allows to account for the fact that traits

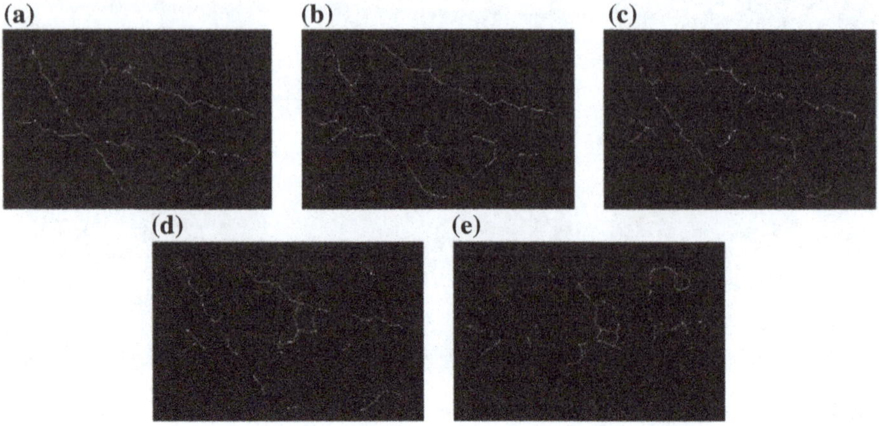


**Fig. 3** 2D templates corresponding to images extracted at: **a** 0.05–0.10 mm, **b** 0.15 mm, **c** 0.20 mm, **d** 0.25 mm, **e** 0.30 mm

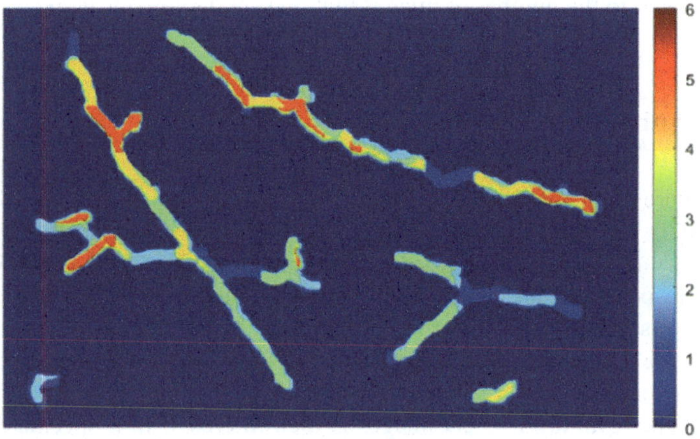


**Fig. 4** Color scale representation of the 3D template: red pixels corresponds to deepest lines

not necessarily develop along a direction orthogonal to the XY plane while the AND operation acts a first filter for secondary or small traits. A graphical representation of a 3D template is shown in Fig. 4.

## 4 Recognition Results

In order to provide a first evaluation of the proposed palmprint recognition system, experiments of verification have been performed exploiting a database of 148 samples acquired from 24 different volunteers. The matching formula adopted to calculate the matching score $R_{2D}$ between two 2D templates is based on a classical pixel-to-area comparison:

$$R_{2D} = \frac{2\sum_{X=1}^{n}\sum_{Y=1}^{m}(T_d(X,Y) > 0 \oplus T_t(X,Y) > 0)}{\sum_{X=1}^{n}\sum_{Y=1}^{m}T_d(X,Y) + \sum_{X=1}^{n}\sum_{Y=1}^{m}T_t(X,Y)} \quad (1)$$

where $T_d$ and $T_t$ are the database and test binary templates extracted by the first image, respectively, and $m \times n$ is the template dimension. The matching score $R_{3D}$ between two 3D templates was computed by inserting an additional term in Eq. 1:

$$R_{3D} = \frac{2\sum_{X=1}^{n}\sum_{Y=1}^{m}|T_d(X,Y) - T_t(X,Y)| < \alpha \oplus (T_d(X,Y) > 0 \oplus T_t(X,Y) > 0)}{\sum_{X=1}^{n}\sum_{Y=1}^{m}T_d(X,Y) + \sum_{X=1}^{n}\sum_{Y=1}^{m}T_t(X,Y)} \quad (2)$$

where $T_d$, $T_t$ and $\alpha$ assume integer values ranging from 0 to 5. The term $(|T_d(X,Y) - T_t(X,Y)| < \alpha)$ acts as an adaptive filter for secondary or low depth traits. The lower is the value of $\alpha$, the higher is the filtering effect. As is known, genuine scores represent comparisons from different images that belong to the same person. Instead, impostor scores represent comparisons from different images that belong to different persons.

Figure 5 shows the plot of the experimental False Acceptance Rates (FAR) versus False Rejection Rates (FRR), called Detection Error Trade-off curve (DET), for various values of $\alpha$ and for the best 2D template, i.e., the one extracted from the

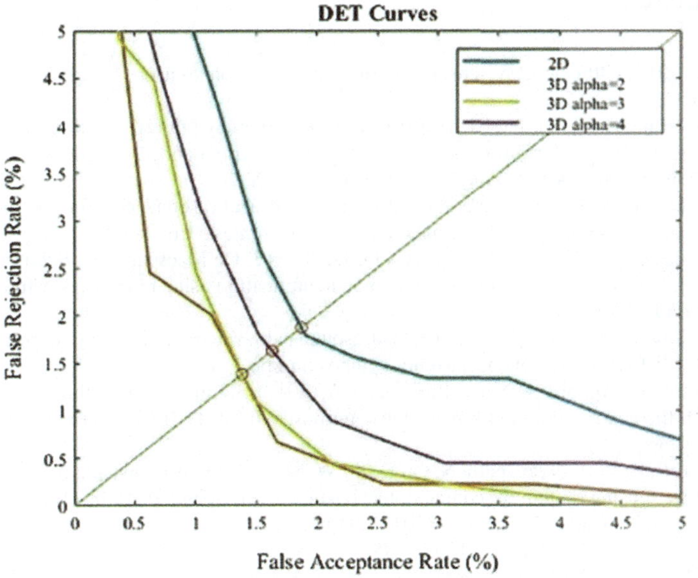


**Fig. 5** DET curves for the various recognition methods

**Table 1** EER for various recognition methods

| Method | 2D | 3D ($\alpha = 2$) | 3D ($\alpha = 3$) | 3D ($\alpha = 4$) |
|---|---|---|---|---|
| EER (%) | 1.87 | 1.37 | 1.37 | 1.63 |

shallowest image. A parameter often used to compare different algorithms is the Equal Error Rate (EER), i.e., the error obtained if the threshold is chosen in such a way that FAR = FRR. As can be seen, for any $\alpha$, the recognition procedure based on the 3D template provides better results than that using a 2D template. Table 1 summarize achieved results.

## 5 Conclusion

A new palmprint recognition procedure for ultrasound images acquired through a system that exploits gel as coupling medium is proposed and experimentally evaluated. Verification experiments have shown that the 3D procedure exhibits very good recognition rates and outperforms the 2D one. Future works will be devoted to improve the 2D features extraction procedure also by applying methods based on deep learning [18, 19].

## References

1. Iula A (2019) Ultrasound systems for biometric recognition. Sensors 19(10):2317. https://doi.org/10.3390/s19102317
2. Jiang X, Tang HY, Lu Y, Ng EJ, Tsai JM, Boser BE, Horsley DA (2017) Ultrasonic fingerprint sensor with transmit beamforming based on a PMUT array bonded to CMOS circuitry. IEEE Trans Ultrason Ferroelectr Freq Control 64:1401–1408
3. Iula A, De Santis M (2011) Experimental evaluation of an ultrasound technique for the biometric recognition of human hand anatomic elements. Ultrasonics 51:683–688
4. Narayanasamy G, Fowlkes J, Kripfgans O, Jacobson J, De Maeseneer M, Schmitt R, Carson P (2008) Ultrasound of the fingers for human identification using biometrics. Ultrasound Med Biol 34:392–399
5. Iula A, Savoia A, Caliano G (2012) 3D Ultrasound palm vein pattern for biometric recognition. In: 2012 IEEE international ultrasonics symposium, pp 1–4
6. De Santis M, Agnelli S, Nardiello D, Iula A (2017) 3D ultrasound palm vein recognition through the centroid method for biometric purposes. In: 2017 IEEE international ultrasonics symposium (IUS)
7. Wang Y, Li Z, Vu T, Nyayapathi N, Oh K, Xu W, Xia J (2018) A robust and secure palm vessel biometric sensing system based on photoacoustics. IEEE Sens J 18:5993–6000
8. Bai X, Gao N, Zhang Z, Zhang D (2017) 3D palmprint identification combining blocked ST and PCA. Pattern Recogn Lett 100:89–95
9. Fei L, Lu G, Jia W, Wen J, Zhang D (2018) Complete binary representation for 3-D palmprint recognition. IEEE Trans Instrum Meas 67:2761–2771

10. Iula A, Hine GE, Ramalli A, Guidi F, Boni E, Savoia AS, Caliano G (2013) An enhanced ultrasound technique for 3D palmprint recognition. In: 2013 IEEE international ultrasonics symposium (IUS), pp 978–981
11. Iula A, Savoia AS, Caliano G (2014) An ultrasound technique for 3D palmprint extraction. Sens Actuators A 212:18–24
12. Iula A, Nardiello D (2016) Three-dimensional ultrasound palmprint recognition using curvature methods. J Electron Imaging 25:033009
13. Iula A, Nardiello D (2019) 3D ultrasound palmprint recognition system based on principal lines extracted at several under skin depths. IEEE Trans Instrum Meas. https://doi.org/10.1109/TIM.2019.2900177
14. Nardiello D, Calia M, Iula A (2016) An enhanced ultrasound technique for 3D palmprint recognition. In: 2016 IEEE international ultrasonics symposium (IUS)
15. Tortoli P, Bassi L, Boni E, Dallai A, Guidi F, Ricci S (2009) ULA-OP: an advanced open platform for ultrasound research. IEEE Trans Ultrason Ferroelectr Freq Control 56:2207–2216
16. Kaur A, Ranade S (2014) Review of speckle noise reduction techniques for ultrasound imaging. Biomed Eng Lett 4:32–40
17. Nardiello D, Iula A (2019) A new recognition procedure for palmprint features extraction from ultrasound images. Lect Notes Electr Eng 512:113–118
18. Bensid K, Samai D, Laallam F, Meraoumia A (2018) Deep learning feature extraction for multispectral palmprint identification. J Electron Imaging 27:033018
19. Guo X, Zhou W, Zhang Y (2017) Collaborative representation with hm-lbp features for palmprint recognition. Mach Vis Appl 28:283–291

# Innovative Lab-on-Disk Technology for Rapid and Integrated Analysis of Pathogen Nucleic Acids

**Emanuele L. Sciuto, Salvatore Petralia and Sabrina Conoci**

**Abstract** Conventional nucleic acids (NA) analysis in diagnostics requires centralized laboratories and specialized staffs that are limiting towards the miniaturization and integration typical of "genetic point-of-care" (PoC) technologies. Moreover, all steps for the analysis need to be performed separately using different instruments and procedures far away from the *sample-in-answer-out* approach. For these reasons, we introduce an easy-to-use PoC technology based on a Lab-on-Disk miniaturized system for pathogen genomes analysis. It is composed by a polycarbonate disk, a silicon chip and a customized reader, integrating electronic and optical modules for the disk driving and the genome detection by Real Time PCR (qPCR). For testing the extraction and detection performances, we used samples of Hepatitis B Virus (HBV) genome.

**Keywords** Point-of-care · Genome extraction · Genome amplification · Microfluidics · Silicon chip

## 1 Introduction

The development of portable systems able to perform the complete nucleic acids (NA) analysis is one of the challenging fields of the research in biosensing. Yields of both NA extraction and detection steps can be influenced by the complexity of the biological matrices employed (blood, urine, saliva etc.), which means the need of a sophisticated systems architectures and protocols. Therefore, the integration of an effective sample preparation module with a NA detection module is a key point for the development of portable device.

This integration can be achieved combining the performances of microfluidics technology with the physical properties of silicon materials.

---

E. L. Sciuto (✉)
CNR-IMM Headquarters, VIII Strada, 5, 95121 Catania, Italy
e-mail: emanueleluigi.sciuto@imm.cnr.it

S. Petralia · S. Conoci
STMicroelectronics, Stradale Primosole, 50, 95121 Catania, Italy

 
G. Di Francia et al. (eds.), *Sensors and Microsystems*, Lecture Notes in Electrical Engineering 629, https://doi.org/10.1007/978-3-030-37558-4_32

Microfluidics introduces a series of advantages in the diagnostic systems (compared to the macroscopic equivalent) such as the small volumes required for the experimental setup and the possibility of a miniaturization together with a higher surface-to-volume ratio [1–3].

On the other hand, silicon materials show physical properties, in terms of biocompatibility, low heat capacity, good thermal conductivity and possible patterned structures to increase the surface–area ratio, which can help improving both the NA extraction and detection [4–8].

In this sense, we introduce an integrated microfluidic biosensor platform based on silicon-plastic hybrid disk able to perform both extraction and detection of nucleic acids through Real-Time PCR (qPCR) technology. Both extraction and amplification data are presented and discussed.

## 2 Materials and Methods

### *2.1 Chemicals and Instruments*

Hepatitis B virus (HBV) clone complete genome (ref product 05960116), consisting of the HBV genome 3.2 kbps and a plasmid PBR322 vector 3.8 kb in TE (Tris 10 mM, EDTA 1 mM, pH_8), and the HBV real time PCR kit (ref product FO2 HBV MMIX KIT 48) were from CLONIT and used for the experiments. Biological sample processing and NA collection were performed by Qiagen QIAamp DNA Mini Kit (Ref. 51306), according to the provided protocol.

### *2.2 NA Extraction Strategy*

The proposed strategy is based on the NA isolation from cell debris by electrostatic interactions between genomic molecules and silica paramagnetic beads in a microfluidic approach [9]. Thanks to the structural layout of the lab-on-disk, the biological sample can be loaded as is (Fig. 1a-1) and then processed in lysis-binding-washing-elution steps.

Disk rotations allow the sample to move through the reaction chambers (Fig. 1a-2 to a-4); in this phase, a specific buffer (1.25 M sodium chloride and 10% polyethylene glycol) creates cation bridges between the NA and the beads surface (Fig. 1b), so that the genomic material can be isolated from cell debris. In the last chamber (Fig. 1a-5), instead, a specific elution buffer removes the cation bridges and separates the NA from beads for the final collection.

Disk rotations are performed by the reader reported in Fig. 2a, b. The disk is fixed on top of a holder and a dedicated software manages the time and speed of rotations.

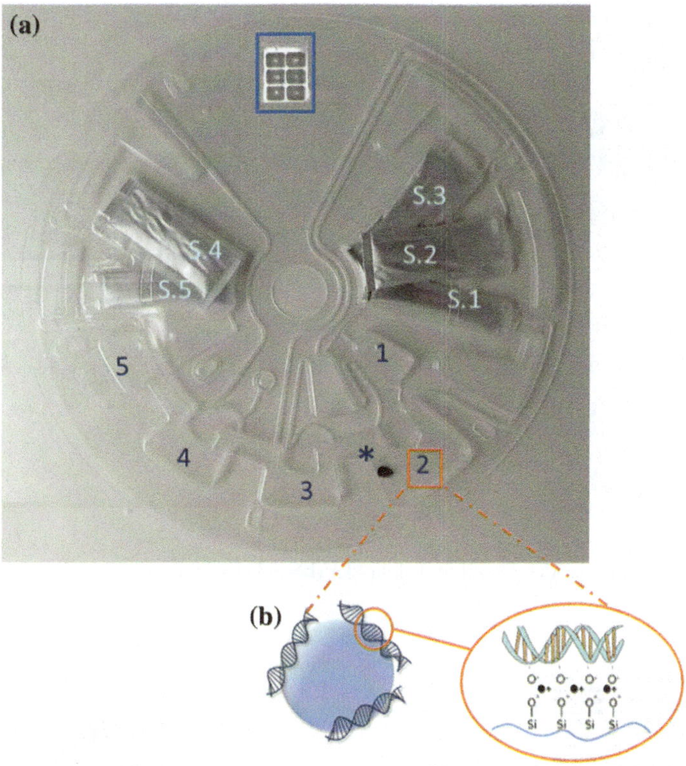


**Fig. 1** **a** Lab-on-Disk: sample (1), lysis (2), binding (3), washing (4) a.nd elution (5) chamber; paramagnetic beads (*); buffer stickpacks (S.1 to S.5). **b** Detail of paramagnetic DNA adsorption and isolation from biological sample

## *2.3 NA Detection Strategy*

The detection strategy is based on the NA (isolated from biological sample) amplification through an integrated silicon chip for the Real-Time PCR reaction. This was developed by STMicroelectronics [10] and integrates temperature sensors and heaters in the rear part and for the thermal control of amplification reaction (Fig. 3). On top, six micro-chambers are covered by a polycarbonate mask to load the NA samples and perform the amplification.

The optical detection is performed by a module mounted below the lid of lab-on-disk reader (Fig. 2c). It consists in a CCD optical detector collecting the fluorescent signal during the NA amplification. The optical module contains of 4 independent optical channels, for multiple Fluorescent reporters (FAM, VIC®). A thermal module completes the system, driving the temperature sensor and heater of the chip to reach the temperature values for qPCR.

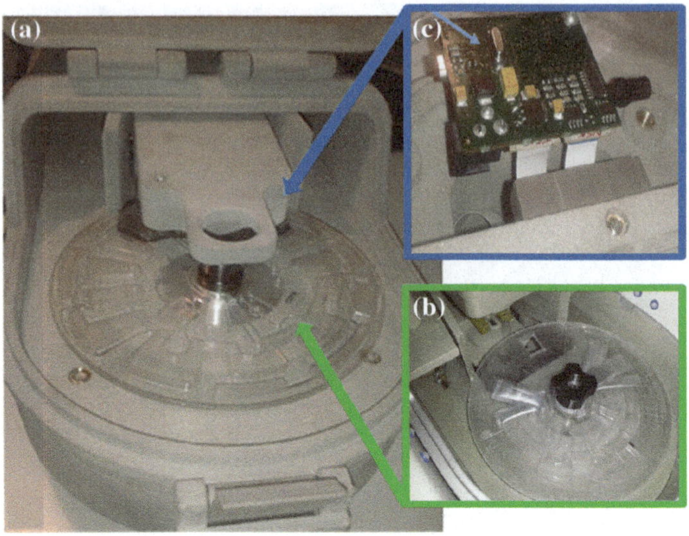


**Fig. 2** **a** Lab-on-disk reader. Detail of disk holder (**b**) and electro-optical detection module (**c**)

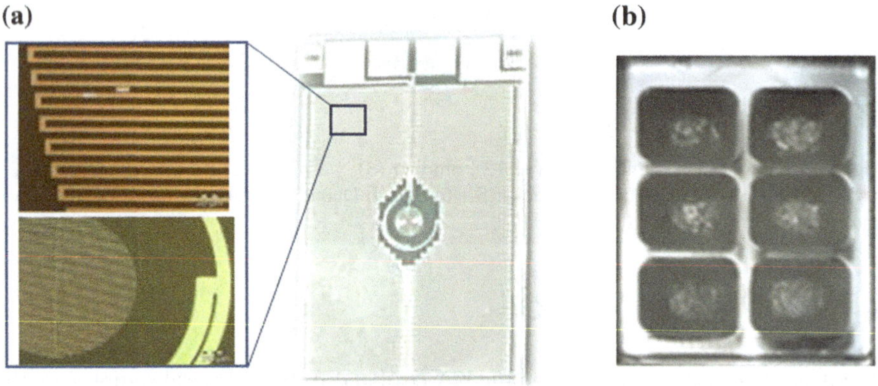


**Fig. 3** Silicon qPCR chip for the NA detection in lab-on-disk: **a** rear part containing the resistors and heaters (blue line); **b** front part containing the reaction wells

# 3 Discussion

The performances of the lab-on-disk hybrid system, in terms of extraction efficiency and detection sensitivity, were characterized by using a HBV synthetic clone complete genome suspended in tap water.

For the extraction yield analysis, a $10^4$ copies/μL HBV sample was prepared. A volume of 200 μL of sample was loaded on hybrid disk and analyzed. Once collected, the starting and purified NA samples were loaded inside two wells of the integrated silicon chip, for the final qPCR amplification and quantification.

The extraction experiment of the HBV sample was performed, also, with a conventional QIAGEN kit, as comparison.

Results of HBV NA extraction yield evaluation are reported in Fig. 4. The data show that the lab-on-disk is able to purify the NA sample with a yield that almost reproduces the one of conventional QIAGEN method (Ct value of 30 and 29.6, respectively). A detail of silicon chip after qPCR cycles is also reported in Fig. 4, with a detail of loaded samples fluorescence appearing from the silicon chip wells.

For the detection sensitivity test, a negative sample (1 μL of water + 14 μL of HBV master mix) and three positive controls ($10^6$, $10^5$ and $10^3$ copies/μL) of HBV genome were loaded on chip. The same samples were amplified with an Applied Biosystems equipment, as comparison.

Results of the detection module sensitivity analysis are reported in Table 1.

For all HBV NA concentration used, data show a gain for the on-board detection module of about 0.5 Ct, respect to the commercial equipment. This enhancement is

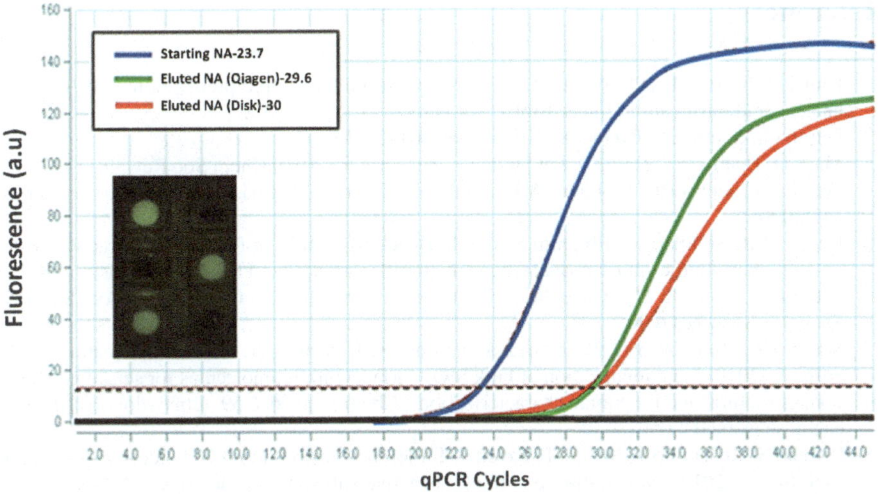


**Fig. 4** HBV NA extraction yield comparison: starting NA

**Table 1** Lab-on-disk optical detection sensitivity compared to the commercial Applied Biosystems

| Lab-on-disk copies/μL: Ct | Applied Biosystems copies/μL: Ct |
|---|---|
| $10^6$: 19.4 | $10^6$: 19.9 |
| $10^5$: 23.9 | $10^5$: 23.9 |
| $10^3$: 28.5 | $10^3$: 29.9 |

the result of a synergy between the properties of the integrated silicon technology and the quality of the electronic detection management involved in the amplification process.

## 4 Conclusion

The Lab-on-Disk technology for pathogen nucleic acids qPCR detection, we propose, reported an optimization of sensitivity of about 1 Ct, thanks to the micro-chambers layout and the reader performances, if compared to the conventional qPCR detection systems (i.e. Applied Biosystem).

This achievement, together with the gain in time and integration of detection process, paves the basis for the development of new Point-of-Care for diagnostic applications on infectious disease.

## References

1. Kim J, Johnson M, Hill P, Gale BK (2009) Microfluidic sample preparation: cell lysis and nucleic acid purification. Integr Biol 1:574–586
2. Petralia S, Sciuto EL, Conoci S (2017) Analyst 142:140
3. Mark D, Haeberle S, Roth G, von Stetten F, Zengerle R (2010) Chem Soc Rev 39:1153
4. Petralia S, Sciuto EL, Di Pietro ML, Zimbone M, Grimaldi MG, Conoci S (2017) Analyst 142:2090
5. Hegab HM, Soliman M, Ebrahim S, Op de Beeck M (2013) In-flow DNA extraction using on-chip microfluidic amino-coated silicon micropillar array filter. J Bionsens Bioelectron 4:1–6
6. Petralia S, Castagna ME, Cappello E, Puntoriero F, Trovato E, Gagliano A, Conoci S (2015) Sens Bio-Sens Res 6:90
7. Sciuto EL, Bongiorno C, Scandurra A, Petralia S, Cosentino T, Conoci S, Sinatra F, Libertino S (2018) Functionalization of bulk $SiO_2$ surface with biomolecules for sensing applications: structural and functional characterizations. Chemosensors 6:59. https://doi.org/10.3390/chemosensors6040059
8. Petralia S, Cosentino T, Sinatra F, Favetta M, Fiorenza P, Bongiorno C, Sciuto EL, Conoci S, Libertino S (2017) Silicon nitride surfaces as active substrate for electrical DNAbiosensors. Sens Actuators B 252:492–502
9. Berensmeier S (2006) Appl Microbiol Biotechnol 73(3):495–504
10. Spata MO, Castagna ME, Conoci S (2015) Image data analysis in qPCR: a method for smart analysis of DNA amplification. Sens Bio-Sens Res 6:79–84

# A Fast Gas Sensing Layer Working at Room Temperature for IOT in Air Quality Scenario

**E. Massera, B. Alfano, M. L. Miglietta, T. Polichetti, S. De Vito, F. Formisano, G. Di Francia, P. Delli Veneri, M. Ferdinandi and M. Molinara**

**Abstract** The work deals with a technique adopted to calibrate in laboratory chemiresistor gas sensing film based on graphene that work at room temperature installed on a micro sensor board for applications in open air and IOT scenario. From the study in controlled environment the beginning of poisoning due to chemisorption can be estimated for the sensing layer and is possible to avoid harmful exposure to the analite during the calibration.

**Keywords** Gas sensors · Graphene · Calibration · Embedded system · IOT

## 1 Introduction

The exceptional sensing properties of graphene are well known [1]. However, after more than 10 years since its discovery, this material has not yet overcome the chronic problems afflicting the gas-sensitive films that operate at room temperature [2]. Electrical stability, repeatability of measurements, desorption of the target gas have a key role for the successful use of a chemoresistive film that can be used in open air in wearable scenario.

Nitrogen dioxide, an important air pollutant, on a sheet of graphene can be physisorbed or chemisorbed. Physisorption involves a fast and weak bond with a good charge exchange between the surface of graphene and the $NO_2$ molecules. Chemisorption at room temperature is a slower and deeper phenomenon that mainly acts on dangling bonds and surface defects. In order to exploit the graphene nanostructure and obtain a rapid and quite reversible response it is essential to find a way to promote physisorption and to avoid chemisorption. Lowering the temporal integral of nitrogen dioxide exposure during calibration seems to be a good way to promote

---

E. Massera (✉) · B. Alfano · M. L. Miglietta · T. Polichetti · S. De Vito · F. Formisano · G. Di Francia · P. Delli Veneri
ENEA, CR-Portici, P.le E. Fermi 1, 80055 Naples, Italy
e-mail: ettore.massera@enea.it

M. Ferdinandi · M. Molinara
Università degli studi di Cassino e del Lazio Meridionale Viale dell'Università—Rettorato (Campus Universitario), Loc. Folcara, 03043 Cassino, FR, Italy



G. Di Francia et al. (eds.), *Sensors and Microsystems*, Lecture Notes in Electrical Engineering 629, https://doi.org/10.1007/978-3-030-37558-4_33

physisorption. The classical laboratory calibration, based on different phases of stationary gas concentration, can lead to a poisoning of the sensor compromising the usability of the sensitive layer. So we adopt a "non-stationary" calibration technique measuring, with high precision, the linearity of the sensor response with sub-ppm concentration of nitrogen dioxide not only in the absorption phase of the target gas but also in the desorption phase. In this way it is possible to estimate the right parameters for a good correlation with the gas concentration for the correct use of the sensor device that minimize chemisorption poisoning.

## 2 Experimental and Results

### 2.1 *Experimental Setup and Calibration Method*

Once prepared, the detection device [3] (Fig. 1a) is connected to the SENSICHIPS board for powering sensor signal conditioning and data acquisition. The SENSICHIPS technology platform (Fig. 1b) is a miniaturized chip with connectivity with external sensing elements. Cross-platform software is based on the java language set and records the electrical parameters including graphene sensor conductance with a high sampling rate (up to 20 Hz).

The sensor system composed by the Sensichip and the chemiresistive graphene sensing device are installed on a steel base with electrical terminals (Fig. 1c) that can be sealed in a 0.4 lt chamber of a Gas Sensor Characterization System (GSCS). Inside the chamber, the air composition (humidity, chemical compounds concentration) can be settled up thanks to an inlet of GAS flux precisely controlled by certified Mass Flow Controllers (MKS 1179 series). The accuracy of the gas chemical composition

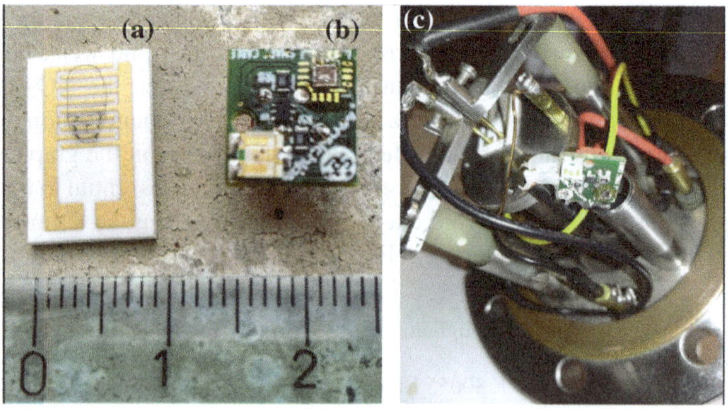


**Fig. 1** **a** The graphene sensing device. **b** The sensing platform SENSICHIPS for the sensor powering and readout. **c** The sensor device and the SENSICHIPS platform connected ready for the laboratory characterization in controlled environment

is ensured by the mixing of certified bottles (Rivoira). Temperature and humidity are recorded with industrial sensors (LSI Pt100). The calibration method consists in injecting in the inlet tube of the chamber a constant flow of the target gas properly diluted at the maximum concentration ($C_0$) with humid synthetic air. The time-rising concentration C(t) of the target gas is precisely predicted by the following exponential law that, in general, describes a transition between two steady states of a physical parameter under a time constant perturbation [3]:

$$C(t) = C_0\left(1 - e^{-\frac{t}{\tau}}\right) \tag{1}$$

At room temperature and for volume under a liter with a chamber of cylindrical geometry the characteristic time ($\tau$) is independent from the gas pollutant and its concentration and can be precisely estimated using a calibrated sensor.

On the other side, also the kinetic response of a sensor to a constant pollutant concentration can be described by the same law with a characteristic time ($\tau_s$). If $\tau_s \gg \tau$ (fast chamber and slow sensor), we can formally exchange the characteristic times $\tau_s$ with $\tau$ and consider the output of a fast sensor in a slow chamber with an equivalent gas concentration ($C_e$) described by

$$C_e(t) = C_0\left(1 - e^{-\frac{t}{\tau_s}}\right) \tag{2}$$

So with this strategy and fitting the output data with the formula (2) is possible to estimate $\tau_s$ and correlate each sampling of the sensor output to the equivalent gas concentration i.e. the gas concentration that the sensor measure in a stable state.

The calibration procedure (run) consists of three time steps: first, synthetic air is injected for the unperturbed state recording of the sensor output (baseline); while in the second step, gas target properly diluted in the gas carrier is injected and the adsorbing phase of the sensor response is recorded; finally the third step is the test chamber washing in a constant flow of synthetic air recording the desorbing phase of the sensor output. With this procedure is possible to verify the sensor output behaviour during the adsorbing and desorbing phase of the chemical compounds on the sensors surface. Sensing hysteresis or poisoning can be detected and measured.

With a gas flow of 0.5 lt/min, the $\tau$ of the chamber is estimated to be 25 $\pm$ 5 s.

## 2.2 *Results*

In Fig. 2a the calibration curve between 0 and 1000 ppm of the sensor 1 exposed to nitrogen dioxide with an integral adsorbent value of 8930 ppb per min. In the adsorption phase (black dots), with a high concentration value, the poisoning effect is clearly visible as a distortion of the response from the linear regression with the concentration of nitrogen dioxide.

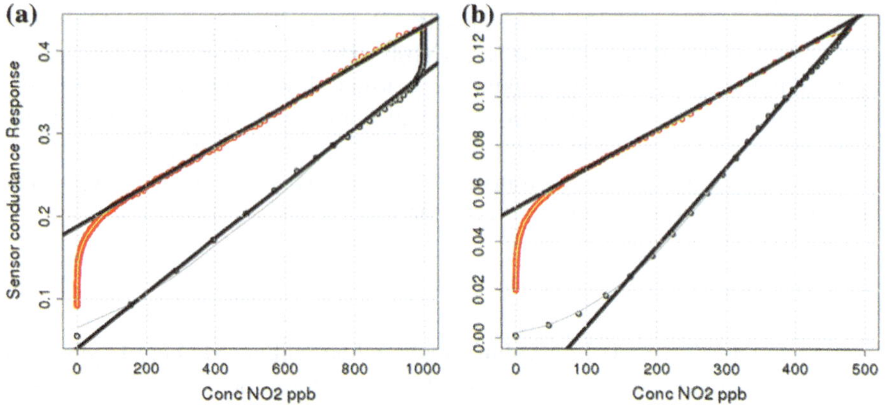


**Fig. 2** **a** Calibration curve of the sensor device 1. **b** Calibration curve of the sensor device 2. Black dots are the experimental points recorded during the adsorbing phase; the red dots are referred to the desorbing phase

The sensitivity value during adsorption is 45%/ppm while in the desorption phase it falls to 27%/ppm. The response time in the adsorption phase is less than 1 min while in the desorption phase it is more than 6 min and the lack of conductance recovery is measured by the intercept of the linear regression of the sensitivity on the y axis. The poisoning effect due to chemisorption accelerates the aging of the sensitive layer. As chemisorption increases, the sensitivity to $NO_2$ decreases and the response time increases.

In Fig. 2b the calibration curve between 0 and 500 ppm of the sensor 2 device this time exposed to a total adsorbing value of nitrogen dioxide of 1800 ppb per minute. This device has a lower sensitivity in the adsorption phase (33%/ppm) and the response time is less than 2 min, but the chemisorption effect is rather reduced. A better performance in the recovery phase can be seen in the lowering of the intercept on the y axis of the desorbing phase linear regression in the test on the sensor 2. All this shows the ability to perform a calibration of the detection device keeping under control the effect of poisoning due to the chemisorption of nitrogen dioxide.

## 3 Conclusion

In this contribution we introduce a different approach for the calibration procedure in laboratory of gas sensing chemiresistor based on graphene that work at room temperature. This non static calibration minimizes critical problems linked to the poisoning effect due to the analite chemisorption.

**Acknowledgements** Thanks to SENSICHIPS srl for providing software and micro-hardware for the chemical sensors readout.

## References

1. Nag A, Mitra A, Mukhopadhyay SC (2017) Graphene and its sensor-based applications: a review. Sens Actuators A Phys
2. Singh E, Meyyappan M, Nalwa HS (2017) Flexible graphene-based wearable gas and chemical sensors. ACS Appl Mater Interfaces 9(40):34544–34586
3. Ricciardella F et al (2014) A calibrated graphene-based chemi-sensor for sub parts-per-million $NO_2$ detection operating at room temperature. Appl Phys Lett 104(18):183502

# Rapid Parallel Calibration for Environmental Bulky Gas Sensor Systems

**E. Massera, B. Alfano, M. L. Miglietta, T. Polichetti, S. De Vito, F. Formisano, G. Di Francia and P. Delli Veneri**

**Abstract** This work explains a technique for the parallel calibration in laboratory of several embedded sensor systems for the air quality monitoring. Thanks to a Large Volume Test Chamber and to a precise injection of the target gas in the chamber is possible to measure with the right precision and accuracy the sensitivity curve of the gas sensors contained in embedded systems for the Internet of Things.

**Keywords** Gas sensors · Calibration · Embedded system · IOT

## 1 Introduction

In the solid state sensor universe, gas sensors stand out for the difficulties that arise in the calibration procedure. Sensors show non homogeneity, aging and drift, their response is affected by interference of environmental parameters such as temperature, humidity as well as by non-target gases and this compromise their predictability [1]. Today the Laboratory Calibration remains the mainstream solution to obtain a certified calibration curve that can ensure the correct usage of a gas sensor in its application scenario. Long time procedures are needed to obtain a reliable calibration curve. Actually it is necessary to expose the gas sensor to several known levels of the target gas concentration waiting a steady state to be reached both for the gas concentration and for the gas sensor output. In the past, a time consuming procedure for gas sensor calibration was not considered a bottleneck because the application scenarios did not request a large number of sensors. Today, thinking to wearable devices and IOT, time consumption in a gas sensor calibration became a fundamental issue for gas sensors manufacturers [2]. To tackle this issue, we have developed a calibration method that with only one calibrated injection of the target gas in a closed chamber can continuously measure and correlate the gas sensors response to the gas concentration with the maximum precision and accuracy achievable by the laboratory

E. Massera (✉) · B. Alfano · M. L. Miglietta · T. Polichetti · S. De Vito · F. Formisano · G. Di Francia · P. Delli Veneri
ENEA, CR-Portici, P.le E. Fermi 1, 80055 Naples, Italy
e-mail: ettore.massera@enea.it

 
G. Di Francia et al. (eds.), *Sensors and Microsystems*, Lecture Notes in Electrical Engineering 629, https://doi.org/10.1007/978-3-030-37558-4_34

instruments. A large volume test chamber is useful to maximize precision and allows calibrating several sensors at the same time.

# 2 Experimental and Results

## *2.1 Experimental Setup and Calibration Method*

A 15 lt "Large Volume Test Chamber" (LVTC, see Fig. 1) has been installed in a state of art Gas Sensor Characterization System (GSCS). In brief the GSCS is composed by a stainless air-tight Test Chamber closed in a adjustable thermal box.

In the LVTC the air composition (humidity, chemical compounds concentration) can be settle up thanks to an inlet of GAS flux precisely controlled by certified Mass Flow Controllers (MKS 1179 series). The accuracy of the gas chemical composition is ensured by the mixing of certified bottles (Rivoira). Temperature and humidity are recorded with industrial sensors (LSI Pt100). The LVTC can sustain the calibration of several complete sensor systems at once. The calibration method consists in injecting in the inlet tube of the LVTC a constant flow of the target gas properly diluted at the maximum concentration ($C_0$) with humid synthetic air. The time-rising concentration C(t) of the target gas is precisely predicted by the following exponential law that, in general, describes a transition between two steady states of a physical parameter under a time constant perturbation [3]:

**Fig. 1** Photo of the 15 lt "Large Volume Test Chamber" (LVTC)

$$C(t) = C_0\left(1 - e^{-\frac{t}{\tau}}\right)$$

The characteristic time (τ) can be precisely estimated using a calibrated sensor, this parameter is proportional to the free space inside the chamber and must be opportunely corrected when several sensors are inside the chamber.

The calibration procedure (run) consists of three time steps: first, synthetic air is injected for the unperturbed state recording of the sensor output (baseline); while in the second step, gas target properly diluted in the gas carrier is injected and the adsorbing phase of the sensor response is recorded; finally the third step is the test chamber washing in a constant flow of synthetic air recording the desorbing phase of the sensor output. With this procedure is possible to verify the sensor output behavior during the adsorbing and desorbing phase of the chemical compounds on the sensors surface. Sensing hysteresis or poisoning can be detected and measured.

With a gas flow of 1 lt/min, the τ of the LVTC is estimated to be 1100 ± 50 s. This means that in 3 h, several sensors with a time response faster than 2 min can be calibrated with maximum precision allowed from 0 to $C_0$.

## 2.2 *Results*

The calibration method is applied to the MONICA wireless sensor systems for air pollution monitoring, equipped with Electrochemical gas sensors [4]. Our LVCT allow to place up to 8 of this sensor systems (Fig. 2).

Sensors calibration is performed versus 0–500 ppb of Nitrogen Dioxide and 0–5 ppm of Carbon Monoxide at a controlled and constant temperature and humidity. The right part of Fig. 2 shows a graph of the time log for a sensor output during a calibration run with injection of 5 ppm of carbon Monoxide. It's easy to distinguish the three steps of the calibration run, the red line underline the adsorbing phase while the blue line the desorbing phase. As a result of the calibration run, a sensitivity curve is estimated by the sensors output log using a script in R language [5]. Once estimated the sensitivity (Fig. 2 left part) with a Linear regression of the data, is possible to explore the precision of the sensor output in all the range of calibration as illustrated in Fig. 3 (right) where is reported the relative error of the sensor estimated gas concentration versus the gas concentration.

# 3 Conclusion

The illustrated work indicates a non-conventional calibration procedure for embedded gas sensor systems in laboratory that overcome critical problems linked to the adoption of a large volume test chamber necessary to contain a complete gas sensor system.

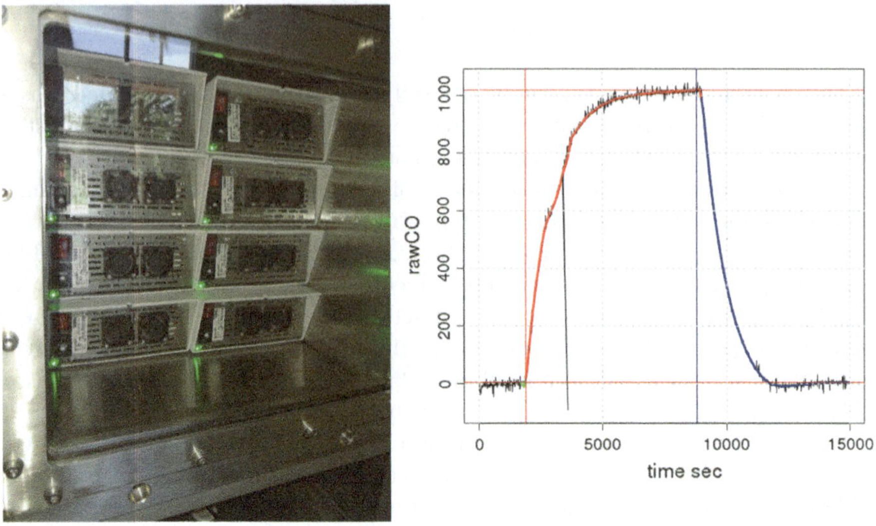


**Fig. 2** On the left, 8 MONICA wireless sensor systems for air pollution monitoring, during the calibration run. On the right, the time log graph for one sensor output during an injection of 5 ppm of carbon monoxide. The red line is the adsorbing phase, the blue line is the desorbing phase of the sensor output

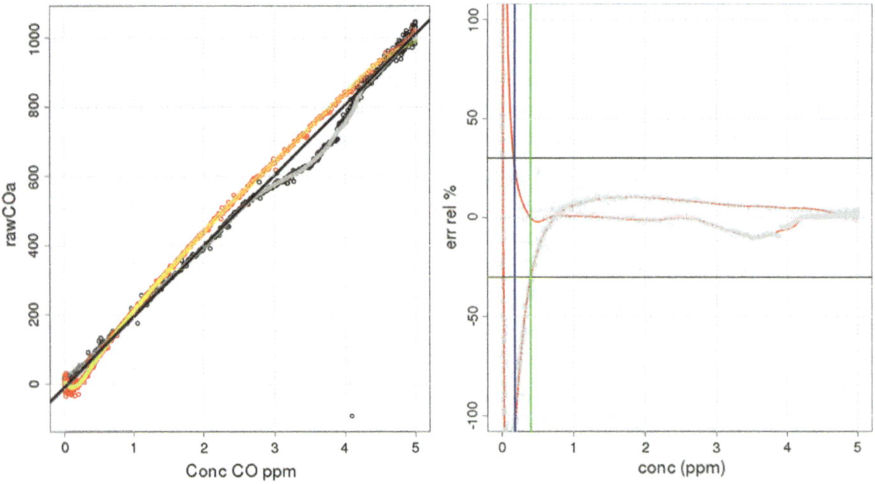


**Fig. 3** On the left the sensitivity curve and the linear regression for the sensor output in the range 0–5 ppm during the adsorbing (black dots) and desorbing (red dots). On the right, the graph of the relative error for the estimated gas concentration by the sensor output with the calculated sensitivity versus the gas concentration

## References


1. Spinelle L et al (2017) Field calibration of a cluster of low-cost commercially available sensors for air quality monitoring. Part B: NO, CO $CO_2$ Sens Actuators B: Chem 238:706–715
2. Cebolla RO et al (2018) Test methodologies for hydrogen sensor performance assessment: chamber vs. flow-through test apparatus. Int J Hydrog Energy 43(45):21149–21160
3. Moxnes E, Davidsen P (2016) Intuitive understanding of steady-state and transient behaviors. Syst Dyn Rev 32(2):130–155
4. De Vito S et al (2017) A crowdfunded personal air quality monitor infrastructure for active life applications. In: 2017 IEEE international workshop on IEEE measurement and networking (M&N)
5. Ripley Brian D (2001) The R project in statistical computing. MSOR Connections Newsl LTSN Maths Stats Netw 1(1):23–25

# Electrochemical Sensors Based on Conducting Polymers: Characterization and Applications

**Giulia Selvolini, Alina Adumitrachioaie, Mariagrazia Lettieri, Oana Hosu, Cecilia Cristea and Giovanna Marrazza**

**Abstract** In this work, we present the development and the characterization of a simple sensing platform based on graphite screen-printed electrodes, which were properly modified with a conducting copolymer nanofilm. The platform was realized by the simultaneous electropolymerization of aniline and anthranilic acid monomers in equal proportion onto the graphite electrodic surface. The obtained copolymer was fully characterized through cyclic voltammetry by using $[Ru(NH_3)_6]^{2+/3+}$ reversible couple as redox probe. The electroactive surface area was calculated by means of the Randles-Sevcik equation and compared with those of the individual polymers (polyaniline and polyanthranilic acid). The selected platform could be then used for different applications, e.g. as a scaffold to immobilize a biological receptor.

**Keywords** Conducting polymer · Sensing platform · Graphite screen-printed electrodes

## 1 Introduction

In the last decade, advanced hybrid materials represent an emerging field in the synthesis of new sensing systems. These are a very large and heterogeneous class of materials, starting with molecular and supramolecular assembled materials, polymers or nano-sized objects to nanostructured and hybrid architectures with inorganic, organic or biological character, having or combining particular properties that cannot be found in other type of materials. Of those synthesized nanomaterials, conducting polymers composites have been widely used in the construction of sensor surfaces. Over the last decades, conductive polymers have emerged as

G. Selvolini · M. Lettieri · G. Marrazza (✉)
"Ugo Schiff" Chemistry Department, University of Florence,
Via della Lastruccia 3, 50019 Sesto Fiorentino, FI, Italy
e-mail: giovanna.marrazza@unifi.it

A. Adumitrachioaie · O. Hosu · C. Cristea
Analytical Chemistry Department, "Iuliu Haţieganu" University of Medicine and Pharmacy, 4 Louis Pasteur Street, 400439 Cluj-Napoca, Romania



G. Di Francia et al. (eds.), *Sensors and Microsystems*, Lecture Notes in Electrical Engineering 629, https://doi.org/10.1007/978-3-030-37558-4_35

an alternative to traditional electrode materials [1]. Conducting polymers i.e. polythiophene, polyaniline, polymethylene blue, polyanthranilic acid, polypyrrole, and poly(o-phenylenediamine) display advantages due to their charge transport properties and electrochemical redox efficiency, which are attributed to the delocalization of π-electrons over the polymeric backbone [2, 3]. For a greater enhancement of the electrochemical sensor performance, different electrode systems have been developed based on surface modification with nanomaterials [4]. The synergy of multifunctional materials, recognition elements, and electrochemical methods is improving the selectivity, stability, and reproducibility, thus promoting the development of sensors for assays and bioassays. Great advancement has been made in the synthesis of novel metal nanomaterials with various sizes, shapes, composition and structure, since all these properties are closely related to their physicochemical properties. In this work, conductive polymers such as polyaniline (PANI) and polyanthranilic acid (PAA), were combined and used to obtain a nano/micropatterned surface at graphite screen printed electrodes (GSPEs). This electrodeposition method combines the portability of screen-printed cells and of a computer-controlled instrument with the low-cost and the easiness of modification. The obtained polymeric films were electrochemically characterized by cyclic voltammetry (CV) to assess their performance as a sensing platform with applications in the biosensors field.

## 2 Materials and Methods

### 2.1 Chemicals

Aniline ($C_6H_7N$), anthranilic acid ($C_7H_7NO_2$), perchloric acid ($HClO_4$), N-(3-dimethylaminopropyl)-N′-ethylcarbodiimide hydrochloride ($C_8H_{17}N_3{\cdot}HCl$, EDC), N-hydroxysuccinimide ($C_4H_5NO_3$, NHS), potassium chloride (KCl), hexaammineruthenium(II) chloride ($[Ru(NH_3)_6]Cl_2$), hexaammineruthenium(III) chloride ($[Ru(NH_3)_6]Cl_3$) were purchased from Merck (Milan, Italy). Deionized water was used for all preparations.

### 2.2 Apparatus

Cyclic voltammetry measurements were carried out with PalmSens portable potentiostat/galvanostat (PalmSens BV, Houten, The Netherlands) and the results were analyzed with PSTrace 5.6 software. The platform was developed by using screen-printed cells based on graphite-working electrodes (3 mm diameter), each one with its own silver pseudo-reference electrode and graphite counter electrode. The screen-printed cells were purchased from EcoBioServices (Florence, Italy).

## 3 Results and Discussion

Conducting polymers have recently attracted a lot of attention in sensing and biosensing, since they show several features (e.g. biocompatibility, flexibility, low cost, process ability) that make them suitable for being applied in this field, as they are a fertile farm for the immobilization of bioreceptors. Among all organic conjugated polymers, polyaniline (PANI) is a very attractive conducting polymer in the field of biosensing. Since PANI shows conductive features only in acidic pH values, which makes its application in biosensing a big deal, it is often doped with something that can increase its conductive properties, such as noble metal (e.g. gold) nanoparticles or conjugate carboxylic acids [5]. PANI doped with gold nanoparticles had been already characterized and successfully applied as an electrochemical platform for the biosensing of pesticides in environmental field [4, 6], while poly-anthranilic acid (PAA), a carboxylated aniline based polymer capable of self-doping, is of interest as a soluble derivative of PANI that had been already used in clinical analysis [7]. The primary surface modification of the graphite screen-printed electrodes was obtained by electrodepositing PANI and PAA via CV. The obtained co-polymer combines conductive properties of aniline with the ability of anthranilic acid to bind the biocomponents through its carboxylic groups [8]. The polymerization process of the monomer mixture (2.5 mM equimolar mixture in 50 mM $HClO_4$) were compared with those of individual aniline and anthranilic acid. The number of cycles for the electropolymerization was optimized taking into account the growth of the current peak height after each cycle. No significant differences in the redox peaks were found by using 10 cycles and thus this value was used for the following experiments. The obtained CV patterns are consistent with previous studies of electropolymerization of PANI [4] and PAA [7], since the redox peaks of the copolymer are located at an intermediate position between the ones of the individual polymers. Table 1 reports the position (e.g. the potential value) of the oxidation peak of the monomers as well as the position of the oxidation and reduction peaks of the polymers.

The observed shift in the potential values confirms that the resulting copolymer has a different structure from the ones obtained by electropolymerizing the individual monomers. The polymers and the copolymer were then characterized through CV at different scan rates (ranging from 25 to 150 mV/s) in presence of 1 mM

**Table 1** Cathodic and anodic potential values for the copolymer poly(aniline-*co*-anthranilic acid (PANI-PAA) and for the individual polymers polyaniline (PANI) and polyanthranilic acid (PAA)

| Platform | | Scan number | $E_{p,a}$ (V) | $E_{p,c}$ (V) |
|---|---|---|---|---|
| PANI | Monomer | 1 | 0.70 | −0.11 |
| | Polymer | 10 | 0.18 | |
| PAA | Monomer | 1 | 0.75 | −0.20 |
| | Polymer | 10 | 0.30 | |
| PANI-PAA | Monomer | 1 | 0.72 | −0.03 |
| | Polymer | 10 | 0.23 | |

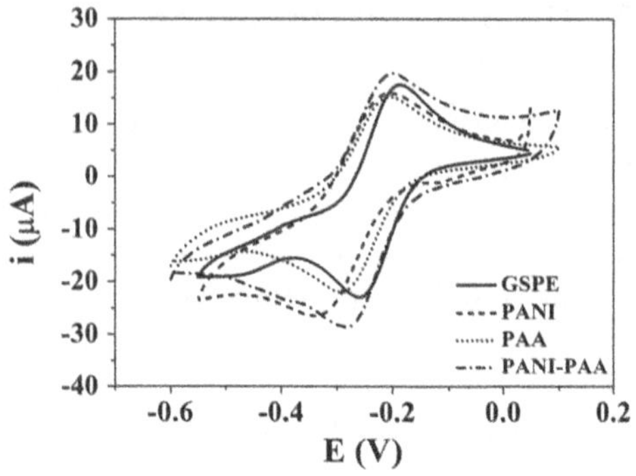


**Fig. 1** Characterization through CV of unmodified GSPE (solid line), PANI-modified GSPE (dashed line), PAA-modified GSPE (dotted line) and PANI-PAA-modified GSPE (dashed-dotted line)

$[Ru(NH_3)_6]^{2+/3+}$ (equimolar mixture in 0.1 KCl) as redox probe and the response was compared with that of an unmodified GSPE. Figure 1 reports the comparative voltammograms of the different electrochemical platforms recorded at 100 mV/s.

The current peak height was plotted against the square root of the scan rate and the electroactive surface area was calculated from the angular coefficient of the obtained linear regression by applying the Randles-Sevcik equation [9]. The results are shown in Fig. 2.

The electroactive area of the unmodified GSPE is higher than electroactive area of modified platforms with PANI (even if its use leads to a more reproducible surface) because there is a charge repulsion between positive charge of $[Ru(NH_3)_6]^{2+/3+}$ and the positive charge of amino groups. In fact, this is not verified in the case of modified platforms with the copolymer and PAA because repulsiveness between charges decrease with presence of negative carboxylic groups provided by anthranilic acid. As a matter of fact, with this positively-charged probe, the electroactive area increases for modified electrodes with PAA and the copolymer.

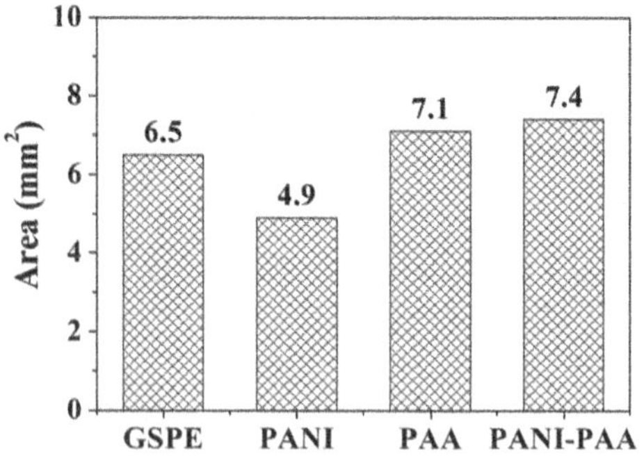


**Fig. 2** Electroactive area values obtained for the different electrochemical platforms: unmodified GSPE, PANI-modified GSPE, PAA-modified GSPE and PANI-PAA-modified GSPE

## 4 Conclusions

The realized platform, based on graphite screen-printed electrodes (GSPEs) modified with poly(aniline-*co*-anthranilic acid) copolymer (PANI-PAA), provides a high surface area and an improved conductivity. These preliminary results encourage its application in the next future for biosensor development, as it can be further functionalized with a biological recognition element in an easy way, e.g. though the use of N-(3-dimethylaminopropyl)-N′-ethylcarbodiimide/N-hydroxysuccinimide (EDAC/NHS) chemistry.

## References

1. Naveen MH, Gurudatt NG, Shim YB (2017) Applications of conducting polymer composites to electrochemical sensors: a review. Appl Mater Today 9:419–443
2. Barsan MM, Ghica ME, Brett CMA (2015) Electrochemical sensors and biosensors based on redox polymer/carbon nanotube modified electrodes: a review. Anal Chim Acta 881:1–23
3. Ravalli A, Rossi C, Marrazza G (2017) Bio-inspired fish robot based on chemical sensors. Sens Actuators B: Chem 239:325–329
4. Selvolini G, Băjan I, Hosu O, Cristea C, Săndulescu R, Marrazza G (2018) DNA-based sensor for the detection of an organophosphorus pesticide: profenofos. Sensors 18:2035–2046
5. Saberi RS, Shahrokhian S, Marrazza G (2013) Amplified electrochemical DNA sensor based on polyaniline film and gold nanoparticles. Electroanalysis 25:1373–1380
6. Rapini R, Cincinelli A, Marrazza G (2016) Acetamiprid multidetection by disposable electrochemical DNA aptasensor. Talanta 161:15–21
7. Taleat Z, Ravalli A, Mazloum-Ardakani M, Marrazza G (2013) CA125 immunosensors based on poly-anthranilic acid modified screen-printed electrodes. Electroanalysis 25:269–277
8. Selvolini G, Lettieri M, Tassoni L, Gastaldello S, Grillo M, Maran C, Marrazza G (2019) Electrochemical enzyme-linked oligonucleotide array for aflatoxin $B_1$ detection. Talanta 203:49–57
9. Elgrishi N, Rountree KJ, McCarthy BD, Rountree ES, Eisenhart TT, Dempsey JL (2018) A practical beginner's guide to cyclic voltammetry. J Chem Educ 95:197–206

# The Best Approach for Early Detection of Fungi in Tomato Sauce

**Domenico Palumbo, Luigi Quercia, Antonella Del Fiore, Patrizia De Rossi and Annamaria Bevivino**

**Abstract** The detection of fungal contaminations, specifically moulds, in tomato sauce stored in the refrigerator is of great importance and very attractive in smart emerging applications. Using an electronic nose (e-nose) and a Fourier transform infrared (FTIR) spectrometer, we examined two sampling methods to early detect fungal contamination: the first method looks at the accumulated headspace while the second one at the actual headspace. Interestingly, we found that we can use only one sensor to detect the moulds even before their visual development.

**Keywords** Mould early detection · E-nose · Machine learning

## 1 Introduction

The detection of fungal contaminations, specifically moulds, into jarred tomato sauce for pasta using an electronic sensory system is of great importance not only because Italy is a world leading producer but also because of its usefulness in smart emerging applications very attractive for the consumer, like the smart fridges [1–3]. The main fungal contaminants of vegetables, such as tomato, both in the field and in the post-harvest, belong to the *Penicillium* genus, including species responsible for vegetables rot [4]. Some strains belonging to *Penicillium spp.* are tolerant to low temperatures and, therefore, are able to contaminate and grow on tomato, but also on tomato sauce stored at low temperatures [5].

In the present work, we used a combined approach based on conventional culture-based method and electronic sensory system in order to detect fungal contamination in tomato sauce. We aimed at determining the best experimental approach to early detect fungal contamination in the headspace of tomato sauce stored in the refrigerator in order to evaluate if the sauce has been contaminated once opened.

Two approaches were examined. The first one takes into account the headspace history of the sample: in this case the headspace was not perturbed from the opening till the measurements were performed, therefore, detecting the accumulated headspace.

D. Palumbo (✉) · L. Quercia · A. Del Fiore · P. De Rossi · A. Bevivino
ENEA, Centro Ricerche Casaccia, Via Anguillarese 301, 00123 Rome, Italy
e-mail: domenico.palumbo@enea.it

 
G. Di Francia et al. (eds.), *Sensors and Microsystems*, Lecture Notes in Electrical Engineering 629, https://doi.org/10.1007/978-3-030-37558-4_36

The second one looks at the actual state of the headspace: before each measurement the accumulated headspace was removed by using an inert carrier so that only the actual, and hence, the real headspace was measured.

## 2 Measurement Methods

### 2.1 Measurement Method Which Looks at the Accumulated Headspace

Using a commercial portable e-nose (PEN3.5, **AIR**SENSE Analytics) we analysed tomato sauce headspace of artificially contaminated (psychrophilic *Penicillium spp.*) and not contaminated samples during 14 days of refrigerated storage at 8 °C.

The samples consisted of glass containers filled for 1/4 with tomato sauce. A psychrophilic strain of *Penicillium spp.*, previously isolated from tomato sauce stored at +4 °C, was used to artificially contaminate the samples. Conidial suspension of this fungal strain grown on Potato Dextrose Agar (PDA) slopes was achieved by recovering the conidia with sterile water/Triton X100 solution, followed by filtration through sterile filter paper (Whatman No. 1) to remove fungal hyphae. Conidial concentration was therefore determined by Thoma chamber counting. Contamination of samples was achieved by inoculating 1 ml of *Penicillium spp.* conidial suspension, corresponding to $10^4$ conidia in 400 ml of tomato sauce. Then all samples were stored at 8 °C.

Three biological replicates of fungal treated and not-treated samples were performed. In order to directly verify the fungal growth during storage in refrigerated conditions, microbiological analyses were performed at the 0th, 3rd, 7th, 10th and 14th day from the inoculation, for both contaminated and not contaminated samples, by plating serial dilutions of tomato sauce on PDA plates in triplicates. In detail, at each sampling time, sauce aliquots (5 ml) were taken from the inoculated and not inoculated jars and tenfold serial dilutions ($10^{-1}$, $10^{-2}$, $10^{-3}$, $10^{-4}$) were carried out in triplicate. A volume of 100 μl of each dilution was then plated onto PDA plates containing 300 mg/L and 150 mg/L of streptomycin and neomycin, respectively, to prevent unwanted bacterial growth. After that, plates were incubated in a thermostatic chamber for one week at 25 °C. The fungal counts were carried out by enumerating the number of colony forming units (CFU) developed on agar plates after 5-days incubation.

In parallel, we also measured the same headspaces using a FTIR (Thermo Fisher Nicolet iS10, 2 m ZnSe gas cell). From these spectral measurements we considered the region 600–740 $cm^{-1}$ to calculate, using a multiple linear regression,[1] the $CO_2$ content of the headspaces of all the samples. Since the $CO_2$ content in the headspaces

[1] All the elaborations in this research were done using Matlab 2015b (MathWorks).

is the result of fungi "respiration" in the contaminated samples, its measurement should be a direct indicator of the contamination.

We performed the measurements under a constant aspiration of 60 sccm of nitrogen for 500 s (the effective FTIR spectra were taken at 250 s from the beginning). The e-nose measurements followed a similar procedure but the aspiration was settled to 400 sccm and lasted 300 s. The headspace accumulated was pumped in the measurement system allowing the ambient air to enter the containers through an opening in the lid.

## 2.2 Measurement Method Which Detects the Actual Headspace

To be more confident on the measurements and to detect the development of mould before it is visually evident, it is necessary a measurement method which "photographs" the effective conservation of the sauce. This new measurement method focuses on the actual state of the tomato sauce "looking" inside the sauce itself by removing the headspace history. This is done by means of a constant flux of an inert gas which flows through the headspace of the sample under measure so that, rejecting the first part of the atmosphere which contains the history of the sample, we monitor the sauce VOCs (Volatile Organic Compounds) emitted in real time. In this case, samples consisted of 20 ml sterile vials with 2 ml of tomato sauce taken from tomato sauce jars artificially inoculated with *Penicillium spp.* ($10^4$ conidia in 240 ml of tomato sauce), and from jars with not inoculated sauce. All vials were sealed in a sterile cabinet and stored at 3 °C. Three biological replicates for each condition were performed. The procedure to measure a sample using a FTIR was the following: 2 min of nitrogen flux to remove the accumulated headspace followed by 15 min in oven at 30 °C to generate the headspace, next we performed the measure under a constant flux of 60 sccm of nitrogen for 500 s (the effective FTIR spectra were taken at 250 s from the beginning). The e-nose measurements followed a similar procedure except for the flux settled to 400 sccm and the time settled to 300 s. FTIR spectra measurements were taken at the 5th, 6th, 7th, 10th, 11th and 13th day from the inoculation whereas e-nose measurements on the same samples were taken at 6th, 7th, 8th, 11th, 12th and 14th day from the inoculation for time necessity. Figures 1a, b show two emblematic spectral regions of the FTIR spectra whereas Fig. 1c shows an example of e-nose measurement.

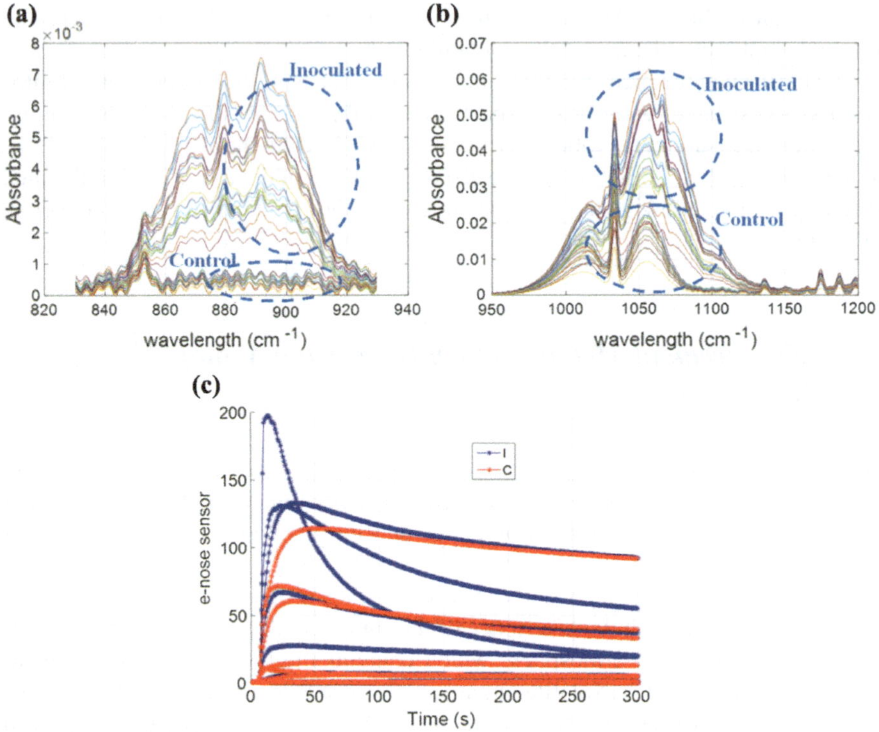


**Fig. 1** **a** and **b** Actual headspace examples of FTIR measured bands, for the tomato sauce at the 5th, 6th, 7th, 10th, 11th and 13th day from the inoculation. **c** Actual headspace e-nose measurements at the 6th day; Control (C) refers to tomato sauce without inoculation and inoculated (I) to tomato sauce inoculated with psychrophilic *Penicillium spp*. The mould is visually evident from the 10th day

# 3 Results

## 3.1 Accumulated Headspace

The e-nose measurements showed a signal enhancement 10 days after the fungal inoculation, when the mould content was visually evident. Correspondingly, the microbiological analysis revealed an increased mycelium growth of *Penicillium spp.* on the 10th day from the inoculation, $2.5 \times 10^4$ CFU/ml, which reached the value of $4.7 \times 10^4$ CFU/ml on the 14th day. The same happened for the FTIR measurements and the estimated ppm of $CO_2$ which rose 10 days after inoculation, although the variance associated with the estimations was high. This was probably due to the measurement method which looked back to all the history of the samples (Fig. 2b).

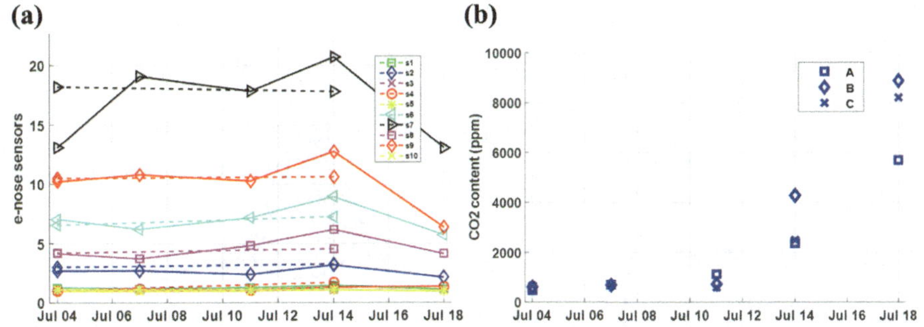


**Fig. 2** Accumulated headspace. Measurements in tomato sauce inoculated with psychrophilic *Penicillium spp.* at the 0th, 3rd, 7th, 10th and 14th day from the inoculation. **a** E-nose signals (dashed-control; Continuous-inoculated); **b** Estimated ppm of $CO_2$ using FTIR spectra; A, B and C are the three biological replicates

## 3.2 *Actual Headspace*

There exist different interesting regions of the FTIR spectra in the case of actual headspace which are useful to distinguish between the inoculated and the control samples. An interesting one is the band around 1050 $cm^{-1}$ fully compatible with the sensible hypothesis that FTIR discrimination could be related to the increase of alcohols following fermentation of contaminated tomato sauce [6]. Actually the two methods look at different compositions (see Fig. 3): the accumulated headspace essentially reveals the $CO_2$ (gas) whereas the method which looks at the actual sauce conservation reveals a more complex headspace composition (VOCs) useful to easily distinguish between contaminated and not contaminated samples through the use of either FTIR or e-nose measurements and allows an early mould detection before its visual evidence.

Figure 4 shows principal component analysis (PCA) done on FTIR and e-nose measurements in the case of actual headspace. Differently from the case of accumulated headspace, a great separation in both cases is evident between control samples and inoculated samples as early as the 5th day, even before mould growth even with one e-nose sensor.

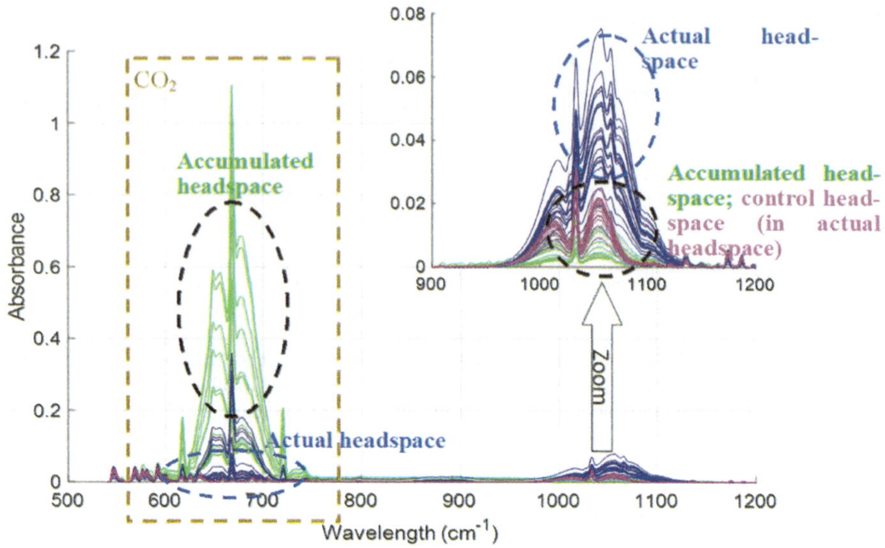


**Fig. 3** All the FTIR spectra measurements of the actual headspace. In the band 600–740 $cm^{-1}$, apart some outliers, we revealed the $CO_2$ concentration whereas around 1050 $cm^{-1}$ we distinguished between the inoculated and the control samples in the actual headspace whereas the accumulated headspace was not useful

## 4 Conclusion

By using the method that examines the actual sauce conservation and by using specific wavelengths or few, even just one, sensor we have detected the differences between food samples before the mould development occurred. This method could therefore be useful for monitoring the qualitative decay of tomato sauce after the eventual contamination of the tomato jar, during the domestic consumption and storage, allowing a correct management and prevention of waste production.

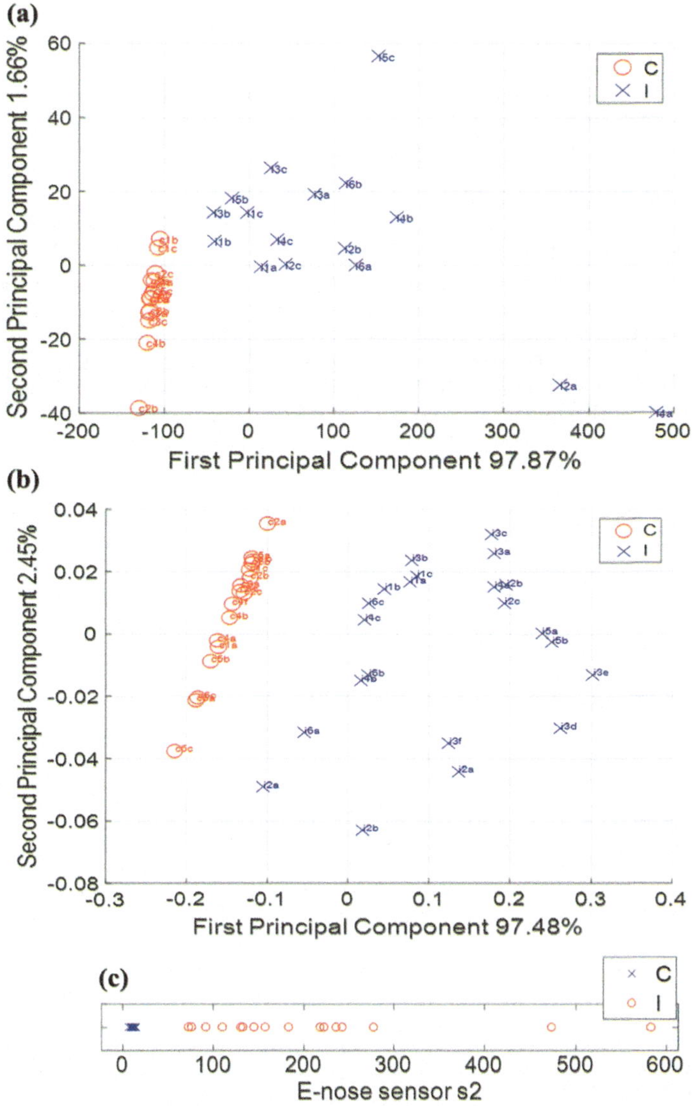


**Fig. 4** PCA analysis done on FTIR and e-nose measurements in the case of actual headspace. **a** PCA done on FTIR wavelength 1050 $cm^{-1}$; **b** PCA done on e-nose measurements; **c** E-nose values of the sensor n° 2 of the e-nose

**Acknowledgements** This work was partially supported by Safe & Smart, Nuove tecnologie abilitanti per la food safety e l'integrità della filiera agro-alimentare in uno scenario globale, project N. CTN01_00230_248064, funded by the Italian Ministry of Education, Universities and Research (MIUR).

## References

1. Wilson AD, Baietto M (2009) Applications and advances in electronic-nose technologies. Sensors 9:5099–5148
2. Tang KT, Chiu SW, Pan CH, Hsieh HY, Liang YS, Liu SC (2010) Development of a portable electronic nose system for the detection and classification of fruity odors. Sensors 10:9179–9193. https://doi.org/10.3390/s101009179
3. Sberveglieri V, Falasconi M, Gobbi E, Núñez CE, Zambotti G, Pulvirenti A (2014) Candida milleri detected by electronic nose in tomato sauce. Procedia Eng 87:584–587
4. Pitt JI, Hocking AD (1999) Fungi and food spoilage. Aspen Publishers Inc., Gaithersburg, Maryland
5. Wani AH (2011) An overview of the fungal rot of tomato. Mycopath 9(1):33–38
6. Quercia L, Capuano R, Khomenko I, Catini A, Martinelli E, Paolesse R, Biasioli F, Di Natale C (2019) PTR-ToF-MS/QMB-electronic-nose synergies exploration: a case study. In: Poster at 8th international PTR-MS conference, 4–7 Feb 2019

# UV Autofluorescence Spectroscopy for Cyanobacteria Monitoring and Discrimination in Source Water

**Gianluca Persichetti, Genni Testa, Romeo Bernini, Emanuela Viaggiu, Roberta Congestri, Victor Dumas and Laurent Labbe**

**Abstract** An analytical technique allowing a distinction between cyanobacteria and other microscopic life forms that exploits autofluorescence in the deep ultraviolet has been developed. The proposed approach is based on the amplitude of relative fluorescence peaks of natural pigments or metabolites in unicellular microorganisms commonly present in the waters. The experimental results showed a clear distinction between cyanobacteria and other planktonic species. This approach has been applied to an aquaponics system receiving input water from the Drennec lake, in France, correctly detecting the presence of cyanobacteria.

**Keywords** Fluorescence spectroscopy · Cyanobacteria · Autofluorescence · Portable sensor

## 1 Introduction

The monitoring of cyanobacteria is of great importance in prevention or intervention in the case of algal blooms that often affect water resources [1]. Usually water quality monitoring requires rather complex techniques or expensive equipment. In some cases, devices for water monitoring are unsuitable for in situ measurements or not able to provide results in real-time (e.g. counting, image analysis [2]).

Autofluorescence spectroscopy is becoming an increasingly used method in water monitoring. Compared to conventional approaches, this method of analysis can measure samples without preliminary operations such as extraction or separation and its advantages lie in simplicity, rapidity as well as extreme sensitivity [3].

G. Persichetti (✉) · G. Testa · R. Bernini
Institute for Electromagnetic Sensing of the Environment (IREA) National Research Council (CNR), 80124 Naples, Italy
e-mail: persichetti.g@irea.cnr.it

E. Viaggiu · R. Congestri
Department of Biology, University of Rome 'Tor Vergata', Via Cracovia 1, 00133 Rome, Italy

V. Dumas · L. Labbe
INRA, UE 0937 PEIMA (Experimental Fish Farming Facility INRA of Monts D'Arrée), 29450 Sizun, France



G. Di Francia et al. (eds.), *Sensors and Microsystems*, Lecture Notes in Electrical Engineering 629, https://doi.org/10.1007/978-3-030-37558-4_37

## 2 Fluorescence Analysis

A portable system based on autofluorescence that exploits ultraviolet excitation at 270 nm has been developed. The elements of the setup have been selected by considering also their portability, therefore this device is suitable for in situ measurements. The elements composing the portable sensor consist an UV LED source emitting at 270 nm, a system for a water jet waveguide approach [4] that is useful for containing the solution and for providing enhancement in the fluorescence collection and a mini-spectrophotometer used for the detection. A laptop manages the mini-spectrophotometer and the LED driver via USB.

Eleven representative microorganisms were selected for this procedure: nine cyanobacteria, one diatom and one green alga. Their spectra, acquired with the developed sensor, are shown in Fig. 1. In the same figure, spectral ranges attributable to the specific autofluorescent pigments are also reported [5–7].

More specifically, the LED emitting at 270 nm allowed to excite the fluorescence of chlorophylls, phycobiliproteins such as phycocyanin (PC) and phycoerythrin (PE) but

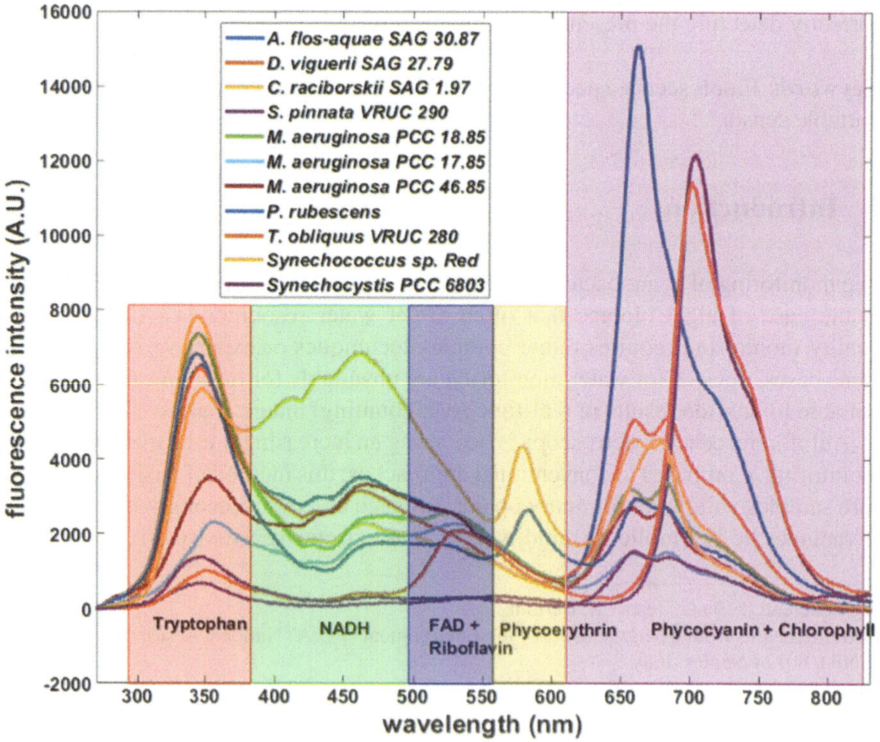


**Fig. 1** Fluorescence spectra acquired with the portable sensor. The transparent bands indicate the spectral ranges corresponding to the specific autofluorescent pigments

also the cellular metabolites that exhibit autofluorescence such as tryptophan, nicotinamide adenine dinucleotide [NADH or NAD(P)H], flavins [derivative of riboflavin such as flavin mononucleotide (FMN) and flavin adenine dinucleotide (FAD)].

The wavelength ranges corresponding to the fluorescence of these pigments (defined in Fig. 1) have been used in an original procedure for cyanobacteria discrimination. This procedure, already described in [8] it is reported here for convenience. The discrimination approach is based on the four following steps:

(1) Calculation of the fluorescence peaks areas subtended by the wavelength range corresponding to specific pigments.
(2) Calculation of the ratio between the areas of the peaks and the area of the peak corresponding to the tryptophan.
(3) At each tested organism is attributed a numerical set of values corresponding to the peak ratios.
(4) All possible ratio combinations have been considered and the ratios showing greater separation/grouping capacities are shown in Fig. 2

The arrangement of the points thus associated allows a grouping of cyanobacteria clearly distinct from that of the green alga (*Tetradesmus obliquus*) and that of the diatom (*Staurosirella pinnata*). The proposed method allows a straightforward implementation in systems for the real-time detection of cyanobacteria that are based on autofluorescence. At the same time, this approach provides an interpretation of the results that is directly related to the presence of specific pigments possessed by the analyzed species.

The previously described method was applied to test water coming from an aquaponics system within the facilities of INRA-PEIMA located in Sizun (France) during a potential algal bloom event in the Drennec lake. PEIMA is one of the largest salmonid experimental station in Europe, this facility uses the Drennec lake to supply

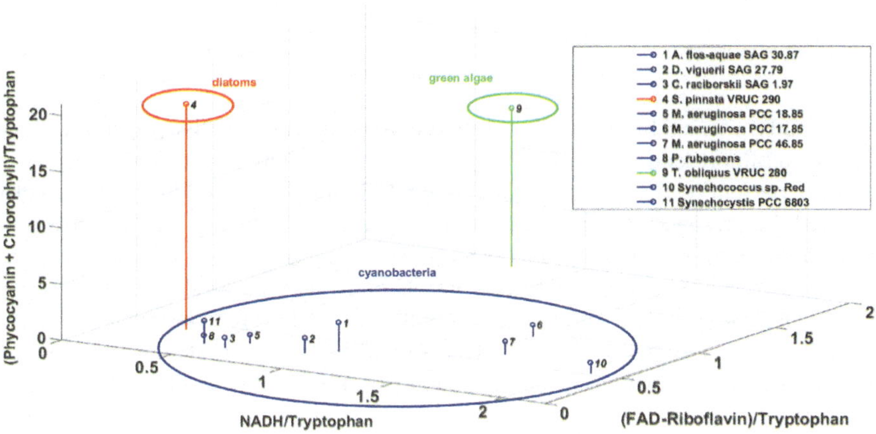


**Fig. 2** Ratio of peaks (phycocyanin + chlorophyll)/tryptophan—NADH/tryptophan—(FAD-riboflavin)/tryptophan

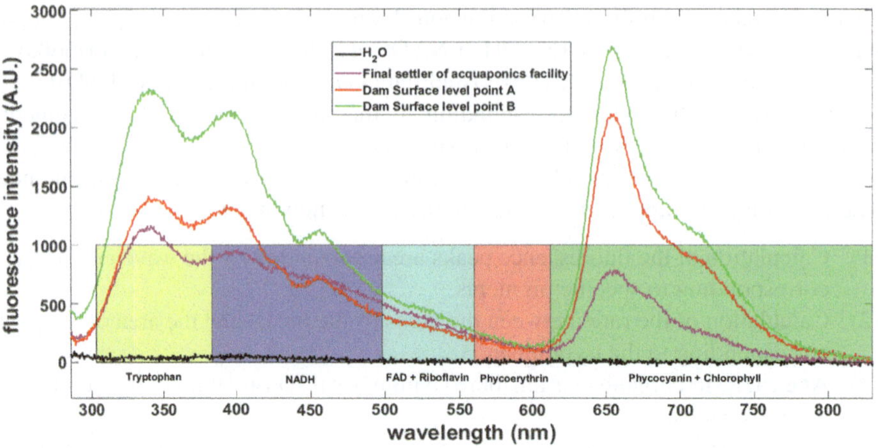


**Fig. 3** Spectra of samples coming from different points of an aquaponic farm during a period of suspect presence of cyanobacteria

water to the farm and its aquaponics system. Water samples were taken at different points in the aquaponics farm, and then sent to our laboratories in an adiabatic container kept at low temperature and in dark conditions. The plot in Fig. 3 shows some of the spectra, acquired with the developed sensor, relative to two sampling points placed near the dam in proximity to the plant, and in a point corresponding to the final settler of the facility. Then, the procedure, described in Sect. 2, was applied to the data and the results are shown in Fig. 4.

Subsequent measurements, carried out at an independent laboratory, showed that algal bloom had originated from the cyanobacteria *Planktotrix isotrix*.

## 3 Conclusion

In this study, a sensor based on UV autofluorescence and an effective procedure based on UV autofluorescence have been applied to the recognition of unknown microorganisms present in freshwater during an algal bloom. The procedure has correctly detected the presence of cyanobacteria in an aquaponics system. The effectiveness of the approach, in terms of cyanobacteria distinction from other microalgae in the phytoplankton, has been demonstrated in spite of the use of a single excitation wavelength. Obviously, further improvements of this technique are expected with the use of multiple excitation wavelengths. This should in fact guarantee a more intense fluorescence of compounds which are more weakly excited at the wavelength of 270 nm.

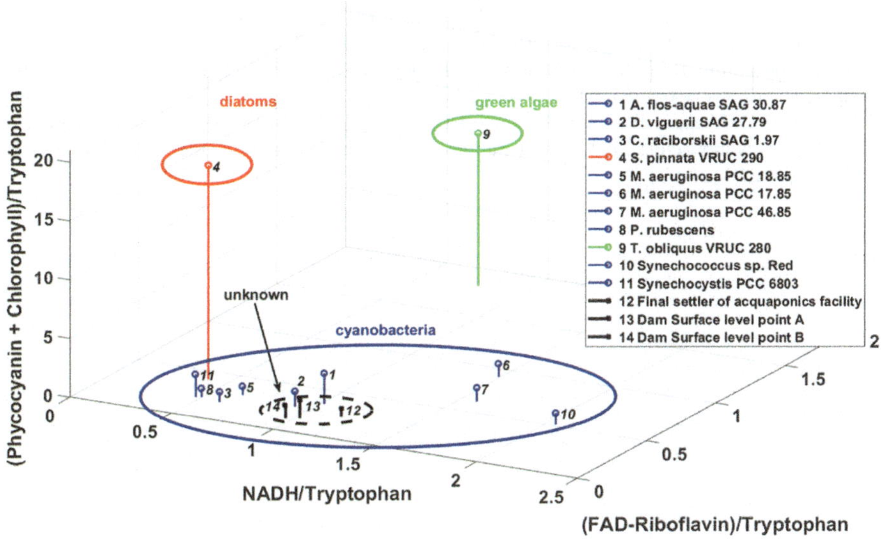


**Fig. 4** Results of the cyanobacteria recognition procedure, applied to unknown samples (circled by a black dotted ellipse) and compared with known results

**Acknowledgements** This research has been partially supported by the ERA-NET Cofund WaterWorks2015 project SMARTECOPONICS, "On-Site Microbial Sensing For Minimising Environmental Risks From Aquaponics To Human Health".

## References


1. He X, Liu YL, Conklin A, Westrick J, Weavers LK, Dionysiou DD, Lenhart JJ, Mouser PJ, Szlag D, Walker HW (2016) Toxic cyanobacteria and drinking water: impacts, detection, and treatment. Harmful Algae 54:174–193
2. Gandola E, Antonioli M, Traficante A, Franceschini S, Scardi M, Congestri R (2016) ACQUA: automated cyanobacterial quantification algorithm for toxic filamentous genera using spline curves, pattern recognition and machine learning. J Microbiol Methods 124:48–56
3. Bourgeois W, Burgess JE, Stuetz RM (2001) On-line monitoring of wastewater quality: a review. J Chem Technol Biotechnol 76:337–348
4. Persichetti G, Testa G, Bernini R (2013) High sensitivity UV fluorescence spectroscopy based on an optofluidic jet waveguide. Opt Express 21:24219–24230
5. Lakowicz JR (2006) Principles of fluorescence spectroscopy. Springer, Baltimore
6. Billinton N, Knight AW (2001) Seeing the wood through the trees: a review of techniques for distinguishing green fluorescent protein from endogenous autofluorescence. Anal Biochem 291:175–197

7. Zhang D, Muller JP, Lavender S, Walton D, Dartnell LR (2012) Fluorescent analysis of photosynthetic microbes and polycyclic aromatic hydrocarbons linked to optical remote sensing. Int Arch Photogramm Remote Sens Spat Inf Sci ISPRS Arch 39:555–559. https://doi.org/10.5194/isprsarchives-xxxix-b8-555-2012
8. Persichetti G, Viaggiu E, Testa G, Congestri R, Bernini R (2019) Spectral discrimination of planktonic cyanobacteria and microalgae based on deep UV fluorescence. Sens Actuators B: Chem 284:228–235

# Application of Polyvinylidene Fluoride Interdigital Capacitors as Parasitic Temperature-Sensing Loads in Passive HF RFID Transponders

**Riccardo Miscioscia, Carmela Borriello, Giuseppe Pandolfi, Giovanni De Filippo, Tommaso Fasolino, Bruno Lanza, Giovanna Zappa and Carla Minarini**

**Abstract** In this work, the introduction of small variations in the electrical impedance of HF (High Frequency) RFID (Radio Frequency Identification) tags has been exploited in the proximity sensing of temperature by applying Polyvinylidene Fluoride (PVDF) interdigital capacitive sensors as a modification of the transponder's circuitry. The adopted electrical parasitic sensing load introduces a shift in the primary resonance peak of the tag. The impedance spectra have been acquired by a dedicated laboratory setup allowing to apply simultaneously thermal forcing and RF signals to the transponder itself. The sensitivity of the modified tag has been assessed after taking into account the thermal response of the transponder chip which is relevant in the formulation of thermal-sensing transponders.

**Keywords** RFID · PVDF · Thermal sensors · Resonant sensors

## 1 Introduction

RFID (Radio Frequency Identification) [1] is a well-established technology utilized to uniquely identify tagged objects. Passive RFIDs have the main advantage to operate wirelessly at short distance (usually about 5–6 cm) with inexpensive transponders and without requiring a continuous power supply. A passive High-Frequency (HF) RFID tag harvests power from the carrier coming from the reader, appearing from the electrical point of view as an inductively coupled resonant RLC circuit tuned at 13.56 MHz. In the recent years, remarkable works have been carried out in the field of RFID systems for remote sensing [2] and temperature monitoring of foods [3] by applying custom integrated circuits equipped with signal processing and data acquisition hardware [4]. However, the adoption of dedicated chips to acquire the transduced

R. Miscioscia (✉) · C. Borriello · G. Pandolfi · G. De Filippo · T. Fasolino · B. Lanza · C. Minarini
ENEA, C. R. Portici, p. le E. Fermi 1, 80055 Naples, Italy
e-mail: riccardo.miscioscia@enea.it

G. Zappa
ENEA, C. R. Casaccia, via Anguillarese 301, 00123 Rome, Italy

 
G. Di Francia et al. (eds.), *Sensors and Microsystems*, Lecture Notes in Electrical Engineering 629, https://doi.org/10.1007/978-3-030-37558-4_38

physical quantities, increases the cost of tags and, in many cases, requires also specific readers to gather the sensed data. Therefore, it could be convenient to experiment tag formulations which rely solely on standard RFID chips (e.g. MIFARE© Classic – 1 K) and moving to reader-side equipment the task of acquiring the sensed quantity.

Given the specifications of common RFID systems [5], drifts from the ideal tuning conditions could be tolerated if a performance loss in terms of reading range is considered acceptable. Therefore, a sensor can be introduced into the transponder circuit and the sensed quantity be remotely acquired by measuring the impedance spectrum of the tag and still keeping the capability of a standard chip to interface itself to an RFID reader/system. Such impedance-spectrum acquisition can be performed in laboratory by the means of Vector-Network Analyzers (VNAs) and dedicated setups [6] or by cheaper (and simpler) portable devices as demonstrated in our patented prototype [7]. Flexible Polyvinylidene Fluoride (PVDF) capacitive sensors have already demonstrated their feasibility in resonant transducer circuits for temperature sensing [8] but, at the best of our knowledge, the integration of such sensors as parasitics in an passive chip-provided HF RFID tag/proximity integrated circuit card has still not been attempted.

In the present work, a temperature-sensing Interdigital Capacitor (IDC) has been experimented as parasitic load in passive HF RFID tags in order to sense temperature changes the reader and the tag are coupled in proximity conditions. In order to interpret the measured data more clearly, the variation of the electric capacitance of the RFID chip and of an IDC having PVDF as dielectric were also measured as a function of the temperature. A customized test-fixture was used to impose thermal variations during the tag impedance spectrum acquisition.

## 2 Experiment Setup

A passive RFID transponder has been carried out on an ISO/IEC 7810 ID-1 shaped Printed Circuit Board (PCB—size about: 85 × 54 mm) 1.6 mm thick FR4 substrate covered by a 35 μm copper conductor. In detail, the layout of an inductive antenna connected to an IDC has been manufactured on the PCB as shown in Fig. 1a by mechanical milling. The IDC had 4 finger per electrode, 7 gaps, 200 μm of gap distance between fingers, 200 μm of finger width and 1600 μm of finger length. Then, a MIFARE© Classic MF1K-compliant chip has been soldered to the antenna pads and commercially-available PVDF dielectric (Sigma Aldrich, average Mn ~ 71,000) has been deposited on the IDC electrodes by drop-casting from a solution of PVDF in Dimethylformamide (DMF). Therefore, in this approach, the IDC has been introduced as electrical load in a passive HF RFID tag by connecting it in parallel to the chip in order to acquire the tag's temperature in the form of impedance spectrum deviation.

A customized test-fixture has been applied in order to impose thermal variations to the capacitor and the chip while measuring the impedance spectrum of the tag. The characterization fixture (see Fig. 1a, b) presents both a round antenna coil (inductance

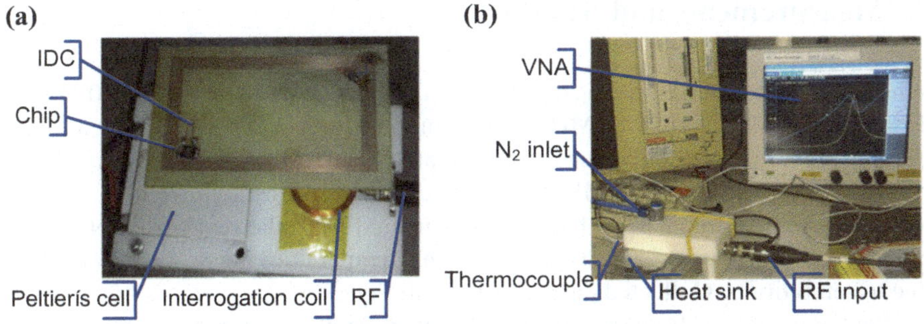


**Fig. 1** A picture of the HF RFID tag (including the IDC sensor) placed on the top of the fixture (**a**); a detail of experimental setup (**b**)

L = 7 μH) and a 20 W Peltier cell to the back surface of the tag. The temperature of the top surface of the tag has been measured by a k-type thermocouple and the RF stimulus has been applied to the interrogation coil by a VNA through a BNC connector (see Fig. 1). The impedance spectra of the system were acquired in a frequency range from 12 to 15 MHz by the means of an Agilent E5061B network analyzer.

In order to evaluate the electrical and thermal characteristics of the main components included in the tag circuitry, the thermal response of the chip has been acquired separately through an impedance analyzer (4192A LF Impedance Analyzer) in a nitrogen-saturated chamber on an ESPEC thermal chuck embedded in a Cascade Summit 11,000 manual probe station at the base frequency of 100 kHz in a shielded four-terminal configuration.

The same experimental setup adopted for the Capacitance/temperature measurements performed on the chip has been employed to measure the capacitance response to temperature of a single IDC carrying drop-casted PVDF as dielectric.

Physical characteristics of this sample are reported in Fig. 2

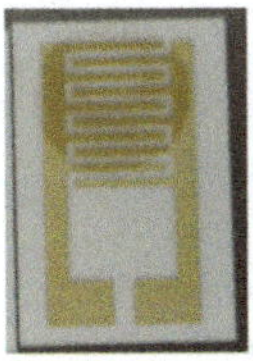

| | |
|---|---|
| Finger material: | Gold |
| Finger width: | 400μm |
| Gap between fingers: | 300μm |
| Finger to Finger faced length: | 4200μm |
| Finger thickness: | 80μm |
| Side contacts width: | 1000μm |
| Number of fingers: | 10 (5 per electrode) |

**Fig. 2** Picture and layout dimensions of the reference IDC sample

## 3 Measurements and Results

When coupled to the fixture, the tag showed a resonance frequency of 13.40 MHz at 25 °C and a bandwidth of 1.81 MHz. The communication channel of the tag, after the introduction of the capacitive load is still capable to interface the chip to commercial readers. A commercial NXP MFRC522 reader has been used to prove that.

The resonance frequencies have been estimated from the real part [Re(Z)] plot of electrical impedance Z versus frequency in a temperature range +20 °C/+ 40 °C, due to setup limits. Collected data are shown in Fig. 3.

The tag resonance frequency decreases linearly with the increase of temperature with a sensitivity S = −5190 Hz/ °C.

This decrease is probably due to an increase of load capacitance. If we consider the load capacitance ($C_{tot}$) to be the resultant of substrate ($C_{substrate}$), chip ($C_{chip}$) and sensor ($C_{sensor}$) capacitances connected in parallel, the load capacitance can be expressed as the sum of such contributions $C_{tot} = C_{chip} + C_{substrate} + C_{sensor}$.

After neglecting antenna resistance, the sum of sensor and substrates capacitance $C_{ss} = C_{sensor} + C_{substrate}$, can be extracted as follows:

$$C_{ss}(T) = \frac{1}{L_{ant}(2\pi f_r(T))^2} - C_{chip}(T) \quad (1)$$

where $f_r$ is the resonance frequency of the tag; $L_{ant}$ is the tag's antenna inductance being approx. $L_{ant} = 5.402$ μH and $C_{chip}$ (the variation of chip capacitance vs temperature). Both $L_{ant}$ has and $C_{chip}$ have been acquired through the impedance analyzer setup as described in the experimental part. By fitting measured data (here not reported) of $C_{chip}(T)$ versus T, its variations could be empirically modeled as $C_{chip} = 14,22$[pF] + 5,96 [fF/°C] * T[°C]. Then, by applying this model to Eq. 1, an estimation of the variation of sensor's capacitance can be graphed versus temperature as in Fig. 4.

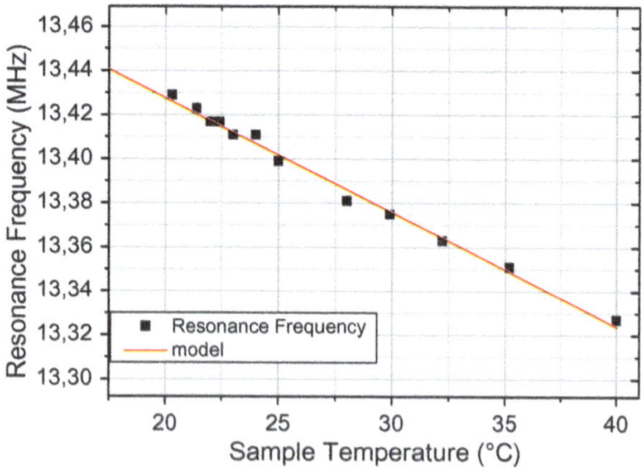


**Fig. 3** Graph of tag resonance frequencies versus temperature

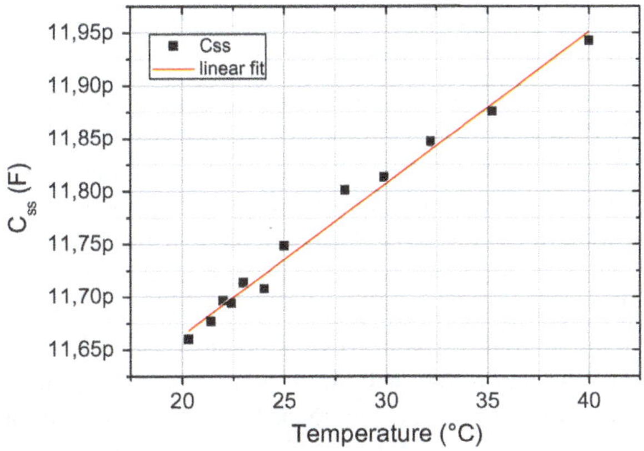


**Fig. 4** Sum of the PVDF capacitance of the sensor and substrate capacitance ($C_{SS}$) plotted versus temperature (b) in the temperature-sensitized RFID tag

The observed behavior is coherent with the increase in capacitance of PVDF capacitors with temperature reported in literature [8] and shows a sensitivity of 14.4 fF/°C. In order to compare the data for the capacitance $C_{SS}(T)$ estimated from the resonance frequency of the tag to the intrinsic behavior of the PVDF capacitor, a reference sample (see Fig. 2) has been characterized at the base frequency of 100 kHz in the same temperature range of the tag. The reference IDC has been found to have a capacitance rest value at +20 °C of about 2 pF. Its variations have been normalized at the value measured at +20 °C and compared to $C_{SS}(T)$ (see Fig. 5) extracted from the above mentioned procedure.

As it can be stated by the graph in Fig. 5, $C_{SS}$ versus temperature plots are in agreement with the reference capacitor too. Sensor curve and tag curve differ for a scale factor due to different IDC extension/geometries and because of the presence of substrate capacitances (from the coil to the PCB).

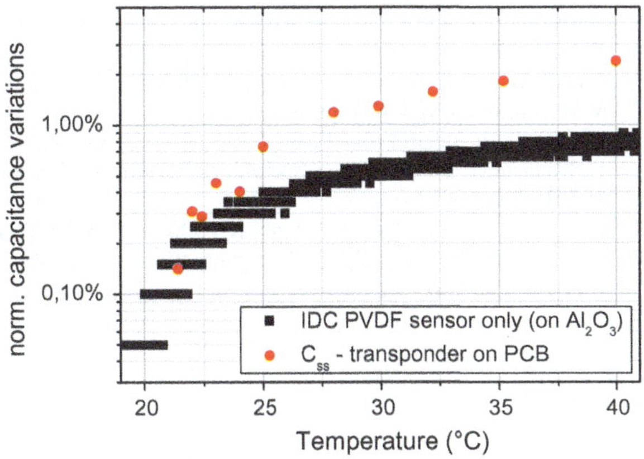


**Fig. 5** Relative variations of capacitance referred to T = 20 °C. Squares: Capacitance of the IDC on Alumina (reference sensor). Circles: $C_{SS}$ extracted from the resonance frequency of the tag

## 4 Conclusions

The feasibility of PVDF-based capacitors as parasitic sensing loads in passive HF RFID tags has been demonstrated. The resonance frequency of the tag decreases as the temperature increases because the load capacitance increases with temperature. The sensitivity of the tag to temperature has been estimated however, for a correct evaluation of the sensor's status, the thermal response of the capacitance of the chip had to be taken into account. Even if experimental setup has still to be improved to characterize tags at temperatures below ambient conditions, these preliminary results point out to the possibility of application of PVDF to temperature sensing RFID tags once unintentional drifts introduced by the chip and other tag components are sort out. Interestingly, the adopted kind of chip has been found to contribute in the sensing of the temperature. Beyond the specific context, the presented information could be useful also generically in the integration of non-thermal transducers as capacitive loads in RFID tags.

## References

1. Finkenzeller K (2010) RFID handbook. 3rd edn. Wiley, New Jersey, ISBN 978-0-470-69506-7
2. Azzarelli JM, Mirica KA, Ravnsbæk JB, Swager TM (2014) Wireless gas detection with a smartphone via RF communication. PNAS 111 (51):18162–18166. 23 Dec 2014; published ahead of print 8 Dec 2014. https://doi.org/10.1073/pnas.1415403111
3. Law M, Bermak A, Luong H (2010) A sub-μW embedded CMOS temperature sensor for RFID food monitoring application. Solid-State Circuits IEEE J 45(6):1246–1255
4. Ussmueller T, Brenk D, Essel J, Heidrich J, Fischer G, Weigel R (2012) A multistandard HF/UHF-RFID-tag with integrated sensor interface and localization capability. In: 2012 IEEE international conference on RFID (RFID), Orlando, FL, pp 66–73. https://doi.org/10.1109/rfid.2012.6193058
5. ISO/IEC 14443-2:2016. Identification cards—Contactless integrated circuit cards—Proximity cards—Part 2: Radio frequency power and signal interface
6. Potyrailo RA, Morris WG, Sivavec T, Tomlinson HW, Klensmeden S, Lindh K (2009) RFID sensors based on ubiquitous passive 13.56-MHz RFID tags and complex impedance detection. Wirel Commun Mob Comput 9:1318–1330
7. Miscioscia R, Citarella A, Pascarella F, Pandolfi G, Loffredo F, Villani F, Fasolino T, Quercia L, Minarini C Dispositivo e Sistema di Misura. ENEA patent N. 837, Italian application N. 102016000079657
8. Khan N, Omran H, Yao Y, Salama KN (2015) Flexible PVDF ferroelectric capacitive temperature sensor. In: 2015 IEEE 58th international midwest symposium on circuits and systems (MWSCAS), Fort Collins, CO, pp 1–4. https://doi.org/10.1109/mwscas.2015.7282063

# Stability of Hydrogenated Amorphous Silicon Diodes as Thin Film Temperature Sensors

**N. Lovecchio, G. de Cesare, A. Nascetti, A. Buzzin and D. Caputo**

**Abstract** This work reports on the characterization of stability of amorphous silicon diodes used as temperature sensors in lab-on-chip systems. We found that under constant forward current injection, the voltage drop over the diode changes depending on the values of current and injection time. The optimized operating conditions for practical applications have been established on the base of the obtained experimental data.

**Keywords** Amorphous silicon diodes · Temperature sensors · Stability · Current injection

## 1 Introduction

Lab-on-chip (LoC) are miniaturized systems able to implement the analytical procedures needed to perform a biomolecular analysis in shorter time and with lower reagent consumption than a standard laboratory [1, 2]. The analytical functions include chemical surface treatment [3, 4], preparation and treatment of the sample [5] as well as detection of an electrical [6, 7] or optoelectronic property [8–11] after the biomolecular recognition.

In particular, most of clinical diagnostics techniques, such as DNA amplification [12, 13], require thermal treatments of the analyte. The thermal power can be provided by a bulky metal block [14] thermally coupled with the LoC or by an integrated thin film heater [15, 16]. A very interesting approach in this context is represented by the integration of both thin film heaters and thin film temperature sensors on the same substrate in order to monitor and control the temperature of the target LoC area [17, 18]. To this aim, hydrogenated amorphous silicon (a-Si:H) diodes are appealing

N. Lovecchio (✉) · G. de Cesare · A. Buzzin · D. Caputo
Department of Information Engineering, Electronics and Telecommunications, Sapienza University of Rome, via Eudossiana 18, 00184 Rome, Italy
e-mail: nicola.lovecchio@uniroma1.it

A. Nascetti
School of Aerospace Engineering, Sapienza University of Rome, via Salaria 851/881, 00138 Rome, Italy

 
G. Di Francia et al. (eds.), *Sensors and Microsystems*, Lecture Notes in Electrical Engineering 629, https://doi.org/10.1007/978-3-030-37558-4_39

as thin film temperature sensors since they can be deposited on different kind of substrates such as plastic, metal and in particular on glass, one of the most used material in biological analysis, and because they show a sensitivity greater than those found in crystalline silicon [19].

As it is well known, biasing a diode in constant forward current and reading the voltage drop across it ($V_{drop}$) as a function of temperature (T) brings to a linear behavior of $V_{drop}$ versus T. However, one of the main issue concerning the use of a-Si:H device is its metastability under non-equilibrium conditions, as determined by light illumination or current injection [20]. Therefore, it is worth to understand and characterize the behavior of a-Si:H temperature sensors under working conditions. Within this framework, here we present a study of stability of a-Si:H diodes under different values of injected current and injection times. The optimized driving conditions for the a-Si: temperature sensors are derived from the experimental data.

## 2 Temperature Sensor Structure and Fabrication

The a-Si:H temperature sensors are metal/amorphous silicon carbide (a-SiC:H) p-type/a-Si:H intrinsic/a-Si:H n-type/metal structure deposited on a Borofloat glass substrate provide by Zaot s.r.l. (Vittuone, Milan, Italy). A structure cross section is reported in Fig. 1, which shows also the thickness of each layer.

The microelectronic fabrication foresees 4 photolithographic steps:

a. vacuum evaporation of 30/150/30 nm-thick Cr/Al/Cr stacked layers, acting as bottom contact;
b. patterning of the metal layer by conventional photolithography and wet etching process (mask #1);
c. deposition by Plasma Enhanced Chemical Vapor Deposition (PECVD) of the a-Si:H stacked structure.
d. deposition by vacuum evaporation of a 50 nm-thick Cr layer, behaving as top contact;
e. wet etching of the 50 nm-thick Cr layer and dry etching of the a-Si:H layers for the mesa patterning of the diodes (mask #2);

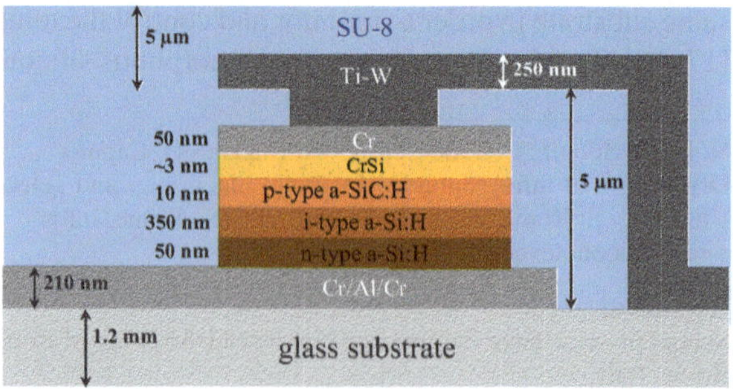


**Fig. 1** Schematic cross section of the fabricated temperature sensors. Thicknesses of the deposited layers are also reported (not in scale)

**Table 1** Deposition parameters of the PECVD process

| Parameters | N-type | Intrinsic | *P*-type | δn-type |
|---|---|---|---|---|
| $SiH_4$ (sccm) | 40 | 40 | 40 | 40 |
| $B_2H_6$ (sccm) | | | 5 | |
| $CH_4$ (sccm) | | | 60 | |
| $PH_3$ (sccm) | 10 | | | 10 |
| RF Power (W) | 4.6 | 4.8 | 4.6 | 4.6 |
| Tdep (°C) | 300 | 280 | 210 | 300 |
| Time (s) | 180 | 2100 | 70 | 3 |

f. deposition by spin coating of a 5 μm-thick SU-8 3005 (from MicroChem, MA, USA) passivation layer and its pattering for opening via holes over the diodes (mask #3);
g. deposition by sputtering of a 250 nm-thick Ti/W alloy layer and its patterning for the definition of the top contacts and of the connection to the pad contacts, which are located on the edge of the glass (mask #4);
h. deposition by spin coating of a 5 μm-thick SU-8 3005 passivation layer.

Table 1 reports the deposition parameters (gas flux, temperature, chamber pressure, power density and time) of the a-Si:H layers. The PECVD process ends with a very thin (2–3 nm) δn-type film, which enhances the formation of a chromium silicide and in turn the achievement of an ohmic contact on the p-type layer [21].

## 3 Temperature Sensor Stability Study

As preliminary characterization step, the diode current-voltage curves in dark condition have been measured. Results are reported in Fig. 2, which shows in forward bias voltage two distinct conduction zones.

Below 0.5 V, which corresponds to a current of about 2 nA, the device behavior reflects the ideal diode law $I_D = I_S[\exp(V_D/\eta kT)-1]$, where $I_D$ is the diode current, $I_S$ the reverse saturation current, $V_D$ the diode drop voltage, $\eta$ the quality factor, k the Boltzmann constant and T the absolute temperature. Above 0.5 V, the dependence of $I_D$ versus $V_D$ is still exponential but with a much higher $I_S$ and much larger $\eta$. Even though the conduction mechanism controlling this higher-current regime can be related to high injection conditions, its detailed model is out of the scope of this work and will be the object of future works.

Stability of the fabricated temperature sensor has been investigated by monitoring the variations of the diode characteristics in different operative conditions, varying both the value of the forward current and its application time. In particular, the variations of the voltage across the diodes have been recorded at different injection currents (1, 5 and 50 nA) around the knee of the forward IV curve for different

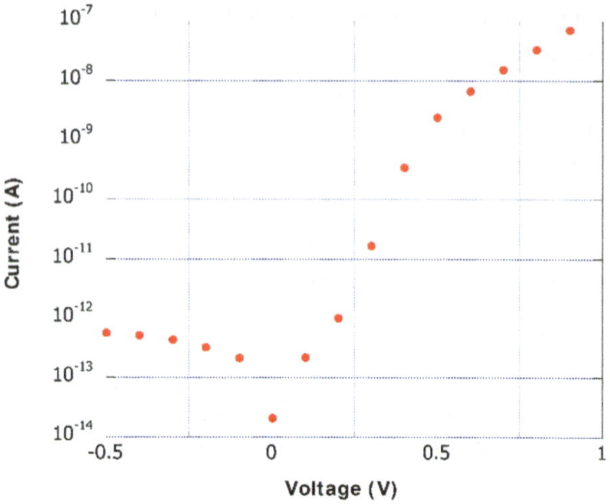


**Fig. 2** Current-voltage temperature sensor curve measured in dark condition

measurement times (1, 2, 5 and 10 h). The current has been provided by a Source Measure Unit Keithley 236 in current source mode. Careful attention has been paid in the control of the environmental conditions to keep constant the sensor temperature at 25 °C.

Results are summarized in Fig. 3, which reports, for each injected current the variation of the diode voltage drop as a function of time.

The voltage variation related to the first experiments (performed at 1 nA) is equal to 2.5 mV. Considering that the slope of the voltage/temperature characteristic of our devices is about 3.6 mV/°C [3], this means that, in this case, the temperature measurement error is about 0.7 °C. For the experiments performed at 5 nA, the voltage variation is 7 mV, with a related temperature error of 1.9 °C. A significant degradation of the device characteristic is recorded only for the highest injection current (50 nA), where the voltage variation is 42 mV, corresponding to an error of 11.7 °C.

Taking into account the current-voltage characteristics reported in Fig. 2, we can state that biasing the a-Si:H temperature sensors in the forward region, whose conduction mechanism is dominated by the ideal diode law, leads to stable behavior of the diode. Indeed, the error in temperature monitoring related only to the intrinsic behavior of the sensor is below 1 °C, which fully satisfies the requirements for biomedical applications in lab-on-chip systems.

## 4 Conclusion

This research has investigated the stability of amorphous silicon diodes acting as temperature sensors. Monitoring the diode voltage drop under constant current injection

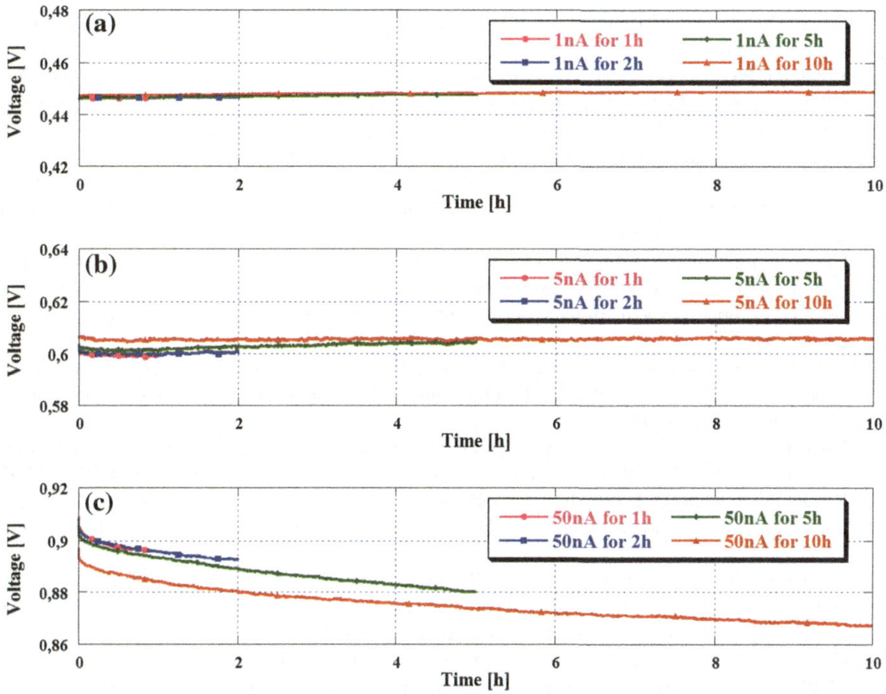


**Fig. 3** Time evolution of diode voltage drop under **a** 1 nA, **b** 5 nA and **c** 50 nA constant current injection at room temperature

in forward bias in dark conditions, we have established that the optimized operating conditions are achieved when the devices are biased in the conduction regime described by the ideal diode law. Polarization outside this region leads to an error on the temperature measurement not acceptable in practical biomedical applications.

## References

1. Whitesides GM (2006) The origins and the future of microfluidics. Nature 442:368–373
2. Lafleur J, Jonsson A, Senkbeil S, Kutter J (2016) Recent advances in lab-on-a-chip for biosensing applications. Biosens Bioelectron 76:213–233
3. Fasolato C, Giantulli S, Silvestri I, Mazzarda F, Toumia Y, Ripanti F, Mura F, Luongo F, Costantini F, Bordi F, Postorino P, Domenici F (2016) Folate-based single cell screening using surface enhanced Raman microimaging. Nanoscale 8(39):17304–17313
4. Fasolato C, Domenici F, Sennato S, Mura F, De Angelis L, Luongo F, Costantini F, Bordi F, Postorino P (2014) Dimensional scale effects on surface enhanced Raman scattering efficiency of self-assembled silver nanoparticle clusters. Appl Phys Lett 105(7):073105
5. Kim J, Maitra R, Pedrotti KD, Dunbar WB (2013) A patch-clamp asic for nanopore-based dna analysis. IEEE Trans Biomed Circuits Syst 7(3):285–295

6. Yue H, Mason AJ (2013) Lab-on-CMOS integration of microfluidics and electrochemical sensors. Lab Chip 13(19):3929–3934
7. Sanghavi BJ, Moore JA, Chavez JL, Hagen JA, Loughnane NK, Chou CF, Swami NS (2016) Aptamer-functionalized nanoparticles for surface immobilization-free electrochemical detection of cortisol in a microfluidic device. Biosens Bioelectron 78:244–252
8. Costantini F, Tiggelaar RM, Salvio R, Nardecchia M, Schlautmann S, Manetti C, Gardeniers H, de Cesare G, Caputo D, Nascetti A (2017) An all-glass microfluidic network with integrated amorphous silicon photosensors for on-chip monitoring of enzymatic biochemical assay. Biosensors 7(4):58
9. Liu R, Ishimatsu R, Yahiro M, Adachi C, Nakano K, Imato T (2015) Fluorometric flow-immunoassay for alkylphenol polyethoxylates on a microchip containing a fluorescence detector comprised of an organic light emitting diode and an organic photodiode. Talanta 134:37–47
10. Robbins H, Sumitomo K, Tsujimura N, Kamei T (2017) Integrated thin film Si fluorescence sensor coupled with a GaN microLED for microfluidic point-of-care testing. J Micromech Microeng 28(2):024001
11. Novo P, Chu V, Conde JP (2014) Integrated fluorescence detection of labeled biomolecules using a prism-like PDMS microfluidic chip and lateral light excitation. Lab Chip 14:1991–1995
12. Petrucci G, Caputo D, Lovecchio N, Costantini F, Legnini I, Bozzoni I, Nascetti A, de Cesare G (2017) Multifunctional system-on-glass for lab-on-chip applications. Biosens Bioelectron 93:315–321
13. Bruijns BB, Costantini F, Lovecchio N, Tiggelaar RM, Di Timoteo G, Nascetti A, de Cesare G, Gardeniers JGE, Caputo D (2019) On-chip real-time monitoring of multiple displacement amplification of DNA. Sens Actuators B: Chem 293:16–22
14. Gulliksen A, Solli LA, Drese KS, Sorensen O, Karlsen F, Rogne H, Hovig E, Sirevag R (2005) Parallel nanoliter detection of cancer markers using polymer microchips. Lab Chip 5(4):416–420
15. Martinez-Quijada J, Caverhill-Godkewitsch S, Reynolds M, Gutierrez-Rivera L, Johnstone RW, Elliott DG, Sameoto D, Backhouse CJ (2013) Fabrication and characterization of aluminum thin film heaters and temperature sensors on a photopolymer for lab-on-chip systems. Sens Actuators A 193:170–181
16. Ohlander A, Zilio C, Hammerle T, Zelenin S, Klink G, Chiari M, Bock K, Russom A (2013) Genotyping of single nucleotide polymorphisms by melting curve analysis using thin film semi-transparent heaters integrated in a lab-on-foil system. Lab Chip 13(11):2075–2082
17. Costantini F, Petrucci G, Lovecchio N, Nardecchia M, Nascetti A, de Cesare G, Tedeschi L, Domenici C, Ruggi A, Placidi P et al (2018) Integrated sensor system for dna amplification and separation based on thin film technology. IEEE Trans Compon Packag Manuf Technol 8(7):1141–1148
18. Mirasoli M, Bonvicini F, Lovecchio N, Petrucci G, Zangheri M, Calabria D, Costantini F, Roda A, Gallinella G, Caputo D et al (2018) On-chip lamp-bart reaction for viral dna real-time bioluminescence detection. Sens Actuators B: Chem 262:1024–1033
19. Lovecchio N, Petrucci G, Caputo D, Alameddine S, Carpentiero M, Martini L, Parisi E, de Cesare G, Nascetti A (2015) Thermal control system based on thin film heaters and amorphous silicon diodes. In: 2015 6th international workshop on advances in sensors and interfaces (IWASI). IEEE, pp 277–282
20. Caputo D (1999) Degradation and annealing of amorphous silicon solar cells by current injection: experiment and modeling. Sol Energy Mater Sol Cells 59:289–298
21. Caputo D, de Cesare G, Ceccarelli M, Nascetti A, Tucci M, Meda L, Losurdo M, Bruno G (2008) Characterization of chromium silicide thin layer formed on amorphous silicon films. J Non-Cryst Solids 354(19-25):2171–2175

# Piezoelectric Multi-Frequency Nonlinear MEMS Converter for Energy Harvesting from Broadband Vibrations

**M. Baù, M. Ferrari and V. Ferrari**

**Abstract** This paper proposes a MEMS piezoelectric converter for energy harvesting from vibrations which exploits nonlinear effects to broaden the operating bandwidth. The converter is composed of an array of cantilevers with different geometric dimensions. Piezoelectric layer and electrodes have been deposited on the cantilevers by a custom low-curing temperature post process. Nonlinearity is achieved by the magnetic interaction of a magnet and ferromagnetic particles deposited on the cantilever tips. Preliminary results show that the converter behaves like a nonlinear system and a downshift of the resonant frequency of the cantilevers with respect to the linear resonant frequency is observed, as expected.

**Keywords** MEMS · Piezoelectric converter · Nonlinear energy harvesting · Low-curing piezoelectric ink

## 1 Introduction

One of the most challenging issues of future smart devices will be the trade-off between dimension downscaling and power consumption [1]. Besides traditional solutions based on batteries, other strategies are being extensively investigated, such as RFID or contactless techniques [2–5]. In both cases, the energy and the information is exchanged through an electromagnetic link, but in the former case the smart device has active electronic on board, while in the latter can be completely passive [6, 7].

Alternatively, an emerging strategy is based on energy harvesting from ambient sources, such as vibrations, movements, and thermal gradients [8]. In the last decade several strategies have been adopted to improve the effectiveness of energy harvesters from broadband vibrations based on piezoelectric conversion principle [9–11]. The

M. Baù (✉) · M. Ferrari · V. Ferrari
Department of Information Engineering, University of Brescia, Via Branze 38, 25123 Brescia, Italy
e-mail: marco.bau@unibs.it

M. Ferrari · V. Ferrari
INO-CNR (National Research Council), Via Branze 45, 25123 Brescia, Italy



G. Di Francia et al. (eds.), *Sensors and Microsystems*, Lecture Notes in Electrical Engineering 629, https://doi.org/10.1007/978-3-030-37558-4_40

aim was to overcome the limitations of linear energy harvesters which best operate at their resonant frequency. One option is to combine the output from different converters to implement a Multi-Frequency Converter Array (MFCA). Differently, nonlinear conversion principles have been demonstrated effective in this regard. In particular, frequency-up techniques, like the one based on impact [12, 13], have been considered, as well as the adoption of configurations with nonlinear mechanisms induced by magnets [14] or other physical principles [15]. As a further step, the possibility to combine the nonlinear and multi-frequency array approaches has been already demonstrated at macro-scale level [16–18].

In this paper, a piezoelectric multi-frequency nonlinear MEMS converter for energy harvesting from broadband vibrations is presented.

## 2 Background Theory and System Description

Figure 1a shows a sketch of the architecture of the converter. The device is composed of an array of tapered piezoelectric cantilevers clamped along a circular frame. A magnet is fixed at the center of the structure and can interact with ferromagnetic masses fixed to the cantilever tips. Figure 1b shows a schematic diagram of the working principle of the proposed converter considering the simplified case with only two cantilevers.

The magnet interacts with the cantilevers through the vertical forces $F_{\mathrm{Av1}}$ and $F_{\mathrm{Av2}}$, which depend on the displacements $x_1$ and $x_2$, while the horizontal forces $F_{\mathrm{Ah1}}$ and $F_{\mathrm{Ah2}}$ are balanced by the clamps. Both the vertical and horizontal magnetic forces depend also on the distances $d_1$ and $d_2$. When the base undergoes mechanical excitation, the dynamical behavior of the cantilevers is determined by the magnetic forces $F_{\mathrm{Av1}}$ and $F_{\mathrm{Av2}}$ and the elastic forces $F_{\mathrm{el1}}$ and $F_{\mathrm{el2}}$ due to the flexural stiffness of the cantilevers. In the considered configuration, the mechanical behavior of each cantilever is independent from the others, i.e. the cantilevers do not interact. Depending on the distances $d_1$ and $d_2$ each cantilever behaves like a nonlinear monostable or a bistable system [18].

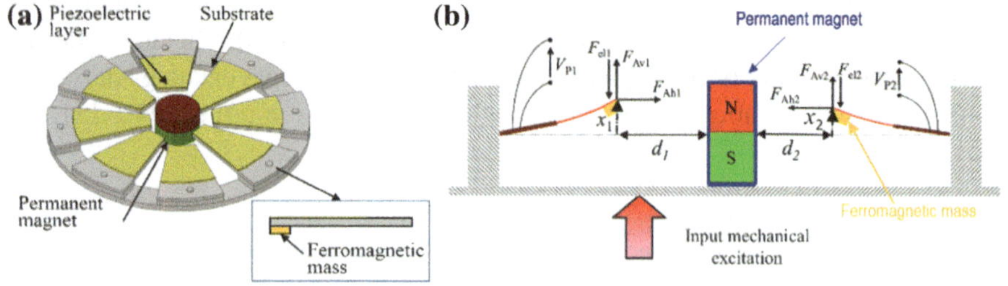


**Fig. 1** Sketch of the nonlinear multi-frequency converter array (**a**). Model of the nonlinear system for the simplified case with only two cantilevers (**b**)

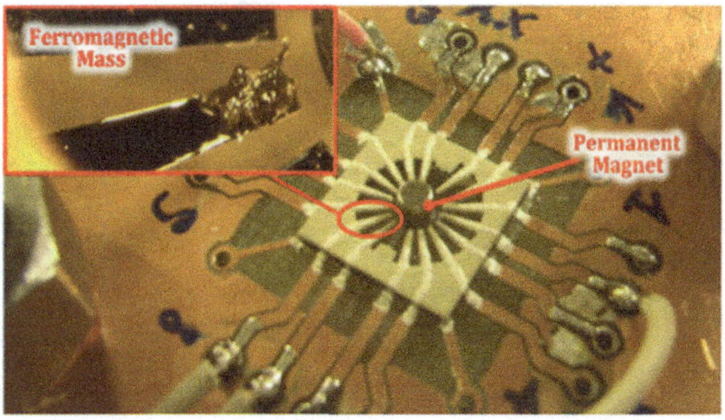


**Fig. 2** Microfabricated harvester with a permanent magnet held in a fixed position and an enlarged view of the deposited ferromagnetic mass (inset)

Figure 2 shows a picture of the fabricated prototype with an enlarged view of the deposited ferromagnetic mass in the inset. The MEMS device has been fabricated in Bonded and Etched-back Silicon On Insulator (BESOI) technology at the Centro Nacional de Microelectrónica (CNM) of Barcelona, Spain. The sixteen silicon cantilevers have thickness of 15 μm and exhibit increasing lengths in the range 1800–2500 μm with a step of 50 μm.

The width at the tip is 260 μm and the width of the clamped edge is in the range of 500–700 μm. A hollow cylinder with an inner diameter of 2 mm has been fabricated in the center of the die to easily align the permanent magnet. A layer of low-curing temperature Ag conductive paste has been screen printed on the top surface of the MEMS as the bottom electrode. The piezoelectric layer and the top electrodes have been deposited by post-processing procedures using the printing system SonoPlot GIX Microplotter Desktop equipped with a custom-developed extrusion tool. In particular, the piezoelectric layer has been obtained by exploiting an ink composed of PZT (Lead Zirconate Titanate) commercial powders (Piezokeramika APC856) dispersed in a low-curing-temperature binder [19–21]. The layer and the electrodes have been cured at 150 °C for 10 min. Subsequently, the piezoelectric layer has been poled with an electric field of 5 MV/m at 130 °C for 10 min.

## 3 Experimental Results

The equivalent capacitance $C_p$ and the parallel resistance $R_p$ of the piezoelectric layer have been measured under no mechanical excitation with a HP4194A impedance analyzer at 100 Hz, resulting in about 10 pF and 10 MΩ, respectively. The MEMS array has been characterized both in the linear regime, obtained without the magnet, and nonlinear regime with the fixed magnet.

Preliminarily impulsive response tests have been carried out to determine the resonant frequency of the cantilevers, measuring the open-circuit output voltages generated by the converters as a result of suitable small impulsive mechanical excitations. Figure 3 shows the waveform for the cantilever named in the following M1

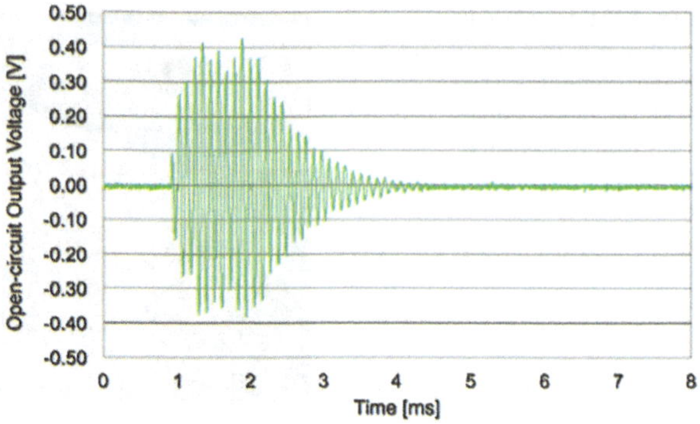


**Fig. 3** Open-circuit output voltage of M1 under impulsive mechanical excitation

where the measured frequency is 10.136 kHz. Subsequently, the microfabricated device has been mounted on the electrodynamic shaker and a stationary sinusoidal excitation has been applied. Similar results, not shown, have been obtained for the cantilevers named M2 and M3, measuring resonance frequencies of 10.352 kHz and 10.413 kHz, respectively.

Figure 4 reports the typical open-circuit output voltages of the cantilevers M1, M2 and M3 when the excitation frequency equals the resonant frequency of the converter M2. In this case, the sinusoidal acceleration imparted to the device has a peak value of about 1 g. The measured output voltage of the converter M2 is larger than the outputs of M1 and M3 as expected. Similar results have been obtained for the other cantilevers excited at their own resonant frequency.

Preliminary results on the nonlinear behavior have been achieved by placing a permanent magnet in the central hollow cylinder of the MEMS converter and depositing ferromagnetic masses on the bottom side of the cantilever tips, as shown in the inset of Fig. 2. Considering the intensity of the magnetic field and the distance from the cantilever tips and the magnet, the system behaves as a nonlinear monostable system. Figure 5 compares the open circuit output voltages generated by converter M3$^*$, i.e. the converter M3 with the added ferromagnetic tip mass, due to a suitable

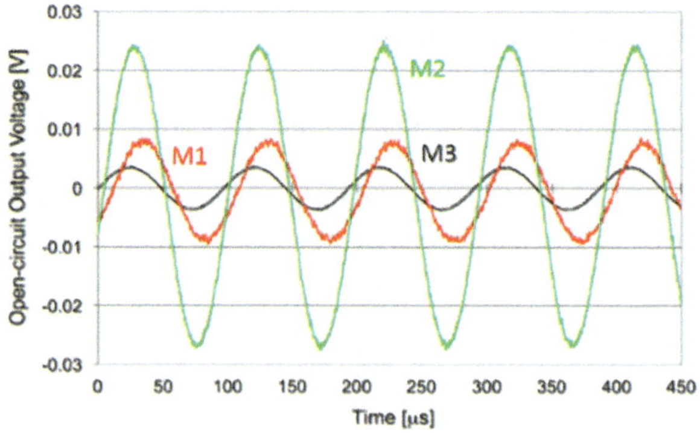


**Fig. 4** Open-circuit output voltages of M1, M2 and M3 under sinusoidal excitation with frequency equal to the resonance of M2

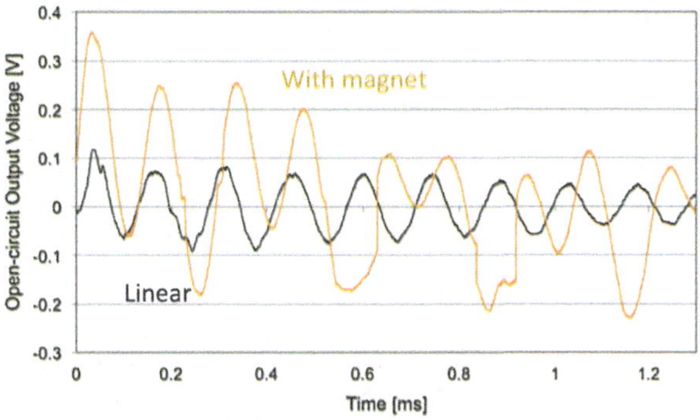


**Fig. 5** Response of M3* for the linear (without magnet) and the nonlinear conditions (with magnet) under small impulsive mechanical excitation

small impulsive mechanical excitation in the linear regime (without magnet) and in the nonlinear regime (with magnet).

It can be observed that the effect of the interaction between the magnet and ferromagnetic mass is to lower the resonant frequency from 6.877 kHz (linear case) to 6.218 kHz (nonlinear case), as expected.

The output voltages from the converters can be rectified and combined with suitable circuits to accumulate energy in storage capacitors for powering an electronic load, exploiting, depending on the voltage levels, series-like or parallel-like combination of the converter outputs [22–24].

## 4 Conclusion

A nonlinear multi-frequency piezoelectric MEMS energy converter array has been presented. The MEMS device has been fabricated in a BESOI process and post-processed with custom procedures to deposit both the top and bottom electrodes and the piezoelectric layer adopting low-temperature curing inks. Ferromagnetic particles have been deposited at the tips of the cantilevers to allow for the interaction with a magnet located at the center of the MEMS device. The device has been experimentally characterized showing a monostable nonlinear behavior when it undergoes base vibrations. The proposed approach is aimed to decrease the cantilever sensitivities to the amplitude of the mechanical excitation, therefore increasing the overall effectiveness of the converter array. In the future, the possibility to combine the outputs of the array converters to power an electronic load will be investigated.

## References


1. Čolaković A, Hadžialić M (2018) Internet of things (IoT): a review of enabling technologies, challenges, and open research issues. Comput Netw 144:17–39
2. Baù M, Ferrari M, Ferrari V (2019) Magnet-less electromagnetic contactless interrogation technique for unwired conductive resonators. Elec Lett 55(11):642–644
3. Demori M, Baù M, Ferrari M, Ferrari V (2018) Electronic technique and circuit topology for accurate distance-independent contactless readout of passive LC sensors. AEU Int J Electron Commun 92:82–85
4. Baù M, Ferrari M, Ferrari V (2017) Analysis and validation of contactless time-gated interrogation technique for quartz resonator sensors. Sensors 17(6):1264
5. Demori M, Baù M, Dalola S, Ferrari M, Ferrari V (2019) Low-requency RFID signal and power transfer circuitry for capacitive and resistive mixed sensor array. Electronics 8(6): 675
6. Demori M, Baù M, Ferrari M, Ferrari V (2018) Interrogation techniques and interface circuits for coil-coupled passive sensors. Micromachines 9(9):449
7. Ferrari M, Baù M, Tonoli E, Ferrari V (2013) Piezoelectric resonant sensors with contactless interrogation for mass-sensitive and acoustic-load detection. Sens Actuators Phys 202:100–105
8. Yildirim T, Ghayesh MH, Li W, Alici G (2017) A review on performance enhancement techniques for ambient vibration energy harvesters. Renew Sustain Energy Rev 71:435–449
9. Wei C, Jing X (2017) A comprehensive review on vibration energy harvesting: modelling and realization. Renew Sustain Energy Rev 74:1–18
10. Tang L, Yang Y, Soh CK (2010) Toward broadband vibration-based energy harvesting. J Intell Mater Syst Struct 21(18):1867–1897
11. Zhu D, Tudor MJ, Beeby S (2010) Strategies for increasing the operating frequency range of vibration energy harvesters: a review. Meas Sci Technol 21:1–29
12. Ferrari M, Baù M, Cerini F, Ferrari V (2012) Impact-enhanced multi-beam piezoelectric converter for energy harvesting in autonomous sensors. Procedia Eng 47:418–421
13. Alghisi D, Dalola S, Ferrari M, Ferrari V (2015) Triaxial ball-impact piezoelectric converter for autonomous sensors exploiting energy harvesting from vibrations and human motion. Sens Actuators Phys 233:569–581
14. Andò B, Baglio S, Baù M, Bulsara AR, Ferrari V, Ferrari M, L'Episcopo G (2012) A nonlinear energy harvester by direct printing technology. Procedia Eng 47:933–936
15. Demori M, Ferrari M, Bonzanini A, Poesio P, Ferrari V (2017) Autonomous sensors powered by energy harvesting by von Karman vortices in airflow. Sensors 17(9):2100
16. Ferrari M, Alghisi D, Baù M, Ferrari V (2012) Nonlinear multi-frequency converter array for vibration energy harvesting in autonomous sensors. Procedia Eng 47:410–413
17. Gu L (2011) Low-frequency piezoelectric energy harvesting prototype suitable for the MEMS implementation. Microelectron J 42:277–282
18. Kaajakari T, Mattila A, Oja A, Seppä H (2004) Nonlinear limits for single-crystal silicon microresonators. J Microelectromech Syst 13(5):715–724
19. Baù M, Ferrari M, Tonoli E, Ferrari V (2011) Sensors and energy harvesters based on piezoelectric thick films. Procedia Eng 25:737–744
20. Ferrari M, Ferrari V, Guizzetti M, Marioli D (2010) Piezoelectric low-curing-temperature ink for sensors and power harvesting. Lect Notes Electr Eng 54:77–81
21. Andò B, Baglio S, Bulsara AR, Marletta V, Ferrari V, Ferrari M (2015) A low-cost snap-through-buckling inkjet-printed device for vibrational energy harvesting. IEEE Sens J 15(6):3209–3220
22. Alghisi D, Ferrari V, Ferrari M, Crescini D, Touati F, Mnaouer AB (2017) Single-and multi-source battery-less power management circuits for piezoelectric energy harvesting systems. Sens Actuators Phys 264:234–246

23. Alghisi D, Ferrari V, Ferrari M, Touati F, Crescini D, Mnaouer AB (2017) A new nano-power trigger circuit for battery-less power management electronics in energy harvesting systems. Sens Actuators Phys 263:305–316
24. Ferrari M, Ferrari V, Guizzetti M, Marioli D (2010) Investigation on electrical output combination options in a piezoelectric multifrequency converter array for energy harvesting in autonomous sensors. In: Proceedings 1st international conference on sensor device technologies and applications, pp 258–263

# Automatic Compensation of Parallel Capacitance of TPoS MEMS Resonator for Accurate Frequency Tracking with PLL-Based Oscillator Circuit

**M. Baù, M. Ferrari, V. Ferrari, A. Ali and J. E.-Y. Lee**

**Abstract** This paper proposes an oscillator circuit based on a phase-locked loop which automatically compensates for the parasitic capacitance of Thin-film Piezoelectric-on-Silicon (TPoS) contour-mode resonators. The circuit enables accurate tracking of the mechanical parameters of the resonators in demanding sensor applications by advantageously combining their favorable Q-factor with the electrical resonance enhancement arising from the automatic cancellation of parallel capacitance. Preliminary results on tracking the resonant frequency of a TPoS resonator due to temperature variations show a temperature coefficient of frequency of 53.4 ppm/°C.

**Keywords** Piezoelectric resonators · Thin-film Piezoelectric-on-Silicon resonator · Phase-locked loop · Automatic capacitance compensation · Oscillator circuit

## 1 Introduction

Piezoelectric MEMS resonators vibrating in contour mode are extensively investigated as promising tools for measurements in air and liquid environments, for their favorable Q-factor also in full immersion in liquid media [1, 2]. The monitoring of the resonant frequency of piezoelectric resonators can be achieved through impedance measurement techniques, which allow also multiple-harmonic analysis [3].

M. Baù (✉) · M. Ferrari · V. Ferrari
Department of Information Engineering, University of Brescia, Via Branze 38, 25123 Brescia, Italy
e-mail: marco.bau@unibs.it

M. Ferrari · V. Ferrari
INO-CNR (National Research Council), Via Branze 45, 25123 Brescia, Italy

A. Ali · J. E.-Y. Lee
Department of Electronic Engineering, City University of Hong Kong, Kowloon, Hong Kong

J. E.-Y. Lee
State Key Laboratory of Terahertz and Millimeter Waves, City University of Hong Kong, Kowloon, Hong Kong



G. Di Francia et al. (eds.), *Sensors and Microsystems*, Lecture Notes in Electrical Engineering 629, https://doi.org/10.1007/978-3-030-37558-4_41

In addition, piezoelectric resonators have been demonstrated to be suitable for contactless operation with readout techniques in frequency [4] and time [5] domains, as is the case of Quartz Crystal Resonators (QCRs) and Resonant-Piezo-Layer (RPL) sensors [6].

Oscillator circuits which track the resonant frequency of the resonator can be adopted but they can be challenging to implement due to possibly low signal-to-noise ratio coming from the variations of parallel capacitance, especially when dielectric loading is present [6–9]. This demands for compensation techniques of the parallel capacitance. One option is the adoption of dummy resonators [10].

As an alternative, this work presents an oscillator circuit with Automatic parallel Capacitance Compensation (ACC) for the accurate tracking of the series resonant frequency and Q-factor of Thin-film Piezoelectric-on-Silicon (TPoS) contour-mode resonators. The circuit is based on a Phase-Locked Loop (PLL) which automatically compensates for the parasitic capacitance [11, 12].

The proposed circuit allows accurate tracking of the mechanical parameters of TPoS resonators (e.g. resonant frequency and Q-factor) in demanding sensor applications. It exploits the favorable Q-factor of the TPoS resonators and combines it with the electrical resonance enhancement arising from the cancellation of parallel capacitance implemented automatically.

## 2 System Description

Figure 1a shows a picture of the adopted MEMS resonator to be embedded in the circuit. The MEMS resonator is an Aluminum Nitride (AlN) TPoS disk with a diameter of 200 μm, which is suspended by four T-shaped tethers. The two Al/Cr top electrodes $P_1$ and $P_2$, and the bottom ground electrode $P_S$ allow the operation of the device as a two-port resonator or, by shorting $P_1$ and $P_2$, as a single-port resonator between the connections A and B, as shown in Fig. 1b.

Figure 2 shows the block diagram of the proposed oscillator circuit. The resonator is represented by its modified Butterworth-Van-Dyke (BVD) equivalent circuit [13].

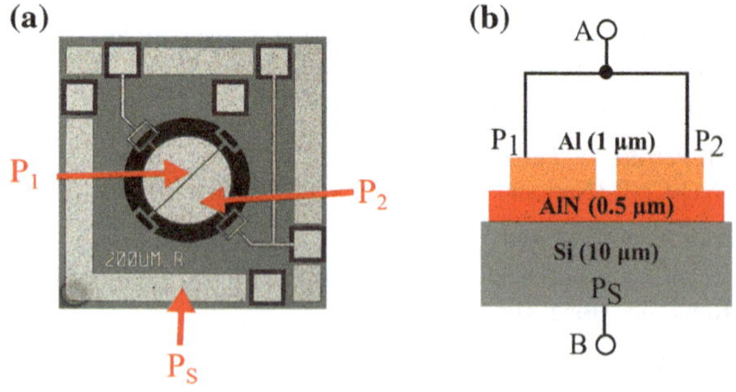


**Fig. 1** Picture (**a**) and sketch (**b**) of the TPoS resonator

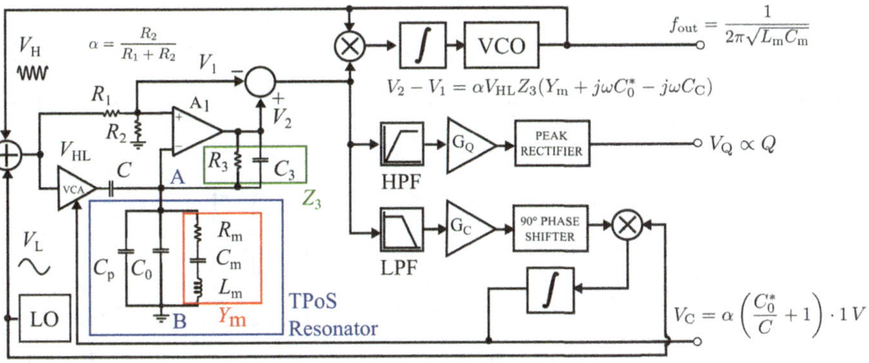


**Fig. 2** Block diagram of the ACC oscillator with the output relevant signals

The motional arm is composed of $R_m - L_m - C_m$, while $C_0^* = C_0 + C_p$ represents the total parallel capacitance, where $C_0$ is the parallel electrical capacitance while $C_p$ represents any additional parasitic capacitance. The TPoS resonator has mechanical admittance $Y_m = R_m + j\omega L_m + 1/j\omega C_m$, while the mechanical resonant frequency and the quality factor are $f_s = [2\pi(L_m C_m)^{1/2}]^{-1}$ and $Q = (L_m/C_m)^{1/2}/R_m$, respectively.

The working principle of the ACC oscillator circuit is to simultaneously excite the resonator at two frequencies, i.e. at $f_{out}$ and at an auxiliary frequency $f_L$ generated by a Local Oscillator (LO). The signal $V_L$ at $f_L$ is processed in a low-frequency feedback loop which measures and automatically compensates $C_0^*$. In particular, the integrator DC output voltage $V_C$ adjusts the gain $G$ of the Voltage-Controlled Amplifier (VCA), which, in turn, modifies the equivalent negative compensating capacitance $C_C$. Therefore, $V_C$ is proportional to the compensated parallel capacitance $C_0^*$ according to the following expression:

$$V_c = \alpha\left(1 + \frac{C_0^*}{C}\right) \cdot 1V \tag{1}$$

where $\alpha = R_2/(R_1 + R_2)$. The signal $V_H$ at $f_{out}$ is processed by a PLL resulting in the locked condition $f_{out} = f_s$. Additionally, the circuit provides the auxiliary signals $V_Q$ and $V_C$ related to $Q$ and $C_0^*$, respectively.

## 3 Experimental Results

Figure 3 shows the normalized magnitude of the signal $(V_2 - V_1)/V_{HL} = \alpha Z_3 Y$ versus frequency, i.e. proportional to the admittance spectrum $Y = Y_m + j\omega C_0^* - j\omega C_C$ of the compensated resonator, for different values of $-C_C$, both in the over- and under-compensated cases. The compensated case, i.e. $C_C = C_0^*$, corresponds to the symmetric curve.

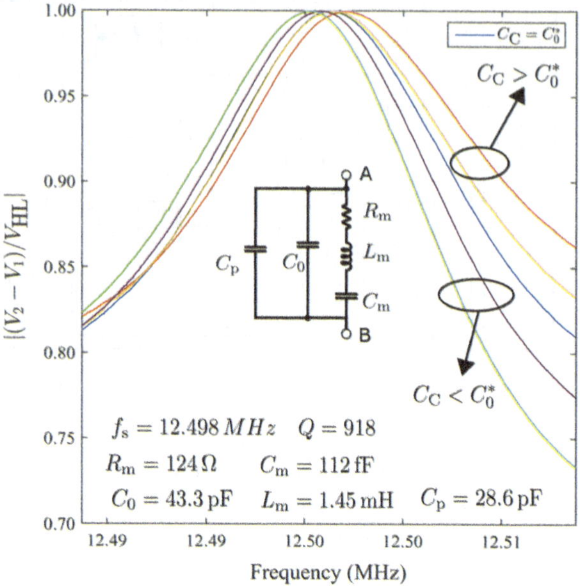


**Fig. 3** Normalized frequency responses of the signal $|(V_2\text{-}V_1)/V_{HL}|$ for different values of the compensating capacitance $C_C$ near the resonance

To assess the circuit effectiveness, tests have been performed by varying the temperature of the resonator by means of a heater. Figure 4 shows the time record of the relevant output signals when the resonator undergoes a temperature variation of about 20 °C measured by a Pt1000 sensor close to the resonator.

The signal $f_{out}$ follows the temperature variations with an opposite correlation as expected, while $C_0^*$ is nearly constant at about 72 pF, accounting for both $C_0 = 43.4$ pF and the total parasitic capacitance $C_p$. Interestingly enough, this suggests that $f_{out}$ variations can be ascribed to variations in the mechanical parameters, while dielectric properties, i.e. $C_0$, are unaffected. Moreover, a slight increase temperature-dependent trend of $V_Q$ is observed.

Figure 5 shows the measured $f_{out}$ versus temperature. On the same plot, reference values of $f_s$ measured by an impedance analyzer (HP4194A) at selected temperatures show a good agreement with the values measured by the oscillator circuit. A Temperature Coefficient of Frequency (TCF) of about −53.4 ppm/°C has been estimated. The results qualify the circuit as suitable for measurements where compensation of $C_0^*$ is a key issue to improve accuracy in tracking the $f_s$ of the resonator, like real-time liquid-phase measurements.

## 4 Conclusion

This work has presented the design and experimental validation of an oscillator circuit based on a PLL for the automatic compensation of the electrical parallel

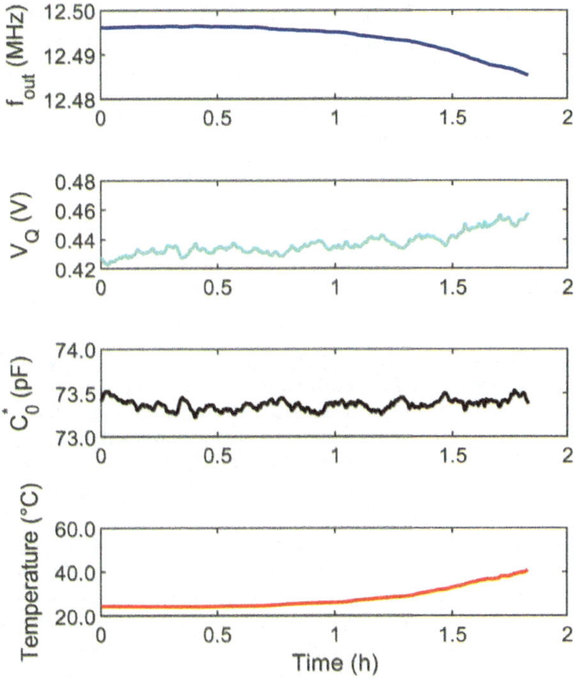


**Fig. 4** Output signals of the circuit obtained with the TPoS resonator subjected to a temperature variation of about 20 °C

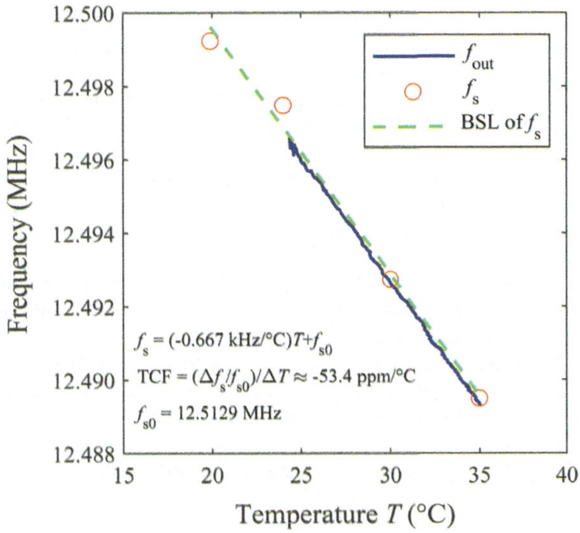


**Fig. 5** Measured $f_{out}$ versus temperature. Red circles represent reference values of $f_s$ measured with an impedance analyzer (HP4194A)

capacitance of TPoS contour-mode MEMS resonators. The circuit advantageously generates an output signal related to the Q factor and to the equivalent compensated capacitance. The circuit has been applied to track the temperature-induced variations of the resonant frequency of a test TPoS device, allowing to estimate a TCF of −53.4 ppm/°C. Further developments will aim to apply the circuit for mass-loading applications or liquid-phase measurements.

## References


1. Zuniga C, Rinaldi M, Piazza G (2009) Quality factor of MEMS and NEMS AlN Contour Mode Resonators in liquid media. In: IEEE international ultrasonics symposium, pp 2568–2571
2. Vignola JF, Judge JA, Jarzynski J, Zalalutdinov M, Houston BH, Baldwin LW (2006) Effect of viscous loss on mechanical resonators designed for mass detection. Appl Phys Lett 88:041921
3. Ferrari M, Ferrari V, Marioli D (2010) Interface circuit for multiple-harmonic analysis on quartz resonator sensors to investigate on liquid solution microdroplets. Sens Actuators B Chem 146(2):489–494
4. Demori M, Baù M, Ferrari M, Ferrari V (2018) Interrogation techniques and interface circuits for coil-coupled passive sensors. Micromachines 9(9):449
5. Baù M, Ferrari M, Ferrari V (2017) Analysis and validation of contactless time-gated interrogation technique for quartz resonator sensors. Sensors 17(6):1264
6. Ferrari M, Baù M, Tonoli E, Ferrari V (2013) Piezoelectric resonant sensors with contactless interrogation for mass-sensitive and acoustic load detection. Sens Actuators A Phys 202:100–105
7. Seo J, Brand O (2008) High Q-factor in-plane-mode resonant microsensor platform for gaseous/liquid environment. Microelectromech Syst J 17:483–493
8. Ali A, Lee JE-Y (2017) Single device on-chip feedthrough cancellation for enhanced electrical characterization of piezoelectric-on-silicon resonators in liquid. Sens Actuator A 260:131–138
9. Ali A, Lee JE-Y (2018) Piezoelectric-on-silicon square wine glass mode resonator for enhanced electrical characterization in water. IEEE Trans Electron Devices 65:1925–1931
10. Toledo J, Manzaneque T, Ruiz-Díez V, Kucera M, Pfusterschmied G, Wistrela E, Schmid U, Sánchez-Rojas JL (2016) Piezoelectric resonators and oscillator circuit based on higher-order out-of-plane modes for density-viscosity measurements of liquids. J Micromech Microeng 26(8):131–138
11. Ferrari M, Ferrari V, Marioli D, Taroni A, Suman M, Dalcanale E (2006) In-liquid sensing of chemical compounds by QCM sensors coupled with high-accuracy ACC oscillator. IEEE Trans Instrum Meas 55(3):828–834
12. Arnau A, García JV, Jimenez Y, Ferrari V, Ferrari M (2008) Improved electronic interfaces for AT-cut quartz crystal microbalance sensors under variable damping and parallel capacitance conditions. Rev Sci Instrum 79(7):075110
13. Cerini F, Ferrari M, Ferrari V, Russo A, Urquia MA, Ardito R, De Masi B, Sedmik RIP (2017) Electro-mechanical modelling and experimental characterization of a high-aspect-ratio electrostatic-capacitive MEMS device. Sens Actuators A Phys 266:219–231

# Doped Zinc Oxide Sensors for Hexanal Detection

**A. Malara, L. Bonaccorsi, A. Donato, P. Frontera and G. Neri**

**Abstract** Hexanal is a volatile compound considered important for food storage and food quality control. Despite this, hexanal has not been extensively studied yet. In a previous work, sensors based on metal oxides semiconductors have been studied and zinc oxide showed promising sensing properties. In this work, the topic was further investigated by preparing zinc oxide-based resistive sensors, together with the aluminum-doped and indium-doped form, for hexanal monitoring in food applications.

**Keywords** Gas sensor · Hexanal · Humidity · Zinc oxide

## 1 Introduction

Metal oxide semiconductor have attracted a lot of attention over the last decade being commonly employed in resistive gas sensors field. Among these, zinc oxide due to high chemical sensitivity, non-toxicity, and low cost, has been widely use [1, 2]. Moreover, the performance of zinc oxide sensor can be easily improved by addition of dopants [3]. In fact, doping with metals enhances the sensitivity and selectivity of gas sensors by changing the energy-band structure and morphology, increasing the surface-to-volume ratio and creating more centers for gas interaction on the metal oxide semiconductor surface, as well its acidity or basicity [1, 4].

In this study, zinc oxide-based resistive sensors (ZnO), together with the aluminum-doped (ZnO:Al) and indium-doped (ZnO:In) form were developed for the monitoring of hexanal. Indeed, the determination of hexanal concentration is of great importance for food quality control and its detection is being a significant indicator of food quality in packaging [5–7]. For instance, hexanal formation is often monitored as a means of determining the onset of rancidity in meat [8]. Anyway,

A. Malara (✉) · L. Bonaccorsi · A. Donato · P. Frontera
Department of Civil, Energy, Environment and Material Engineering, Mediterranean University of Reggio Calabria, Via Graziella, Loc. Feo Di Vito, 89122 Reggio Calabria, Italy
e-mail: angela.malara@unirc.it

G. Neri
Department of Engineering, University of Messina, Contrada Di Dio, 98166 Messina, Italy



G. Di Francia et al. (eds.), *Sensors and Microsystems*, Lecture Notes in Electrical Engineering 629, https://doi.org/10.1007/978-3-030-37558-4_42

despite its importance as a quality marker in the food industry, very few works deal with its detection. To this regard, $In_2O_3$, ZnO, $SnO_2$ based resistive sensors have been proposed for hexanal sensing and monitoring [9, 10]. Indeed, as previously reported [10], ZnO, among the investigated oxides, showed the best compromise between sensing temperature and response. It resulted highly sensitive but saturated at high hexanal concentration. In this study, zinc oxide-based resistive sensors, ZnO, ZnO:Al and ZnO:In, were synthesized and tested for monitoring of low hexanal concentrations, as low as its limit threshold, in dry and humid air, as to accurately replicate the real operating conditions.

## 2 Experimental

### 2.1 *Synthesis of Sensing Materials*

The starting materials used for the synthesis of doped zinc oxides were zinc nitrate hexahydrate, aluminium acetate and indium nitrate hydrate. All the reagents used were of analytical grade and were used as received without further purification. Powders were synthesized by chemical co-precipitation method. For this, an aqueous solution of aluminium or indium precursor was added to the zinc aqueous solution, keeping the Al/Zn or In/Zn molar ratio equal to 0.01, and stirred till the homogenization. The solution was then hydrolyzed with an aqueous potassium carbonate solution (1 M). The precipitates were then filtered, washed with deionized water, dried at 110 °C for 12 h and then calcined at 500 °C for 2 h in air.

### 2.2 *Characterization*

Samples were characterized by means of complementary investigation techniques. Structural characterization was performed by XRD analysis (Bruker, D2 Phaser) in the 2θ range 10–80°, in steps of 0.02° and a count time of 2 s per step (Cu $K\alpha 1$ = 1.54056 Å). Powders morphology was studied by Scanning Electron Microscopy SEM (Phenom ProX) equipped with an energy-dispersive X-ray (EDX). EDX analysis was used to evaluate the content and the dispersion of dopants [11].

### 2.3 *Sensor Fabrication*

In order to prepare the sensor, a paste was obtained by mixing the oxide powder with a proper quantity of ethanol and deposited on an alumina planar substrate (3 mm × 6 mm) supplied with interdigitated Pt electrodes and a heating element on the back side. The device was positioned in a stainless steel testing cell and sensing tests were performed flowing a mixture of dry/humid air and hexanal vapor at different concentrations. A moisturizer was used to produce water moisture and the relative humidity level was detected by a humidity meter (Humidity and Temperature Transmitters HMT120, Vaisala). The hexanal vapor was obtained by bubbling dry air in liquid hexanal maintained at a controlled temperature by a refrigerated circulating bath (temperature range −5/−15 ± 0.01 °C) [10]. Fluxes were measured by Brooks mass flow controller systems for a total gas stream of 100 sccm. The sensors resistance data were collected in the fourpoint mode by an Agilent 34970A multimeter while a dual-channel power supplier instrument (Agilent E3632A) allowed to control the sensor temperature. Sensor response S to hexanal was defined as the ratio of the resistance registered in air ($R_{air}$) and that in hexanal vapor ($R_{hexanal}$). Hexanal concentrations as low as 5, 15 and 30 ppm were considered, at different operating sensors temperatures (200, 250 and 300 °C) and different relative humidity (RH) conditions at room temperature (10, 40 and 60%).

## 3 Results and Discussion

The XRD spectra of the ZnO, Al- and In-doped ZnO samples are showed in Fig. 1. Patterns show the characteristic diffraction peaks of hexagonal wurtzite zinc oxide (Fig. 1a). The crystallite size for ZnO and doped ZnO samples was determined by Debye-Scherrer equation: $D = K\lambda/(\beta \cos \theta)$, where, D is the crystallite size, K is a dimensionless shape factor assumed 0.9, λ is the wavelength of X-ray, β is the full width at half maximum of the peak and θ is the Bragg angle. The crystallite sizes of the samples are reported in the inset table of Fig. 1b.

The calculated values indicated that crystallite sizes of doped samples changed due to the doping effect. Indeed, the crystal size of In- and Al-doped samples were

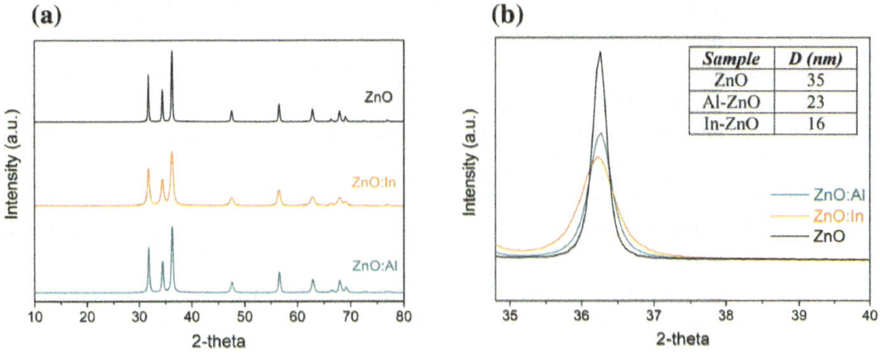


| *Sample* | *D (nm)* |
|---|---|
| ZnO | 35 |
| Al-ZnO | 23 |
| In-ZnO | 16 |

**Fig. 1** XRD spectra of ZnO, ZnO:Al and ZnO:In samples and crystallite sizes determined by Debye-Scherrer equation

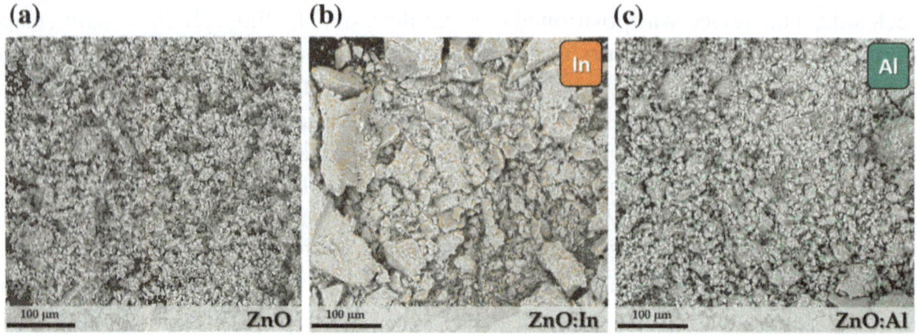


**Fig. 2** SEM micrographs and energy-dispersive X-ray (EDX) maps

smaller than that of the ZnO sample proving that In and Al atoms settled inside the lattice of ZnO and in turn control its nucleation rate, thus a change in the crystal size [12].

From SEM analysis it became clear that the synthesized oxides consisted of grains in the ZnO and ZnO:Al powders (Fig. 2a and c), and composed of fine particles densely aggregated in the case of ZnO:In sample (Fig. 2b). EDX analysis gave the qualitative composition of nanoparticles and also indicated the quantitative presence of indium and aluminum dopants well as their dispersion, homogeneous throughout the zinc oxide matrix (colored points overlapped on the SEM micrographs in Fig. 2b and c).

In Fig. 3, the responses of ZnO, ZnO:Al and ZnO:In as a function of temperature, different hexanal concentrations and relative humidity conditions are shown.

Zinc oxide response to hexanal concentration of 5, 15 and 30 ppm in dry air, at different operating temperatures is shown in Fig. 1a. The sensor is sensitive in all the tested conditions, giving the highest response at 250 °C. In order to study the effect of humidity, the sensor response to different humidity range was registered operating at the temperature of maximum sensitivity. Figure 1b reports the response of ZnO sensor when humid air at 10, 40 and 60% RH (@ 25 °C) was used. As clearly evident, increasing humidity the response has worsened and, moreover, the recovery time has increased as shown in Fig. 4. For a sensor temperature higher than 150 °C, the water vapor in the air stream cannot forms a continuous monolayer covering the oxide surface but is partially chemisorbed by interaction with the surface hydroxyl groups [13].

The chemisorbed water increases the electrical conductivity of the semiconductor oxide mainly for two reasons: the displacement of the adsorbed oxygen species and the interaction of hydronium ions with the lattice oxide [14, 15]. The combined effect of these two mechanisms, however, caused the reduction of the sensor response with humidity observed for ZnO in Fig. 3b. The oxygen species displacement by the adsorbed water molecules lowered the sensitivity of the ZnO sensor because a decrease of oxidation sites available for the hexanal detection [13–15].

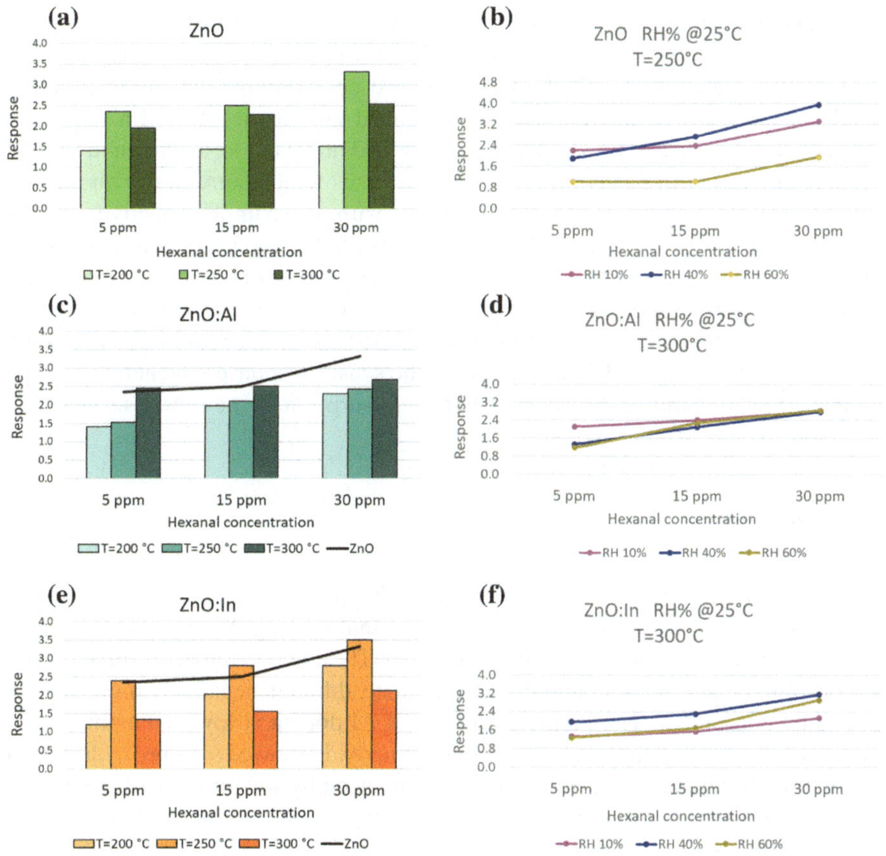


**Fig. 3** Sensors response evaluated as $R_{air}/R_{hexanal}$ as function of temperatures and hexanal concentrations in dry air (**a**, **c**, **e**) and RH (**b**, **d**, **f**)

**Fig. 4** ZnO sensor response at 250 °C and 15 ppm hexanal in dry and wet air (RH 40% @ 25 °C)

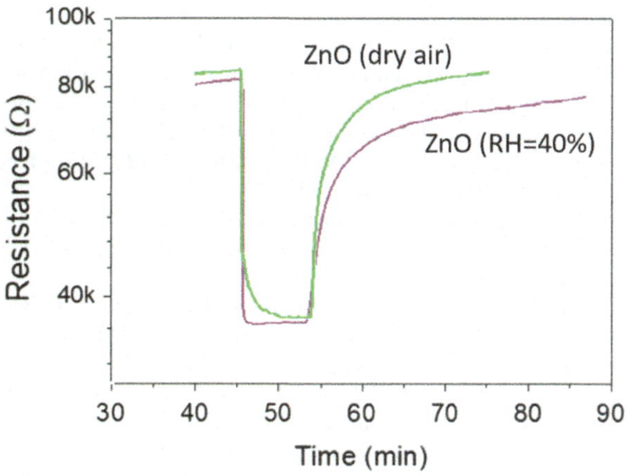

An increasing trend with T, reaching the highest response at 300 °C, was observed for the ZnO:Al sample (Fig. 3b), even if, unless for the higher operating temperature (300 °C), the sensitivity resulted lower than pure ZnO, proving the inefficient effect of Al dopant at low temperatures. The Al doped zinc oxide showed a sensitivity decrease at increasing RH% similar to the undoped metal oxide, as shown in Fig. 3d.

As regards ZnO:In behavior, the response versus temperature in dry air showed a slight increment compared to ZnO (Fig. 3e) with a maximum sensitivity at T = 250 °C. In wet air, the ZnO:In response was higher at the highest operating temperature (300 °C) and, surprisingly, demonstrated to be less sensitive to water interaction even at the maximum tested humidity of 60% RH (@ 25 °C), as evident in Fig. 3f. The reasons for this improved behavior of In doped zinc oxide was ascribed to lower grain size (Fig. 1b) that is know to improve the sensor response in general [2, 3, 14] but that effected in this case also the water chemisorption on the sensor surface [15].

## 4 Conclusion

In conclusion, zinc oxide and doped zinc oxide based sensors, synthesized by chemical co-precipitation method, were used for hexanal sensing. Very low hexanal concentrations both in dry and humid air were tested. The comparison of the three sensors demonstrated that ZnO is an interesting candidate although negatively influenced by a humid environment. The inefficient effect of Al dopant at low temperatures was reported, whereas doping zinc oxide with indium oxide showed to improve the sensor response in dry air and to be only partially influenced by wet air up to 60% RH if an operating temperature of 300 °C is used.

## References

1. Hjiri M, El Mir L, Leonardi SG, Pistone A, Mavilia L, Neri G (2014) Al-doped ZnO for highly sensitive CO gas sensors. Sens Actuators B: Chem 196:413–420
2. Zhang J, Qin Z, Zeng D, Xie C (2017) Metal-oxide-semiconductor based gas sensors: screening, preparation, and integration. Phys Chem Chem Phys 19(9):6313–6329
3. Qi J, Zhang H, Lu S, Li X, Xu M, Zhang Y (2015) High performance indium-doped ZnO gas sensor. J Nanomaterials 16(1):74
4. Angiolini L, Benelli T, Giorgini L, Mauriello F, Salatelli E (2007) Chiroptical and optical thermoplastic acid sensors based on chiral methacrylic polymers containing azoaromatic moieties. Sens Actuators B: Chem 126(1):56–61
5. Spadafora ND, Amaro AL, Pereira MJ, Müller CT, Pintado M, Rogers HJ (2016) Multi-trait analysis of post-harvest storage in rocket salad (*Diplotaxis tenuifolia*) links sensorial, volatile and nutritional data. Food Chem 211:114–123
6. Jayasena DD, Ahn DU, Nam KC, Jo C (2013) Flavour chemistry of chicken meat: a review. Asian-Australas J Anim Sci 26(5):732
7. Schindler S, Krings U, Berger RG, Orlien V (2010) Aroma development in high pressure treated beef and chicken meat compared to raw and heat treated. Meat Sci 86(2):317–323

8. Skibsted LH, Mikkelsen A, Bertelsen G (1998) Lipid-derived off-flavours in meat. Flavor of meat, meat products and seafoods. Blackie Academic, London, pp 217–256
9. Huang K, Zhu C, Yuan F, Xie C (2013) Nanoscale $SnO_2$ flat-type coplanar hexanal gas sensor arrays at ppb level. J Nanosci Nanotechnol 13(6):4370–4374
10. Malara A, Bonaccorsi L, Donato A, Frontera P, Piscopo A, Poiana M, Leonardi SG, Neri G (2018) Sensing properties of indium, tin and zinc oxides for hexanal detection. Convegno Nazionale Sensori. Springer, Cham, pp 39–44
11. Malara A, Paone E, Frontera P, Bonaccorsi L, Panzera G, Mauriello F (2018) Sustainable exploitation of coffee silverskin in water remediation. Sustainability 10(10):3547
12. Hendi AA, Alorainy RH, Yakuphanoglu F (2014) Humidity sensing characteristics of Sn doped zinc oxide based quartz crystal microbalance sensors. J Sol-Gel Sci Technol 72(3):559–564
13. Helwig A, Müller G, Sberveglieri G, Eickhoff M (2009) On the low-temperature response of semiconductor gas sensors. J Sens 2009
14. Farahani H, Wagiran R, Hamidon M (2014) Humidity sensors principle, mechanism, and fabrication technologies: a comprehensive review. Sensors 14(5):7881–7939
15. Wang C, Yin L, Zhang L, Xiang D, Gao R (2010) Metal oxide gas sensors: sensitivity and influencing factors. Sensors 10(3):2088–2106

# An Optical SPR Sensor for Monitoring Accelerated Ageing of Oil-Paper Insulation of Transformers

**L. De Maria, L. Cice, D. Bartalesi, J. Borghetto, A. Tavakoli, F. Scatiggio, D. Gasparini and M. Pesavento**

**Abstract** Degradation of electrical insulation due to the oil-paper insulating system is one of the main factors that affect power transformer service life. Although the end of life failure is inevitable, the life of the insulating paper can be maximized through careful monitoring and maintenance. In this paper an optical SPR sensor is reported for high sensitive detection of a chemical marker of the transformers' insulating paper degradation. Temperature tests were carried out on oil–paper specimens representative of the solid insulating system of transformers, subjected to initial different drying processes and successively exposed to aging temperature. A comparative analysis with standard methods (HPLC—High Performance Liquid Chromatography) confirms that the chemical concentrations detected in oil–paper sampling of different specimens with optical SPR sensors are in good agreement with those measured by standard techniques.

**Keywords** Optical sensor · Surface plasmon resonance · Oil-paper insulation · Power transformers

## 1 Introduction

The power transformer is a key element of the electricity transmission and distribution network. The out of service of a transformer produces serious disservices to users with relevant economic impact. During its operation, the transformer can be subjected to high overloads which can affect windings integrity; for example, under particularly severe operating conditions, an early degradation of the transformers'

L. De Maria (✉) · L. Cice · D. Bartalesi · J. Borghetto · A. Tavakoli
RSE S.p.A, Milan, Italy
e-mail: Letizia.demaria@rse-web.it

F. Scatiggio · D. Gasparini
TERNA Rete Italia S.p.A, Venice, Italy

M. Pesavento
Università di Pavia, via Taramelli 12, 27100 Pavia, Italy

 
G. Di Francia et al. (eds.), *Sensors and Microsystems*, Lecture Notes in Electrical Engineering 629, https://doi.org/10.1007/978-3-030-37558-4_43

solid insulating system (paper and mineral oil) can occur, leading to a drastic reduction of the transformer's life [1]. Actually, the deterioration of the winding's solid insulating system, which generally occurs towards the end of the transformer design life (35–40 years), can be anticipated if undesired thermal, oxidative and hydrolytic processes occur during the service of the component. A sudden and premature failure of the transformer could be avoided by planning suitable maintenance actions to slow down the degradation of the insulating system and to extend the transformer's life. An early detection of physical and chemical parameters involved in these degenerative processes, by means of suitable monitoring techniques, is therefore timely for Utilities and Electrical Operators. It is widely reported that furaldehyde, 2-FAL, is one of the main decomposition by-product of the transformers' insulating kraft paper [2]. The measurements of 2-FAL concentrations in oil sampling is therefore an indirect way to assess the integrity of the windings inside transformers. Generally the standard method used for 2-FAL measurements in oil sampling is based on High Performance Liquid Chromatography (HPLC); this is a complex method which requires specialized operators both for acquisition and data analysis. Recently an optical sensor based on Surface Plasmon Resonance (SPR) was described by the authors in [3, 4] as a potentially suitable for 2-FAL detection directly in standards and in used mineral oil.

In this paper the feasibility of the optical SPR approach to transformer oil control has been investigated on oil samples of specimens, representative of the solid insulating system of transformers, exposed to controlled thermal treatment to simulate aging of the insulating paper.

## 2 Optical SPR Sensor

The optical sensor is based on the excitation of Surface Plasmon Resonance (SPR) [2] at the interface between a thin metal film and a sensing receptor. The SPR optical platform is implemented in a D-shaped plastic optical fibre (POF). The removal of the cladding layer in the POF sensing region is accomplished by hand-polishing the fibre along half the circumference.

A buffer layer of photoresist (Microposit S1813), about 1.5 μm thick, is spin coated on the exposed core; a thin gold film (60 nm thickness) is then deposited on the buffer layer by means of Electron Beam Physical Vapor Deposition (EBPVD) or sputtering techniques. Finally a Molecular Imprinted Polymer (MIP) sensing layer, which behaves as artificial receptor for 2-FAL molecule in the oil sample, is spin coated on the gold film. The procedure for prepolymerix mixture preparation and for the polymer layer formation by thermal polymerization have been previously reported [3]. Figure 1 shows a scheme of the SPR POF-MIP sensor and a picture of the manufactured chemo-sensor.

A white light source (Halogen Lamp) is used for the spectral interrogation of the SPR chemo-sensor. The light spectrum transmitted through the sensor is measured

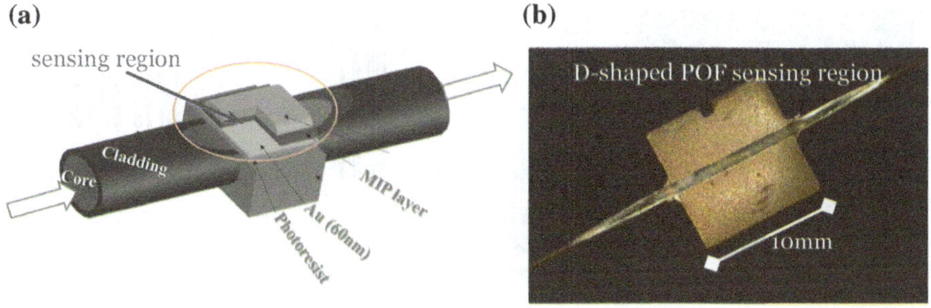


**Fig. 1** The SPR-MIP sensor scheme (**a**) and an example of the manufactured SPR chemo-sensors top view (**b**)

by means of a spectrum analyser for visible light (Ocean Optics USB2000+), controlled by a computer. The measured SPR transmission spectra are normalized to the reference spectrum measured with air as external medium and recorded on the computer. The local binding of the MIP combination sites with 2-FAL molecules, induces a change in the refractive index of the MIP layer at the interface. In SPR spectral interrogation mode the change of MIP refractive index (δn) is read as a wavelength shift δλ (nm) of the SPR resonance. For this type of sensor the sensitivity S is conveniently defined through the following equation:

$$S = \frac{\delta\lambda}{\delta C}\ (\text{nm/M})$$

in which δC represents the change in 2-FAL concentration. The spectral shift δλ (nm) is correlated to the 2-FAL concentration in oil.

## 3 Experimental Setup and Method

A set of four oil-paper specimens with a simplified parallel-plate capacitor geometry (Wafer configuration, W), were used for ageing tests (Fig. 2). Each specimen consists of vertical stacks of six copper plates covered with insulating paper (kraft paper) and clamped together by means of fiberglass rods and bolts. The copper plates are spaced from each other by paper layers or pressboard spacers to allow moisture diffusion through both surfaces. Each specimen was inserted into a stainless steel cylindrical cell filled with about 5 litres of uninhibited mineral oil (Nynas Libra), to simulate the most common mass ratio of oil, paper/pressboard and copper inside of power transformers.

The mineral oil was previously degassed to eliminate moisture contents. Before being inserted into the test cells, the specimens were subjected to different thermal treatments, that is, autoclave treatment for W1, W3 and W4 and vapor-phase for W2, in order to reproduce different initial conditions of moisture content, as reported in

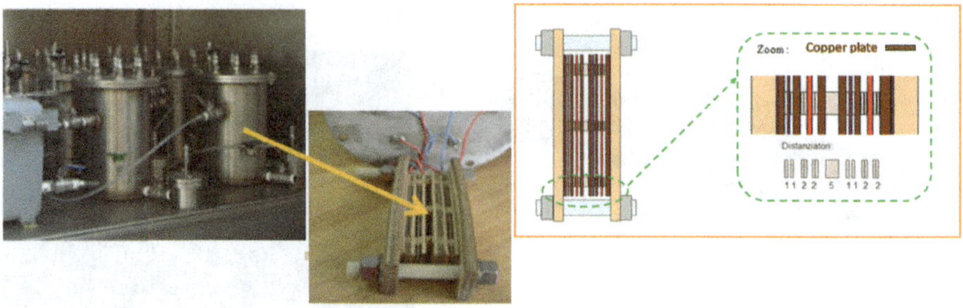


**Fig. 2** Test bench for simulating accelerated aging of oil paper specimens and a detail of the wafer specimen, before being inserted into the stainless steel cylindrical chamber

**Table 1** Expected moisture content of wafer specimens

| Oil-paper specimen | Moisture content (%) |
|---|---|
| Wafer 1 (W1) | 2 |
| Wafer 2 (W2) | 0.1 |
| Wafer 3 (W3) | 2 |
| Wafer 4 (W4) | 4 |

Table 1, which may critically accelerate the aging of the paper. The specimens have been successively exposed to temperature up to 80 degree in a controlled oven. The changes in the insulating oil volume due to temperature inside the steel chambers are compensated by means of an external expansion chamber.

Standard methods have been used to test oil moisture contents (Karl Fisher Titration according to the IEC60814 standard [4]) and 2-FAL concentrations (HPLC analysis according to IEC61198 [5]) in oil sampling of each paper oil specimen, respectively.

During thermal exposure stressed oil samples have been periodically extracted by means of airtight syringes from each steel test chamber. Few microliters (μL) of the oil sampling have been used for 2-FAL optical detection. A set of four SPR optical sensors have been prepared for the 2-FAL analyte detection in the Libra oil samples of whole series of specimens.

## 4 Results

Figure 3a shows the trend of 2-FAL concentrations in the insulating oil of wafers (W1, W2, W3 and W4) exposed to increasing temperature values were measured by HPLC method (left axis, blue line) and optical SPR method (yellow dot). On the right axis the optical response is reported as the shift of the resonance wavelength (nm). As expected, the paper degradation is dramatically influenced by the initial moisture content of the wafers, artificially produced by drying the insulating paper. The lowest

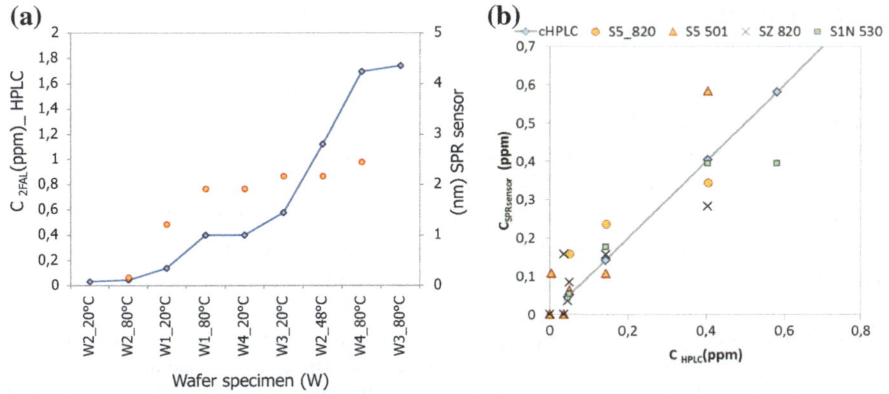


**Fig. 3** **a** Comparison between 2-FAL HPLC analysis (left axis) and optical SPR response (right) assessed on different stressed oil samples of wafer specimens (W) (The left scale is uncorrect), **b** 2-FAL concentration values obtained by Langmuir model from SPR sensor experimental data against values determined by HPLC

2-FAL concentration is for the W2 specimen, which was dried under vapour-phase pre-treatment. Considering samples W1 (at 80 °C) and W4 (at 20 °C), the 2-FAL concentration (0.4 ppm) is the same and it has also been correctly measured by the optical sensor, confirming the efficacy of the method. On the other hand, it appears that the determinable concentration range is smaller in the case of the optical sensing, which is due to the limited number of receptors sites which are present in the MIP at the sensor surface.

The concentrations of 2-FAL in each wafer at different thermal exposure were subsequently assessed by measuring the whole series of Libra oil samples with each of the four SPR-MIP sensors. For each oil sample and for each SPR-MIP sensor, the SPR wavelength shift, $\delta\lambda$ (nm), with respect to white (oil not containing 2-FAL) was evaluated and the Langmuir model [3] was applied for quantification. The parameters were the same for all the oil samples measured with the same SPR-MIP sensor and were those obtained from standardization curves previously determined on fresh Libra oil [3]. Figure 3b shows results of the values of 2-FAL concentration assessed by optical SPR method against the values determined by HPLC. The line with slope $P = 1$ and origin $OO = 0$, corresponding to the perfect equivalence of the two methods, is reported too for comparison. A good agreement is observed between the values determined with the SPR optical sensors compared to HPLC. The observed differences in SPR responses among the various optical sensors, used for the tests, are essentially due to the hand polishing method applied for the preparation of the SPR sensor surface.

Anyway for all the SPR sensors used it clearly emerges that the SPR sensor is able to detect low concentrations, as expected due to the high sensitivity of the SPR transduction method.

# 5 Conclusion

Preliminary results of comparative analysis showed that the optical SPR sensor response is not affected neither by the nature of the specimen nor by its thermal pre-treatment and that the trend of the new SPR sensors' response is in good agreement with that of data acquired by standard methods (HPLC—High Performance Liquid Chromatography). Further tests are on-going on different family of specimens, in order to assess the final performances of the SPR sensors in real samples, in terms of reproducibility and reversibility, on a wider statistical base.

**Acknowledgements** This work has been financed by the Research Fund for the Italian Electrical System under the Contract Agreement between RSE and the Ministry of Economic Development-General Directorate for Energy and Mining Resources stipulated on 29 July 2009 in compliance with the Decree of 19 March 2009. The authors would like to thank N. Cennamo and L. Zeni for helpful discussion.

## References

1. Cigre Brochure n.323 (2007) Ageing of cellulose in mineral oil insulated transformer, 10
2. N'Cho JS, Fofana I, Hadjadj Y, Beroual A (2016) Review of physicochemical-based diagnostic techniques for assessing insulation condition in aged transformers. Energies 9:367
3. Cennamo N, De Maria L, D'Agostino G, Zeni L, Pesavento M (2015) Monitoring of low levels of furfural in power transformer oil with a sensor system based on a POF MIP platform. Sensors 15:8499
4. IEC 60814 Standard (1997) Insulating liquids—oil-impregnated paper and pressboard—determination of water by automatic coulometric Karl Fischer titration
5. IEC 61198 Ed.1.0 (1993) Mineral insulating oils—methods for the determination of 2-furfural and related compounds. International Electrotechnical Commission, Geneva, Switzerland

# Distance-Independent Contactless Interrogation of Quartz Resonator Sensor with Printed-on-Crystal Coil

**M. Ferrari, M. Demori, M. Baù and V. Ferrari**

**Abstract** A novel quartz crystal resonator sensor, which embeds a conductive printed planar coil that enables electromagnetic contactless interrogation techniques is presented. An aerosol-jet process is used to precisely and accurately deposit electronic inks onto a 330 μm-thick bare piezoelectric quartz crystal to print the planar coil and the electrodes. The proposed interrogation technique enables distance-independent operation, and is based on the measurement of the reflected impedance of the quartz resonator sensor through the planar primary coil of the coupled inductors. The resonant frequency, measured without contact using the primary coil connected to an impedance analyzer, results 4.790260 MHz. Contactless operation distances up to 12.2 mm have been obtained. The experimental results have a maximum deviation of about 50 Hz, i.e. 10.5 ppm, with respect to reference measurements taken via contact probes.

**Keywords** Quartz crystal resonator (QCR) · Distance-independent contactless interrogation · Printed coil · Electromagnetic coupling

## 1 Introduction

Acoustic-wave resonators, such as quartz crystal resonators (QCRs), have been widely used for sensing in-air and in-liquid environment as microbalances [1–5]. To track the fundamental resonant frequency, higher harmonics and Q-factor of QCRs, measuring techniques based on oscillator circuits or impedance measurements have been adopted [6–8]. Contactless interrogation of QCRs [9, 10], resonant piezo layer (RPL) devices [11] and mechanical resonators [12, 13] has been demonstrated both in frequency and time domains. Electromagnetic contactless interrogation techniques for LC resonant sensors have been demonstrated [14–19], also exploiting circuits and

---

M. Ferrari (✉) · M. Demori · M. Baù · V. Ferrari
Department of Information Engineering, University of Brescia, Via Branze 38,
25123 Brescia, Italy
e-mail: marco.ferrari@unibs.it



G. Di Francia et al. (eds.), *Sensors and Microsystems*, Lecture Notes
in Electrical Engineering 629, https://doi.org/10.1007/978-3-030-37558-4_44

techniques to ensure distance-independent measurements [20–22]. In several applications, in particular for in-liquid operation, classical QCR sensors with contactless interrogation need to be connected to an external coil that could become unpractical.

In this paper, a novel quartz resonator sensor with printed-on-crystal coil is presented. The sensor does not require any additional component or cable connection, as it is completely passive and equipped with a coil that can be electromagnetically coupled with a readout coil to perform contactless interrogation.

## 2 Quartz Resonator Sensor with Printed-on-Crystal Coil

Figure 1a shows the sketch of the proposed quartz resonator sensor. The substrate is a bare AT-cut crystal with diameter and thickness of 25.4 mm and 330 μm, respectively. The aerosol-jet process used to print the coil and the electrodes exploits the aerodynamic focusing by a sheath gas to precisely and accurately deposit electronic inks onto the crystal. The conductive ink (Novacentrix Metalon HPS-108AE1) is placed into the atomizer of the machine (Optomec AJ300-UP), which creates a dense mist of material, and is then delivered to the deposition head where it is focused. The minimum feature sizes and the minimum space between lines are about 10 μm and 50 μm, respectively. After the deposition, a 15-min curing process at 150 °C has been performed. The measured thickness of the obtained conductive path is about 15 μm. The 5.5-mm electrode and the planar coil terminating at Pad A have been printed on the top side, while the 5.5-mm electrode terminating at Pad B has been printed on the bottom side. Figure 1b shows a picture of the fabricated prototype with an enlarged view of the planar coil.

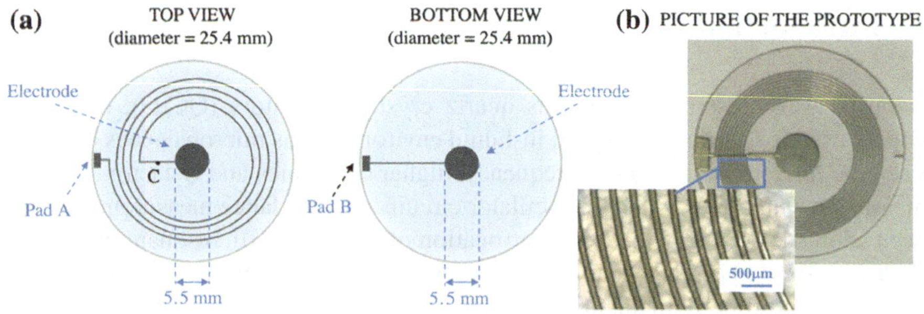


**Fig. 1** Sketch of the quartz resonator sensor with printed-on-crystal coil (**a**). Picture of the prototype with enlarged view of the planar coil deposited by aerosol-jet printing technology (**b**)

## 3 Experimental Results

The impedance $Z_{AC} = R_2 + j\omega L_2$ of the printed planar coil and the admittance $Y_{BC} = G + jB$ of the quartz crystal resonator have been measured using contact probes and an impedance analyzer (HP4194A) by keeping pads A and B in open-circuit configuration. The equivalent circuit and the sensor parameters measured are reported in Fig. 2, where the QCR is represented with the Butterworth-Van Dyke (BVD) equivalent circuit. The series resonant frequency $f_S = (4\pi^2 L_S C_S)^{-1/2}$, measured by taking the frequency where the admittance $G$ reaches its maximum, results 4.790260 MHz.

Pads A and B have been then shorted to allow the contactless interrogation of the sensor. Figure 3 shows the schematic diagram of the working principle and the

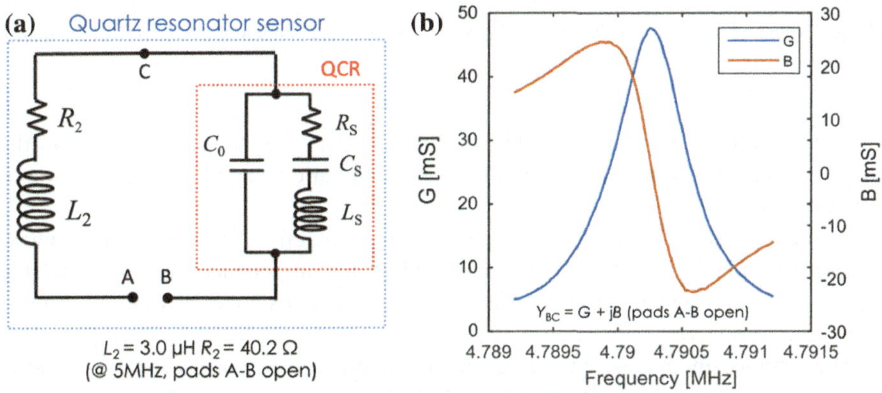


**Fig. 2** Equivalent circuit and measured $Z_{AC} = R_2 + j\omega L_2$ (**a**) and $Y_{BC} = G + jB$ (**b**) of the printed coil and the quartz crystal resonator, respectively (pads A and B in open-circuit configuration)

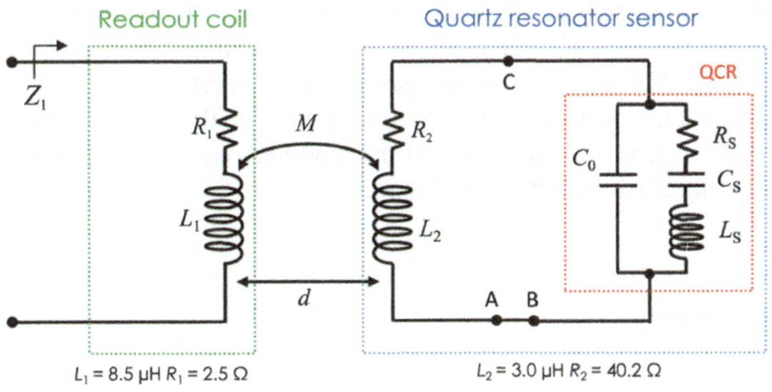


**Fig. 3** Equivalent circuits of the readout coil and quartz resonator sensor (pads A and B shorted)

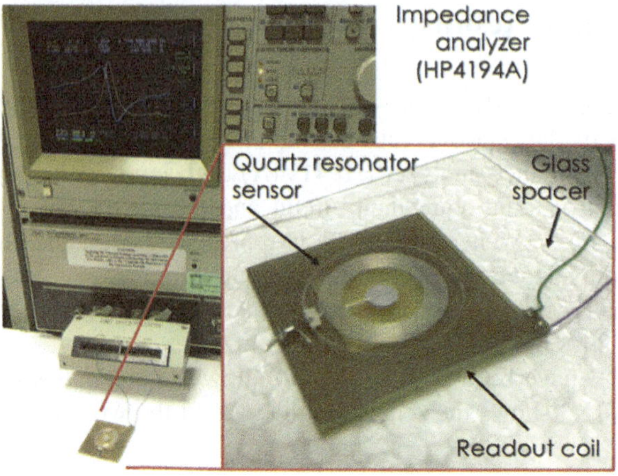


**Fig. 4** Experimental setup for the electromagnetic distance-independent interrogation of the quartz resonator sensor (pads A and B shorted)

measured equivalent parameters. The coil $L_1$ is electromagnetically coupled to the coil $L_2$ of the QCR sensor, enabling the electromagnetic contactless interrogation of the sensor.

Figure 4 shows the experimental setup for the contactless interrogation, in which the primary planar coil has been connected to the impedance analyzer, and the quartz resonator sensor has been placed at different distances $d$ ranging from 1 up to 12.2 mm.

The distance-independent interrogation technique is based on the measurement of the frequency where the real part of the reflected impedance $Z_1$ reaches its maximum [21]. Figure 5 shows the real part $\text{Re}\{Z_1\}$ and imaginary part $\text{Im}\{Z_1\}$ measured with different interrogation distances $d$, maintaining pads A and B shorted, where red unfilled dots represent the frequency at $\max(\text{Re}\{Z_1\})$. Figure 6 shows the frequency and magnitude at $\max(\text{Re}\{Z_1\})$ measured with different interrogation distances $d$, evidencing that the mutual inductance $M$ acts as a scaling factor on the magnitude of $\text{Re}\{Z_1\}$ without affecting the measured readout frequency.

The experimental results show a good agreement with the reference measurements of Fig. 2, with a maximum deviation of the readout frequency of about 50 Hz, i.e. 10.5 ppm, in the considered distance range.

## 4 Conclusion

In this paper, a novel quartz resonator sensor with printed-on-crystal coil has been presented. An aerosol-jet process is used to precisely and accurately deposit the planar coil and the electrodes onto a 330 μm-thick bare piezoelectric quartz crystal. The proposed distance-independent contactless interrogation technique exploits a

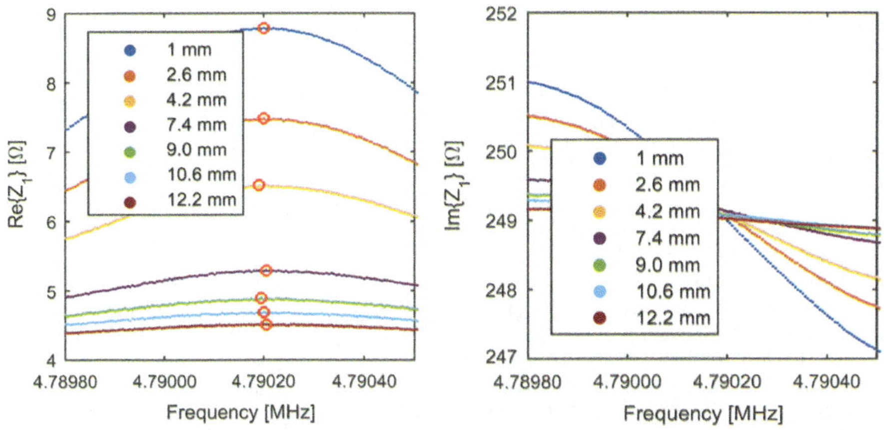


**Fig. 5** Real and imaginary parts of the impedance $Z_1$ measured with different interrogation distances (pads A and B shorted). Red unfilled dots represent the frequency at max(Re$\{Z_1\}$)

**Fig. 6** Frequency and magnitude at max(Re$\{Z_1\}$) measured with different interrogation distances (pads A and B shorted)

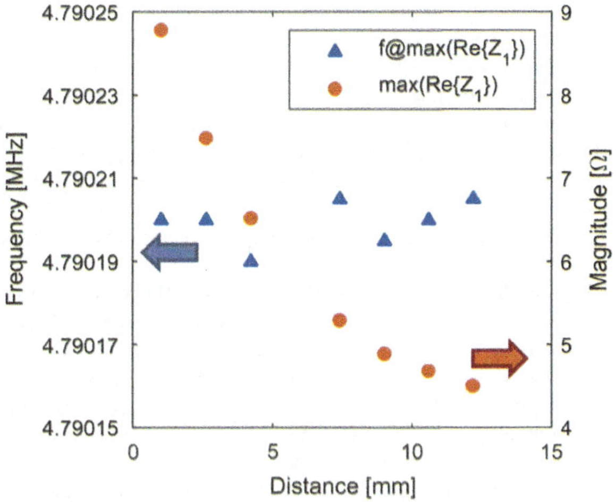


pair of electromagnetically coupled inductors, and it is based on the measurement of the frequency where the real part of the reflected impedance reaches its maximum.

The experimental results confirm that the mutual inductance $M$ acts as a scaling factor for the magnitude of the reflected impedance, without affecting the measured readout frequency for operation distances up to 12.2 mm. A good agreement with the reference measurements taken via contact probes has been obtained, with a maximum deviation of about 50 Hz, i.e. 10.5 ppm,

Further optimization of the coil layout, as well as the use of passivation layers, will be carried out towards the improvement of the maximum interrogation distance and the effectiveness of the sensor for in-air and in-liquid operation.

## References

1. Benes E, Gröschl M, Burger W, Schmid M (1995) Sensors based on piezoelectric resonators. Sens Actuators A Phys 48(1):1–21
2. Ferreira GNM, da-Silva A-C, Tomé B (2009) Acoustic wave biosensors: physical models and biological applications of quartz crystal microbalance. Trends Biotechnol 27(12):689–697
3. Ferrari M, Ferrari V, Marioli D, Taroni A, Suman M, Dalcanale E (2006) In-liquid sensing of chemical compounds by QCM sensors coupled with high-accuracy ACC oscillator. IEEE Trans Instrum Meas 55(3):828–834
4. Ferrari M, Ferrari V, Marioli D, Taroni A, Suman M, Dalcanale E (2004) Cavitand-coated PZT resonant piezo-layer sensors: properties, structure, and comparison with QCM sensors at different temperatures under exposure to organic vapors. Sens Actuators B Chem 103(1–2):240–246
5. Alassi A, Benammar M, Brett D (2017) Quartz crystal microbalance electronic interfacing systems: a review. Sensors 17(12):2799
6. Arnau A, García JV, Jimenez Y, Ferrari V, Ferrari M (2008) Improved electronic interfaces for AT-cut quartz crystal microbalance sensors under variable damping and parallel capacitance conditions. Rev Sci Instrum 79(7):075110
7. Ferrari M, Ferrari V (2009) An oscillator circuit for dual-harmonic tracking of frequency and resistance in quartz resonator sensors. Meas Sci Technol 20(12):124005
8. Ferrari M, Ferrari V, Marioli D (2010) Interface circuit for multiple-harmonic analysis on quartz resonator sensors to investigate on liquid solution microdroplets. Sens Actuators B Chem 146(2):489–494
9. Baù M, Ferrari M, Ferrari V (2017) Analysis and validation of contactless time-gated interrogation technique for quartz resonator sensors. Sensors 17(6):1264
10. Jiang H, Chang Z, Pertijs MAP (2016) A 30 ppm < 80 nJ ring-down-based readout circuit for resonant sensors. IEEE J Solid State Circuits 51:187–195
11. Ferrari M, Baù M, Tonoli E, Ferrari V (2013) Piezoelectric resonant sensors with contactless interrogation for mass-sensitive and acoustic-load detection. Sens Actuators A Phys 202:100–105
12. Baù M, Tonoli E, Ferrari V, Marioli D (2011) Contactless electromagnetic switched interrogation of micromechanical cantilever resonators. Sens Actuators A Phys 172(1):195–203
13. Baù M, Ferrari M, Ferrari V (2019) Magnet-less electromagnetic contactless interrogation technique for unwired conductive resonators. Electron Lett 55(11):642–644
14. Nopper R, Has R, Reindl L (2011) A wireless sensor readout system—circuit concept, simulation, and accuracy. IEEE Trans Instrum Meas 60:2976–2983
15. Huang QA, Dong L, Wang LF (2016) LC passive wireless sensors toward a wireless sensing platform: status, prospects, and challenges. J Microelectromech Syst 25:822–840
16. Salpavaara T, Antniemi A, Hänninen A, Lekkala J, Kellomäki M (2016) Inductively coupled passive resonance sensor for monitoring biodegradable polymers in vitro. Procedia Eng 168:1304–1307
17. Babu A, George B (2016) A linear and high sensitive interfacing scheme for wireless passive LC sensors. IEEE Sens J 16:8608–8616
18. Bhamra H, Tsai JW, Huang YW, Yuan Q, Shah JV, Irazoqui P (2017) A subcubic millimeter wireless implantable intraocular pressure monitor microsystem. IEEE Trans Biomed Circuit Syst 11:1204–1215
19. Siddiqui A, Mahboob MR, Islam TA (2017) Passive wireless tag with digital readout unit for wide range humidity measurement. IEEE Trans Instrum Meas 66:1013–1020
20. Demori M, Baù M, Ferrari M, Ferrari V (2018) Electronic technique and circuit topology for accurate distance-independent contactless readout of passive LC sensors. AEU Int J Electron Commun 92:82–85

21. Demori M, Baù M, Ferrari M, Ferrari V (2018) Interrogation techniques and interface circuits for coil-coupled passive sensors. Micromachines 9(9):449
22. Wu W, Greve DW, Oppenheimer IJ (2018) Inductively coupled sensing using a quartz crystal microbalance. In: Proceedings of the 2008 IEEE ultrasonics symposium, Beijing, China, 2–5 Nov 2008, pp 1018–1021

# "Green" Sensors Based on Bacterial Cellulose

**Giovanna Di Pasquale, Salvatore Graziani, Antonino Pollicino and C. Trigona**

**Abstract** An all-organic Bacterial Cellulose (BC)-PEDOT:PSS composite, impregnated by Ionic Liquids (ILs), working as a vibrating transducer, in cantilever configuration, is here investigated as sensing element. The generating properties of the composite will be presented, demonstrating the possibility of motion measuring. The sensing system is based on cheap, flexible, biodegradable and ecofriendly materials, suitable for the realization of next to come smart ubiquitous sensing systems. The performance of the sensing system are described in the following.

**Keywords** Bacterial cellulose · Ionic liquids · Greener transducers · Biodegradable devices

## 1 Introduction

Green sensors and transducers represent an emerging research area that implies the development of devices based on new materials and innovative technologies. The main aim is obtaining systems having the prerogatives of being: eco-friendly, bio-compatible [1, 2], recyclable, environmentally safe, and biodegradable. In this context, recently, we started investigating the possibility of using Bacterial Cellulose (BC) as a suitable compound for realizing green sensors [3, 4]. BC is a substitute of plant-derived cellulose and it presents the advantage to be much more eco-friendly and green, in the realization and purification processes, as respect the classical cellulose. In literature, manuscripts are available demonstrating the electromechanical

G. Di Pasquale
D.S.C., Dipartimento di Scienze Chimiche, University of Catania, Viale Andrea Doria 6, 95125 Catania, Italy

S. Graziani · C. Trigona (✉)
D.I.E.E.I., Dipartimento di Ingegneria Elettrica Elettronica e Informatica, University of Catania, Viale Andrea Doria 6, 95125 Catania, Italy
e-mail: carlo.trigona@dieei.unict.it

A. Pollicino
D.I.C.A.R., Dipartimento di Ingegneria Civile e Architettura, University of Catania, Viale Andrea Doria 6, 95125 Catania, Italy

 
G. Di Francia et al. (eds.), *Sensors and Microsystems*, Lecture Notes in Electrical Engineering 629, https://doi.org/10.1007/978-3-030-37558-4_45

**Fig. 1** **a** Family of sensors based on BC with several quantities of polymer used as electrodes. **b** SEM micrographs of BC

transduction properties of this family of compounds, however the field of sensor based on BC is somehow new and few papers started to appear in the scientific community [5].

In particular, BC with Ionic Liquids (ILs) and polymeric electrodes has been investigated to realize mass and deformation sensors [3, 4]. Figure 1a shows a family of realized devices with various configurations and thickness of polymeric contacts (notice the different shades of black). Furthermore, Fig. 1b shows a SEM picture of BC, after being impregnated with ILs.

## 2 Setup and Measurement System

### *2.1 Experimental Setup*

In order to study the sensing properties of the composite, the tailored setup, shown in Fig. 2, has been used. It is composed of: a shaker, for mechanically exciting the structure, a signal generator for driving the shaker, an oscilloscope, for data acquiring and two laser sensors, for measuring the displacement of the sensor at the tip and at the anchor. For the validation of the sensing principle, several different mechanical excitations have been applied and various analyses have been pursued.

### *2.2 Experimental Results*

In particular, Fig. 3a shows the output FFT of the sensor for several applied sinusoidal waveforms. The FFT shows the resonant frequency of the BC beam and, in accordance with the theory [4], the spikes decrease in frequency, when the device is driven

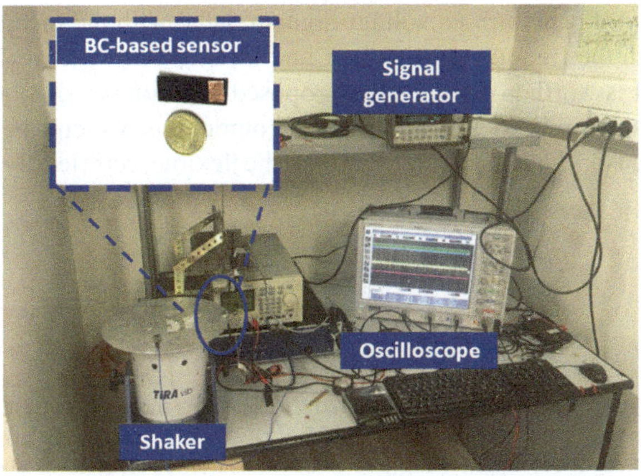


**Fig. 2** Setup used to test the BC-based transducer

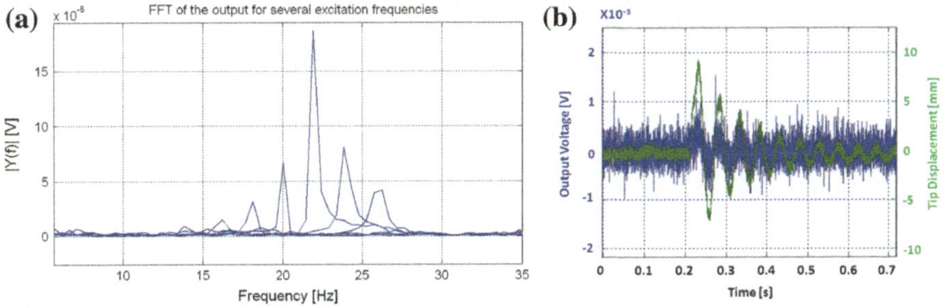


**Fig. 3** **a** FFTs of the output signal, when the system is excited with a sinusoidal motion. Several frequencies have been considered. **b** Step response of the BC-based sensor. The graph includes the output voltage of the device and the tip displacement

by frequencies different than its mechanical resonant frequency. Figure 3b shows the step response. More specifically, the graph includes the output voltage and the tip displacement of the beam. Work is in progress for an exhaustive characterization of BC-based sensors.

## 3 Conclusion

In this paper, the implementation of a bacterial cellulose compound able to be used as a sensing element has been presented. The conceived sensor is based on a nanocomposite consisting of BC, impregnated with ILs, and covered with a conductive polymer. The structure is mounted in cantilever configuration and its flexural behaviour has been studied. The beam deforms as function of the external mechanical source and it capability of generating a voltage signal related with the measurand has been confirmed. Both the frequency behaviour and the step response of the system have been studied. It is worth noting that the proposed solution is a demonstration of the generating sensing properties of this class of compounds, which can be used for the realization of sensors having the prerogative to be flexible, ecofriendly, biodegradable and low-cost.

## References

1. Sabo R et al (2016) Nanocellulose-enabled electronics, energy harvesting devices, smart materials and sensors: a review. J Renew Mater 4(5):297–312
2. Mangayil R et al (2017) Engineering and characterization of bacterial nanocellulose films as low cost and flexible sensor material. ACS Appl Mater Interfaces 9:19048–19056
3. Di Pasquale G, Graziani S, Pollicino A, Trigona C (2019) Paper-based sensor for deformation measurements. In: Proceeding of IEEE-I2MTC 2019, pp 1–4
4. Di Pasquale G, Graziani S, Pollicino A, Trigona C (2019) Green inertial sensors based on bacterial cellulose. In: Proceeding of IEEE-SAS 2019, pp 1–4
5. Wang YH et al (2018) A paper-based piezoelectric accelerometer. Micromachines 9:1–12

# Piezoelectric Beams, Magnets and Stoppers as Fundamental Blocks for Transducers and Autonomous Sensors

**C. Trigona, Bruno Andò and Salvatore Baglio**

**Abstract** This paper reports the development of a family of devices that are all based on fundamental blocks compose of: piezo electric beams, magnets, and mechanical stoppers that act as electric contacts. We have implemented combination of the proposed building blocks in order to realize devices for harvesting kinetic energy from weak and wide spectrum sources. It is worth noting that a proper use of these blocks has been also exploited to demonstrate the realization of devices which can be used as voltage rectifiers and multiplication circuits.

**Keywords** Piezoelectric transducer · Magnetic materials · Mechanical transducers · Stoppers · Fundamental blocks

## 1 Introduction

Transducers for vibration energy harvesting are receiving a considerable amount of interest in the scientific community [1]. It is worth noting that several works focused on autonomous solutions to power small-scale electronic mobile devices also through the adoption of piezoelectric materials. Furthermore, novel configurations of management architectures arouse interest also in vibrationally noisy environments and in presence of weak levels of generated voltage [1, 2]. This specific task has been studied and, in particular, here we address a family of devices which exploit the interaction of piezoelectric materials, flexible beams, magnets and mechanical stoppers together with the study of nonlinear dynamics used to develop architectures for autonomous sensors, smart nodes and measurement systems in vibrating environments [2–4]. We address here a review, starting from a single block (main block), able to work in presence of kinetic sources, to convert this mechanical signal into an electrical power and it is capable to manage the power also in presence of voltage levels <100 mV (voltage levels less than the diodes threshold, which typically, must

C. Trigona (✉) · B. Andò · S. Baglio
D.I.E.E.I., Dipartimento di Ingegneria Elettrica Elettronica e Informatica, University of Catania, Viale Andrea Doria 6, 95125 Catania, Italy
e-mail: carlo.trigona@dieei.unict.it



G. Di Francia et al. (eds.), *Sensors and Microsystems*, Lecture Notes in Electrical Engineering 629, https://doi.org/10.1007/978-3-030-37558-4_46

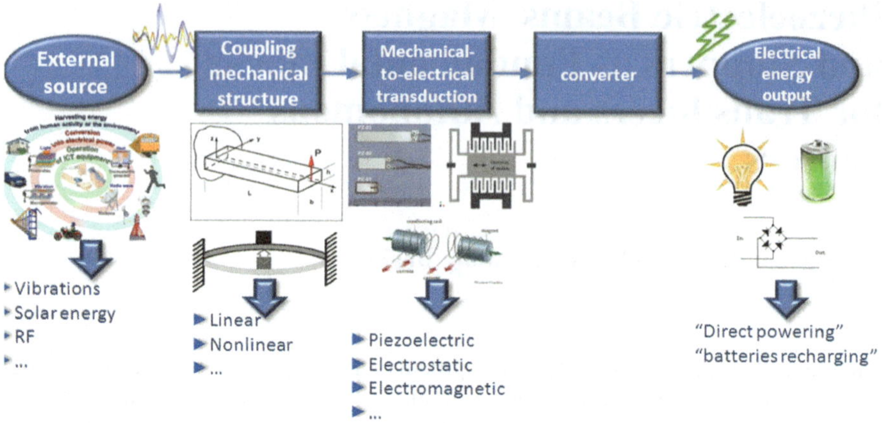


**Fig. 1** Functional blocks in a vibration energy harvesting system

be accounted in "classical" conditioning circuits for harvesting and sensing). Considering the functional blocks in a vibration energy harvesting system, as shown in Fig. 1, a large family of devices has been developed using a main block. Starting from this device, it has been developed a diodeless voltage multiplier AC/DC, DC/DC, filter, harvesters and rectifiers operating at any voltage level without the necessity of diodes.

## 2 Family of Transducers

Figure 2 shows a family of transducers suitable to be used in autonomous sensor nodes as converters and conditioning circuits for signals. In particular from the top of the figure to the bottom, it is possible to note a voltage multiplier [5] realized starting from the main block. It is able to multiply the level of voltage produced by a piezoelectric transducer and it is based on capacitors and mechanical switches (cantilever beams and stoppers) with a specific connection. It is worth noting that no diodes are required nor active circuits. Two blocks coupled with two permanent magnets have been also used to realize nonlinear energy harvesters [6] able to save energy inside a capacitor and useful to supply sensor nodes. The nonlinearities have been exploited in order to increase the bandwidth of response in order to collect more energy from the environment. A similar structure has been also used as diodeless H-bridge voltage rectifier [7] where two piezoelectric bistable cantilevers, that respond anti-symmetrically to environmental mechanical vibrations, are used to store energy, implementing, at the same time, the desired rectifying action. A configuration based on two or three coupled blocks have been also used to implement conditioning circuits based on switched capacitor topologies [8]. Configurations of amplifiers, and filters have been implemented both used as conditioning circuits for sensors and, at the same

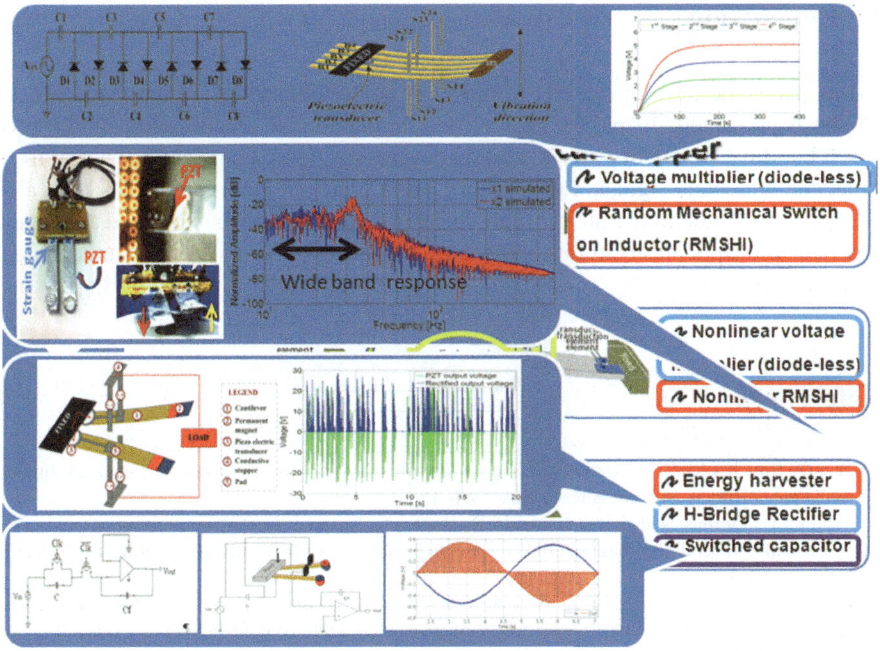


**Fig. 2** Family of transducers for autonomous sensors able to work in vibrationally noisy environments

time, as harvesters. A proper use of these blocks has been also exploited in this review to show the realization of devices which can perform well, in presence of stimulus of external vibrations, as conditioning circuits, sustainers and energy harvesting for autonomous or quasi autonomous sensor nodes. The work is in progress through the implementation of macroscale device and coupled architectures, such as proposed in Fig. 3, with a realization in MEMS technology.

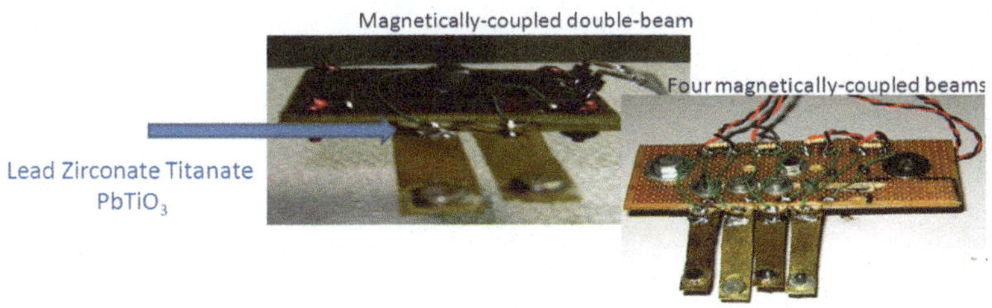


**Fig. 3** Macroscale prototypes

## 3 Conclusion

In this paper the development of a family of devices based on fundamental blocks has been presented. In particular we have implemented combination of the proposed building blocks in order to realize devices for harvesting kinetic energy from weak and wideband signals. Furthermore a proper use of these blocks has been also accomplished to demonstrate the realization of devices which can be used as voltage rectifiers, multiplication circuits and conditioning architectures based on mechanical transducers.

## References

1. Wei C, Jing X (2017) A comprehensive review on vibration energy harvesting: modelling and realization. Renew Sustain Energy Rev 74:1–18
2. Trigona C, Dumas N, Latorre L, Andò B, Baglio S, Nouet P (2011) Exploiting benefits of a periodically-forced nonlinear oscillator for energy harvesting from ambient vibrations. Procedia Eng 25:819–822
3. Penella MT, Gasulla M (2007) A review of commercial energy harvesters for autonomous sensors. In: 2007 IEEE instrumentation & measurement technology conference IMTC 2007, pp 1–5
4. Culler DE, Mulder H (2004) Smart sensors to network the world. Sci Am 290:84–91
5. Giusa F, Maiorca F, Noto A, Trigona C, Andò B, Baglio S (2014) A diode-less mechanical voltage multiplier: a novel transducer for vibration energy harvesting. Sens Actuators A 212:34–41
6. Andò B, Baglio S, Maiorca F, Trigona C (2013) Analysis of two dimensional, wide-band, bistable vibration energy harvester. Sens Actuators A 202:176–182
7. Maiorca F, Giusa F, Trigona C, Andò B, Bulsara AR, Baglio S (2013) Diode-less mechanical H-bridge rectifier for "zero threshold" vibration energy harvesters. Sens Actuators A 201:246–253
8. Trigona C, Andò B, Baglio S (2017) Self-generating SC-conditioning circuit for autonomous sensors. In: Sensors applications symposium (SAS), pp 1–5

# Influence of Oxygen Vacancies in Gas Sensors Based on Metal-Oxide Semiconductors: A First-Principles Study

**Soufiane Krik, Andrea Gaiardo, Matteo Valt, Barbara Fabbri, Cesare Malagù, Giancarlo Pepponi, Davide Casotti, Giuseppe Cruciani, Vincenzo Guidi and Pierluigi Bellutti**

**Abstract** Despite advantages highlighted by MOX-based gas sensors, these devices still show drawbacks in their performances (e.g. selectivity and stability), so further investigations are necessary. $SnO_2$ is the most used semiconductor for chemoresistive gas sensors production due to its broad spectrum of physical-chemical properties, and then it represents the best candidate for the innovative work here proposed. Indeed, among the gaps in research on this material, it is placed the study of oxygen deficiency and its impact on the tin dioxide physicochemical properties. A series of first-principles study was carried out in order to study the impact of oxygen vacancies on the physical-chemical properties of $SnO_2$. The results showed a high electrical conductivity for the samples with oxygen vacancies, which can give a decrease of the operating temperature that sensing material needs to be thermo-activated. The arrangement of the impurity states is one of the important parameters that involve the reactions on the material surface, making the excitation of weakly bound valence electrons into the unoccupied energy levels in the conduction bands.

**Keywords** Oxygen vacancies · $SnO_2$ · Gas sensor · DFT

## 1 Introduction

In order to forecast the performance of a gas sensor based on semiconducting material, it is mandatory to deeply investigate its physical-chemical properties. In this work, we propose a study concerning the influence of oxygen vacancies in Metal-OXide (MOX) sensing materials, widely used for chemical sensors [1].

$SnO_2$ is the most studied semiconductor for chemoresistive gas sensors production [2], and then it represents the best candidate for the innovative work here proposed.

S. Krik (✉) · A. Gaiardo · M. Valt · B. Fabbri · C. Malagù · D. Casotti · G. Cruciani · V. Guidi
Department of Physics and Earth Sciences, University of Ferrara, 44122 Ferrara, Italy
e-mail: soufiane.krik@unife.it

S. Krik · A. Gaiardo · G. Pepponi · P. Bellutti
Center of Materials and Microsystems, Bruno Kessler Foundation, 38128 Trento, Italy



G. Di Francia et al. (eds.), *Sensors and Microsystems*, Lecture Notes in Electrical Engineering 629, https://doi.org/10.1007/978-3-030-37558-4_47

Hitherto, literature presents lacks of studies on how number and arrangement of oxygen vacancies affect the sensing performance.

The physical-chemical properties of materials emerging from the total energy, is the DFT calculations principle. The basis of the approximate exchange-correlation functional within the framework of Density Functional Theory (DFT) [3] are The Local Density Approximation (LDA) [4] and the Generalized Gradient Approximation (GGA) [5], which improve the LDA results taking into account the electron density gradient due to the non-homogeneity of the electron density. Both methods have proved their ability to accurately determine the structural properties of materials [6]. Nevertheless, their applications underestimate the band gap energy of materials compared to experimental results [7, 8]. To improve the results about the band gap energy the Tran-Blaha's modified Becke-Johnson exchange potential model (TB-mBJ) [9], which is a semi local approximation, has been implemented in the Wien2k code and it gave a better concordance with the experimental data.

In the present investigation, a series of first principles study was carried out. It was used the most successful method to solve the Khon Sham equation [10], i.e. The full potential linearized augmented plane wave (FPLAPW) method [11] within the framework of DFT implemented in the Wien2k code [12].

Moreover, in this work the GGA parameterized PBE (Perdew-Burke-Ernzerhof) [13] has been used for structural properties calculations, for the electrical and the electronic ones the TB_mBJ potential has been adopted. To calculate the electrical conductivity of the material studied, we applied the Boltzman's transport theory [14] as implemented in the BoltzTrap package [15] enforced in the wien2k code.

## 2 Computational Details

Cassiterite $SnO_2$ structure (Fig. 1a) has been constructed using literature, the crystal structure has the space group P42/mnm (No. 136) [16]. The initial cell parameters were $a = b = 4.7374$ Å and $c = 3.1864$ Å [16]. The coordinates of tin atoms were: (0, 0, 0), (1/2, 1/2, 1/2) and those of oxygen atoms were: $\pm(u, u, 0)$; $\pm(1/2 + u, 1/2 - u, 1/2)$ with $u = 0.3056$ corresponds to the internal parameter. Then a (1 * 1 * 3) supercell was created (Fig. 1b), which gave us a material containing 18 atoms in total: 6 of tin and 12 of oxygen ($Sn_6O_{12}$), afterwards we created oxygen vacancies with two different concentrations that gave us two different samples of $SnO_{2-x}$ ($x = 0.17$, Fig. 1c and $x = 0.33$, Fig. 1d).

The self-consistent field (SCF) cycles were performed using 2.1 and 1.81 as values of the muffin tin radius for Sn and O respectively. The sampling of the Brillouin zone was carried out according to the Monkhorst-Pack scheme [17] and the number of $k$ points was optimized to obtain a well convergence in the calculations, the set value was 190. The number of plane waves in a Fourier expansion of the potential in the interstitial region (RMT × Kmax) was also optimized for a minimum total energy and the set value used was 7.5. The geometry of the structure was relaxed by minimizing the forces acting on each atom reaching values below 2 mRy/a.u.

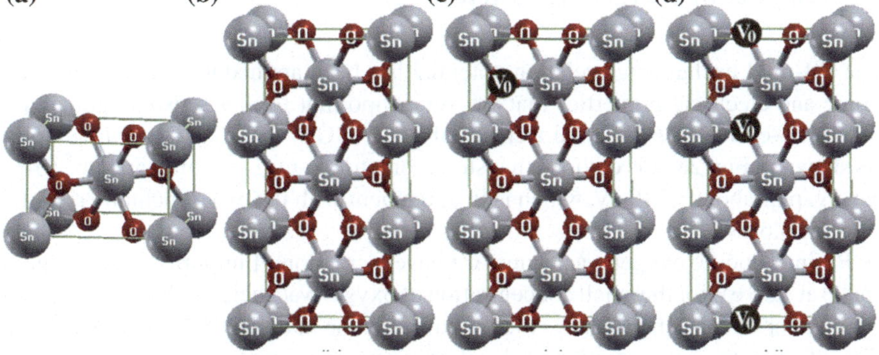


**Fig. 1** The structures used during calculations in this work: **a** cassiterite $SnO_2$, **b** (1 * 1 * 3) supercell "$Sn_6O_{12}$", **c** one oxygen vacancy $V_O$ have been introduced "$Sn_6O_{11}$" and **d** two atoms of oxygen $V_O$ have been removed "$Sn_6O_{10}$"

# 3 Results and Discussions

## *3.1 Stochiometric $SnO_2$*

In a first step we simulated the physical-chemical properties of the stochiometric $SnO_2$ (Fig. 1a) using GGA-PBE. The structural properties calculated in this work and those identified in literature are reported in Table 1. Crystal lattice parameters $a$ and $c$ are optimized by adjusting the total energy to the Murnaghan state equation [18]. The obtained results are close to the experimental [19, 20] and to other simulated [21, 22] ones: the GGA-PPE method gave results for the parameters $a$, $c$ and $u$ in good agreement with the experimental data mentioned in Refs. [19, 20] with a relative error of about $E_r(a) = 0.025$ and $E_r(c) = 0.024$, respectively with $E_r = (E_l - E_s)/E_l$, here $E_l$ is the value identified in literature and $E_s$ is the simulated value on this work.

We can notice an underestimation of the bulk module $B$ and the pressure derivative $B'$ [18], which can be explained taking in account that GGA-PPE method miscalculate the bulk module [23].

**Table 1** The structural parameters obtained in this work compared to those identified in literature

| | | a | c | c/a | u | B | B′ |
|---|---|---|---|---|---|---|---|
| Simulated | Present work | 4.8174 | 3.2399 | 0.671 | 0.3055 | 185.892 | 4.865 |
| | [21] | 4.699 | 3.165 | 0.674 | 0.340 | 244.7 | 4.44 |
| Experimental | [19] | 4.754 | 3.19 | 0.671 | N/A | N/A | N/A |
| | [20] | 4.7367 | 3.1855 | 0.672 | 0.3070 | 205 | 7.4 |

## 3.2 Impact of Oxygen Vacancies

The TB-mBJ exchange-correlation potential has been applied to calculate the electronic and electrical properties that are very important for gas sensing application. Figure 2a–c illustrates the total Density of State (DOS) calculated for our samples. As shown in Fig. 2a, it is obvious that stoichiometric $SnO_2$ is a semiconductor with a band gap of about 3.346 eV, which is in agreement with the experimentally reported value of 3.6 eV.

By introducing oxygen vacancies, we have a creation of impurity states (Fig. 2b and c) at 1.55 eV for the small concentration of oxygen vacancies and at 1.15 eV when the number of oxygen vacancies increased inside the material. We can notice that in Fig. 2d and e the concentration of the charge careers increases in the impurity level by increasing the number of oxygen vacancies, this lead to increase the conductivity of the material and consequently to enhance the reactions on its surface. We can conclude then, the increase of oxygen vacancies concentration gives a decrease in the energy that electrons need to be excited from the defect state to the conduction band. Decreasing of that energy makes the ionization of the dopant level easier and involves the reactions on the material's surface.

Besides, Fig. 3 represents the electrical conductivity over the relaxation time $(\sigma/\tau)$ versus temperature. We can notice an increase in the electrical conductivity with the increase in the number of oxygen vacancies inside $SnO_2$, which can give a decrease in the temperature that sensing material needs to be thermo-activated.

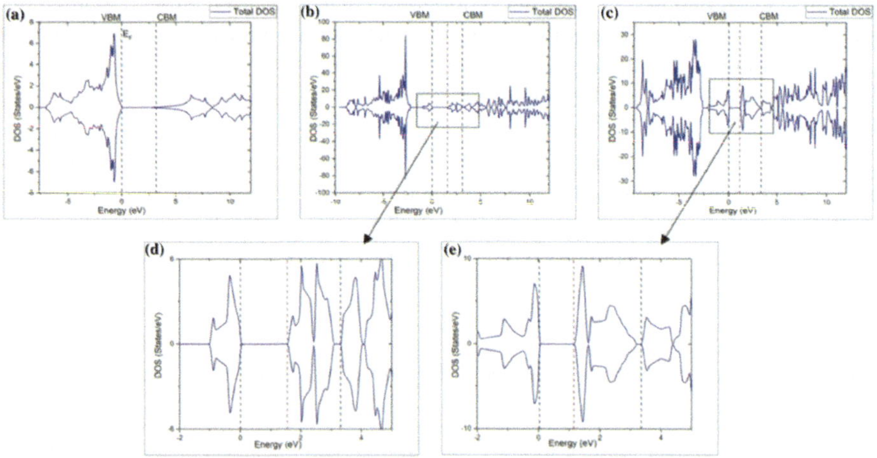


**Fig. 2** Calculated total density of state for: **a** stochiometric $SnO_2$, **b** $Sn_6O_{11}$, **c** $Sn_6O_{10}$, **d** and **e** zoom of the framed area in both figures (**b**) and (**c**), respectively. The Fermi level has been set at 0

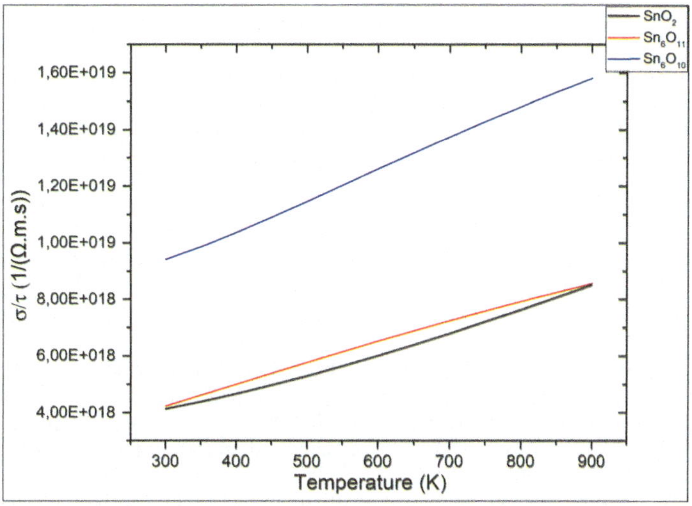


**Fig. 3** The electrical conductivity versus temperature for the stochiometric $SnO_2$ (black curve), $Sn_6O_{11}$ and $Sn_6O_{10}$ (red and blue curves, respectively)

## 4 Conclusion

The impact of oxygen vacancies creation on the physical-chemical properties of tin dioxide was studied from a theoretical point of view. Obtained results show clearly an influence on electrical properties of $SnO_2$, which is one of the most relevant properties of MOX semiconductors for their use as sensing film for chemoresistive gas sensors. This open up to further theoretical and experimental works to deeply investigate the influence of this kind of defects inside MOX semiconductors for sensing applications.

## References

1. Batzill M, Diebold U (2005) The surface and materials science of tin oxide. Prog Surf Sci 79(2–4):47–154
2. Wang C, Yin L, Zhang L, Xiang D, Gao R (2010) Metal oxide gas sensors: Sensitivity and influencing factors. Sensors 10(3):2088–2106
3. Schwarz K (2003) DFT calculations of solids with LAPW and WIEN2k. J Solid-State Chem 176(2):319–328
4. Perdew JP, Wang Y (1992) Accurate and simple analytic representation of the electron-gas correlation energy. Phys Rev B 45(23):13244–13249
5. Langreth DC, Mehl MJ (1983) Beyond the local-density approximation in calculations of ground-state electronic properties. Phys Rev B 28(4):1809–1834
6. Chikr ZC, Mokadem A, Bouslama M, Besahraoui F, Ghaffour M, Ouerdane A, Boulenouar K, Chauvin N, Benrabah B (2013) The investigation of the electron behavior of $SnO_2$ by the

simulation methods GGA and mBJ associated with the eels experimental analysis technique. Surf Rev Lett 20(5):art. no. 1350050
7. Dufek P, Blaha P, Schwarz K (1994) Applications of Engel and Voskos generalized gradient approximation in solids. Phys Rev B 50(11):7279–7283
8. Heyd J, Peralta JE, Scuseria GE, Martin RL (2005) Energy band gaps and lattice parameters evaluated with the Heyd-Scuseria-Ernzerhof screened hybrid functional. J Chem Phys 123(17):art. no. 174101
9. Tran F, Blaha P (2009) Accurate band gaps of semiconductors and insulators with a semilocal exchange-correlation potential. Phys Rev Lett 102(22):art. no. 226401
10. Kohn W, Sham LJ (1965) Self-consistent equations including exchange and correlation effects. Phys Rev 140(4A):A1133–A1138
11. Singh DJ, Nordstrom L (2006) Planewaves, pseudopotentials and the LAPW method, 2nd edn. Springer, New York, NY, pp 1–134
12. Blaha P, Schwarz K, Madsen GKH, Kvasnicka D, Luitz J, Laskowski R, Tran F, Marks LD (2018) WIEN2k, an augmented plane wave + local orbitals program for calculating crystal properties. Karlheinz Schwarz Techn. Universität, Wien, Austria. ISBN 3-9501031-1-2
13. Perdew JP, Burke K, Ernzerhof M (1996) Generalized gradient approximation made simple. Phys Rev Lett 77(18):3865–3868
14. Allen PB (1996) Boltzmann theory and resistivity of metals. In: Chelikowsky JR, Louie SG (eds) Quantum theory of real materials. Kluwer, Boston, pp 219–250
15. Madsen GKH, Singh DJ (2006) BoltzTraP. A code for calculating band-structure dependent quantities. Comput Phys Commun 175(1):67–71
16. Bolzan AA, Fong C, Kennedy BJ, Howard CJ (1997) Structural studies of rutile-type metal dioxides. Acta Crystallogr Sect B: Struct Sci 53(3):373–380
17. Monkhorst HJ, Pack JD (1976) Special points for Brillouin-zone integrations. Phys Rev B 13(12):5188–5192
18. Murnaghan FD (1944) The compressibility of media under extreme pressures. Proc Natl Acad Sci 30(9):244–247. https://doi.org/10.1073/pnas.30.9.244
19. Chetri P, Choudhury A (2013) Investigation of optical properties of $SnO_2$ nanoparticles. Physica E 47:257–263
20. Haines J, Léger J (1997) X-ray diffraction study of the phase transitions and structural evolution of tin dioxide at high pressure: relationships between structure types and implications for other rutile-type dioxides. Phys Rev B: Condens Matter Mater Phys 55(17):11144–11154
21. Zhu B, Liu C-M, Lv M-B, Chen X-R, Zhu J, Ji G-F (2011) Structures, phase transition, elastic properties of $SnO_2$ from first-principles analysis. Physica B: Condens Matter 406(18):3508–3513
22. Godinho Kate G, Walsh A, Watson GW (2009) Energetic and electronic structure analysis of intrinsic defects in $SnO_2$. J Phys Chem C 113(1):439–448
23. Zupan A, Causà M (1995) Density functional LCAO calculations for solids: comparison among Hartree-Fock, DFT local density approximation, and DFT generalized gradient approximation structural properties. Int J Quantum Chem 56(4):337–344

# W-Sn Mixed Oxides: New Materials for Gas Sensing

**Ambra Fioravanti, Maria Cristina Carotta, Sara Morandi, Alessia Amodio, Mauro Mazzocchi and Michele Sacerdoti**

**Abstract** Thick films of tungsten trioxide ($WO_3$), tin dioxide ($SnO_2$), and tungsten-tin (W-Sn) mixed oxides at different nominal Sn molar fraction (0.1 and 0.3) were prepared. The functional materials were synthesized and characterized by SEM, X-ray diffraction, specific surface area measurements (BET), UV-Vis-NIR and IR spectroscopies. The gas sensing measurements highlighted that the mixed oxides based sensors offer a better response towards $NO_2$ with respect to pure $WO_3$ being almost insensitive to CO.

**Keywords** W-Sn mixed oxides · Thick film gas sensors · UV-Vis-NIR and FT-IR spectroscopies

## 1 Introduction

Nowadays there is a variety of structures, materials and working principles exploitable to realize gas sensors. Among them, semiconductor gas sensors using metal oxides (MOX) as gas sensing element are a convenient choice to detect a wide range of gases [1] both in scientific research and in industrial and environmental applications.

Road traffic, industrial processes and fuel combustion utilities strongly influence the air quality in urban atmosphere, above all $NO_x$ and CO in addition to particulate matter. Thereby, the monitoring of the atmospheric pollutants is an important issue.

A. Fioravanti (✉) · M. C. Carotta
Laboratorio Sensori e Nanomateriali CNR—IMAMOTER,
Via Canal Bianco 28, 44124 Ferrara, Italy
e-mail: a.fioravanti@imamoter.cnr.it

S. Morandi · A. Amodio
Dipartimento di Chimica, Università di Torino, Via P. Giuria 7, 10125 Turin, Italy

M. Mazzocchi
CNR—ISTEC, Via Granarolo 64, 48018 Ravenna, Italy

M. Sacerdoti
Dipartimento di Fisica e Scienze della Terra, Università di Ferrara,
Via Saragat 1, 44122 Ferrara, Italy



G. Di Francia et al. (eds.), *Sensors and Microsystems*, Lecture Notes in Electrical Engineering 629, https://doi.org/10.1007/978-3-030-37558-4_48

Portable and versatile devices based on thick film MOX gas sensors, having low requirements in terms of power, consumables, maintenance and installation costs is an effective solution [2].

In this work, sensors to detect atmospheric pollutant gases like $NO_x$ and CO have been taken into account.

Pure oxides $SnO_2$ and $WO_3$ were prepared through sol-gel route. W-Sn mixed oxides at increasing nominal Sn molar fraction (0.1 and 0.3) were synthesized by sol-gel co-precipitation with the aim to join advantages of high sensitivity toward oxidizing gases for $WO_3$ [3] and Sn addition to reduce $WO_3$ grain growth with temperature.

## 2 Materials, Methods and Results

### *2.1 Materials and Methods*

The single oxides $WO_3$ and $SnO_2$, W-Sn mixed oxides were synthesized in from of nanopowders and used to prepare thick film gas sensors. W-Sn mixed oxides (nominal Sn molar fraction of 0.1 and 0.3) were obtained by sol-gel co-precipitation method using required stoichiometric proportions of Tin(II) 2-ethylhexanoate in a proper amount of 1-butanol and $WCl_6$ dissolved in absolute ethanol. Diluted $HNO_3$ solution was added to catalyze the hydrolysis process and the resulting yellowish green powder was calcined at 550 °C for 2 h. The prepared materials were characterized through FE-SEM electron microscopy, BET and XRD analysis. Diffuse reflectance UV-Vis-NIR and absorbance FT-IR spectroscopies, in situ conditions, were employed to characterize the electronic properties of the materials. Screen printing technique was used to deposit the sensing layers and the corresponding thick films (all fired at 650 °C in this work) were tested for sensing in a sealed test chamber using the flow-through technique.

### *2.2 Morphological and Structural Characterizations*

FE-SEM observations confirmed that the addition of Sn to $WO_3$ successfully reduced the coalescence of $WO_3$ grains as shown in Fig. 1 in which the images of all calcined materials are reported.

XRD analysis, performed onto pure $SnO_2$ powder (Fig. 2), attributes a tetragonal crystalline structure (space group $P4_2/mnm$) with an average crystallite size of 12 nm calculated by Scherrer's formula.

XRD patterns of pure $WO_3$, W-Sn(0.1) and W-Sn(0.3) powders correspond to monoclinic pseudo-cubic crystal structure (space group $P2_1/n$) of polycrystalline $WO_3$ without any other phases (Fig. 2). No change in the W-Sn materials unit cell

**Fig. 1** SEM images of pure $SnO_2$, W-Sn(0.3), W-Sn(0.1) and pure $WO_3$ powders calcined at 550 °C

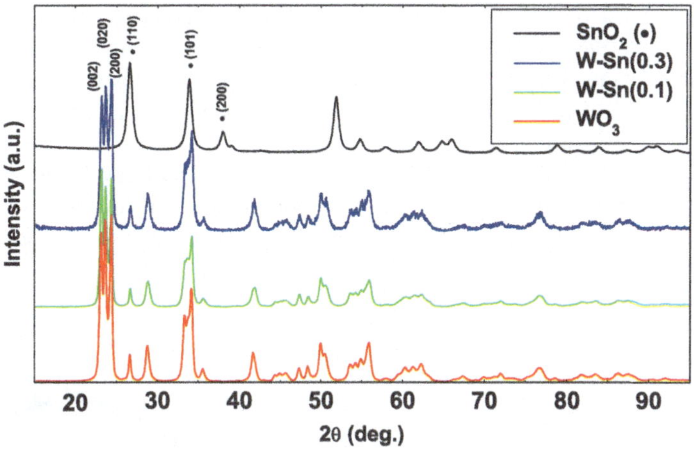


**Fig. 2** XRD pattern of pure $SnO_2$, pure $WO_3$, and W-Sn powders with nominal Sn molar fraction of 0.1 and 0.3

parameters and unit cell volume was observed with respect to pure $WO_3$. The crystallite size of the above powders, evaluated by Scherrer's formula, being 85, 25, 24 nm, respectively.

The trend of the crystallite size is in agreement with the specific surface area measurements (Table 1): coherently, the smaller is the crystallite size the higher is the BET surface area.

**Table 1** Specific surface areas and mean crystallite sizes of the powders

| Sample | Crystallite size (nm) | Specific surface area ($m^2/g$) |
|---|---|---|
| $SnO_2$ | 12 | 27.9 |
| W-Sn(0.3) | 24 | 24.5 |
| W-Sn(0.1) | 25 | 23.9 |
| $WO_3$ | 85 | 9.7 |

## 2.3 Spectroscopic and Electrical Characterizations

All studied materials behave as n-type semiconductors because of lattice defects as highlighted in the Arrhenius plots (not shown here for sake of brevity) in which the semiconductor behavior was modulated by the Schottky barrier formation at gas–semiconductor interface. UV-Vis-NIR and FT-IR spectroscopies (see Fig. 3) highlighted the main electronic defects in the examined materials. In CO reducing atmosphere, $WO_3$ shows polarons formation (band in the near IR region), while $SnO_2$ shows the increase of mono-ionized oxygen vacancies (band in the medium IR). The mixed oxides W-Sn(0.1) and W-Sn(0.3) show polarons formation as in $WO_3$.

In Tables 2 and 3, the electrical responses for all samples to 50 ppm of carbon monoxide and to 1 ppm of nitrogen dioxide obtained at different working temperatures are reported.

It can be observed, that the only studied material really effective toward CO was $SnO_2$. On the other hand, $SnO_2$ was poorly responsive toward $NO_2$ except at very low temperatures, while all other materials were almost insensitive to CO, but exhibited

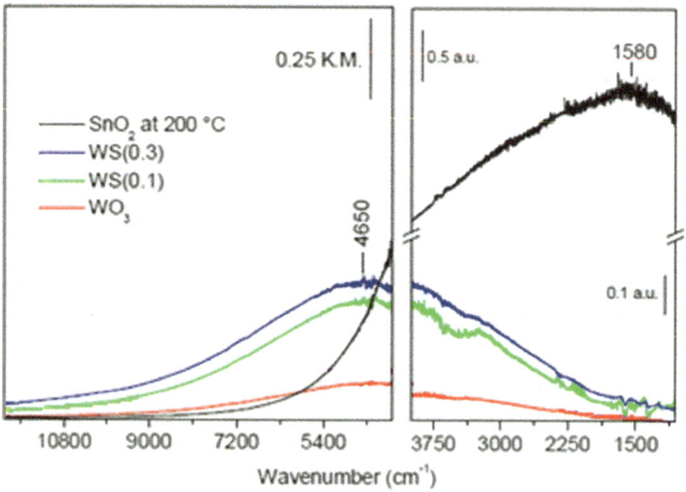


**Fig. 3** Vis-NIR and IR spectra of $SnO_2$, W-Sn(0.1), W-Sn(0.3) and $WO_3$ samples during interaction with $CO/O_2$ (1:5, pCO = 5 mbar) at the temperature of the maximum absorption intensity

**Table 2** Responses of all sensors to 50 ppm of CO reported as $G_{CO}/G_{air}$ at different working temperatures

| Sample | Temperature | | | | |
|---|---|---|---|---|---|
| | 250 °C | 300 °C | 350 °C | 400 °C | 450 °C |
| $SnO_2$ | 1.94 | 2.83 | 6.01 | 4.98 | 3.83 |
| W-Sn(0.3) | 1.05 | 1.09 | 1.08 | 1.09 | 1.05 |
| W-Sn(0.1) | 1.03 | 1.05 | 1.09 | 1.08 | 1.09 |
| $WO_3$ | 1.06 | 1.10 | 1.10 | 1.10 | 1.10 |

**Table 3** Responses of all sensors to 1 ppm of $NO_2$ reported as $G_{air}/G_{NO_2}$ at different working temperatures

| Sample | Temperature | | | | |
|---|---|---|---|---|---|
| | 250 °C | 300 °C | 350 °C | 400 °C | 450 °C |
| $SnO_2$ | 2.49 | 1.81 | 1.72 | 1.43 | 1.14 |
| W-Sn(0.3) | 30.24 | 22.12 | 19.03 | 10.21 | 2.87 |
| W-Sn(0.1) | 24.32 | 16.47 | 14.29 | 8.22 | 2.13 |
| $WO_3$ | 21.85 | 16.15 | 13.54 | 5.12 | 1.64 |

high responses to $NO_2$. In particular, the sensor with nominal Sn molar fraction of 0.3 exhibited the highest response in the dynamical measurement toward 1 ppm of $NO_2$.

# 3 Conclusion

The single oxides $WO_3$ and $SnO_2$ and the W-Sn mixed oxides, with nominal Sn molar fraction of 0.1 and 0.3, were successfully synthesized without the presence of any other phases. The mixed materials W-Sn(0.1) and W-Sn(0.3) achieved the goal to reduce the $WO_3$ grain growth with temperature.

All studied materials, $SnO_2$, $WO_3$, and W-Sn behaved as n-type semiconductors because of lattice defects. In $SnO_2$, oxygen vacancies acted as electron donor levels, while $WO_3$ and W-Sn were characterized by polarons as evidenced by spectroscopic analysis. The gas sensing measurements toward $NO_2$ and CO highlighted that $SnO_2$ was more performant in CO detection. The mixed oxide sensors offered a better response with respect to pure $WO_3$ toward $NO_2$, at the same time maintaining the characteristics of almost complete insensitivity to carbon monoxide.

## References

1. Wang C, Yin L, Zhang L, Xiang D, Gao R (2010) Metal oxide gas sensors: sensitivity and influencing factors. Sensors 10:2088–2106. https://doi.org/10.3390/s100302088
2. Fioravanti A, Bonanno A, Gherardi S, Carotta MC, Skouloudis AN (2016) A portable air-quality station based on thick film gas sensors for real time detection of traces of atmospheric pollutants. IOP Conf Ser: Mater Sci Eng 108:012005. https://doi.org/10.1088/1757-899x/108/1/012005
3. Shendage SS, Patil VL, Vanalakar SA, Patil SP, Harale NS, Bhosale JL, Kim JH, Patil PS (2017) Sensitive and selective $NO_2$ gas sensor based on $WO_3$ nanoplates. Sens Actuators B Chem 240:426–433. https://doi.org/10.1016/j.snb.2016.08.177

# Lossy Mode Resonance Enabling Ultra-Low Detection Limit for Fibre-Optic Biosensors (INVITED)

**F. Chiavaioli, A. Giannetti, S. Tombelli, C. Trono, I. Del Villar, I. R. Matias, P. Zubiate, C. R. Zamarreño, F. J. Arregui and F. Baldini**

**Abstract** The combination of optical fibre-based biosensors with nanotechnologies is providing the opportunity for the development of in situ, portable, lightweight, versatile and high-sensitivity optical sensing platforms. We report on the generation of lossy mode resonances (LMRs) by means of the deposition of nm-thick $SnO_2$ film on optical fibres. This allows measuring precisely and accurately the changes in refractive index of the fibre-surrounding medium with very high sensitivity compared to other optical technology platforms, such as long period grating or surface plasmon resonance. This approach, mixed with the use of specialty fiber structures such as D-shaped fibres, allows improving the light-matter interaction in strong way. Different imaging systems, i.e. SEM and TEM along with X-EDS tool, have been used to study the optical features of the fiber coating. The shift of the LMR has been monitored in real-time thanks to conventional wavelength interrogation system and ad hoc developed microfluidics. A big leap in performance has been attained by detecting femtomolar concentrations in human serum. The biosensor reusability has been also tested by using a solution of sodium dodecyl sulphate.

**Keywords** Lossy mode resonance · In-fiber optical sensing platform · Nanofilm deposition · Femtomolar concentration · Reusability

F. Chiavaioli (✉) · A. Giannetti · S. Tombelli · C. Trono · F. Baldini
Institute of Applied Physics "Nello Carrara", National Research Council, Via Madonna del Piano 10, 50019 Sesto Fiorentino, Firenze, Italy
e-mail: f.chiavaioli@ifac.cnr.it

I. Del Villar · I. R. Matias
Institute of Smart Cities, Public University of Navarra, Campus de Arrosadia, 31006 Pamplona, Spain

P. Zubiate · C. R. Zamarreño · F. J. Arregui
Electrical and Electronic Engineering Department, Public University of Navarra, Campus de Arrosadia, 31006 Pamplona, Spain



G. Di Francia et al. (eds.), *Sensors and Microsystems*, Lecture Notes in Electrical Engineering 629, https://doi.org/10.1007/978-3-030-37558-4_49

# 1 Introduction

Compact, portable, lightweight and high sensitivity devices are more and more demanded by industrial companies as well as novel sensing concepts are continuously studied in fundamental research. Great advantages are offered from fibre-optic devices over other optical technology platforms thanks to the peculiarities of optical fibres [1, 2]. Moreover, the opportunity of depositing nanometre films on optical fibres with a high degree of accuracy, precision and reproducibility has allowed the broadening of the application domains of this technology [3, 4].

One of the most widespread phenomenon related to metallic thin films for sensing is surface plasmon resonance (SPR) [5]. More recently, starting from the studies on semiconductor waveguides [6], the concept of guided mode resonance was also applied to fibre-optic, under the name of lossy mode resonance (LMR) [7]. LMR occurs when the real part of the thin film permittivity is positive and greater in magnitude than both its own imaginary part and the permittivity of the material surrounding the thin film [8]. Therefore, metallic oxides and polymers are used to generate LMRs [9], instead of the metallic materials typically used to generate SPRs [10]. Differently from SPR, both TE- and TM-polarized light can excite LMR. Another important characteristic is the possibility of tuning the spectral position of the LMR by adjusting the thin film thickness [7].

LMR-based sensors have been recently implemented in D-shaped single-mode fibres instead of using multi-mode fibres, where LMRs are broader and more difficult to analyse [11]. Due to the asymmetric shape of the transversal section of the D-shaped fibre, it is possible to excite more than a single LMR, thus it allows tracking the wavelength shifts of the first LMR, the most sensitive LMR, at wavelengths in the infrared region, where the sensitivity is also improved if compared to the visible region [12].

Biosensing with optical fibres should gain a great push from this technology, since the requirement of very low limit of detection (LOD) is high in many applications. This causes in turn a change.

In this paper, we present the performance of $SnO_2$-coated D-shaped fibre-optic LMR biosensor by developing an IgG/anti-IgG immunological assay. The sensing principle is quite simple: when a target analyte interacts with the fibre surface firstly coated with an nm-thick film and then suitably functionalised, this induces a change in the spectral position of the LMR that can be measured accurately and precisely by means of a conventional wavelength interrogation system and an ad hoc integrated microfluidics [12]. The results detail that it is possible to achieve femtomolar concentrations of the analyte.

## 2 Materials and Methods

The fibre-optic used here was a D-shaped single-mode fibre (SMF) purchased from Phoenix Photonics Ltd. (Birchington, UK). This fibre consisted of a standard SMF (SMF-28, Corning®) with a side-polished length of 17 mm (Fig. 1). Afterwards, the polished surface of the fibre, hereafter called sensitive region, was coated with a thin film of $SnO_2$ with a 99.99% of purity and purchased from ZhongNuo Advanced Material Technology Co. The D-shaped SMF was carefully handled and placed on an ad hoc developed substrate in a DC sputter machine (ND-SCS200, Nadetech S.L.). The fabrication parameters were: $9 \times 10^{-2}$ mbar argon partial pressure and 90 mA current intensity. The experimental setup is sketched in Fig. 1. It consisted of a broadband multi-LED light source (FIBRELABS, Inc., SLD-1310/1430/1550/1690), an optical spectrum analyser (OSA, Anritsu MS9030A-MS9701B), an in-line polarizer and a polarization controller, which allowed exciting the selected TE or TM polarized travelling light.

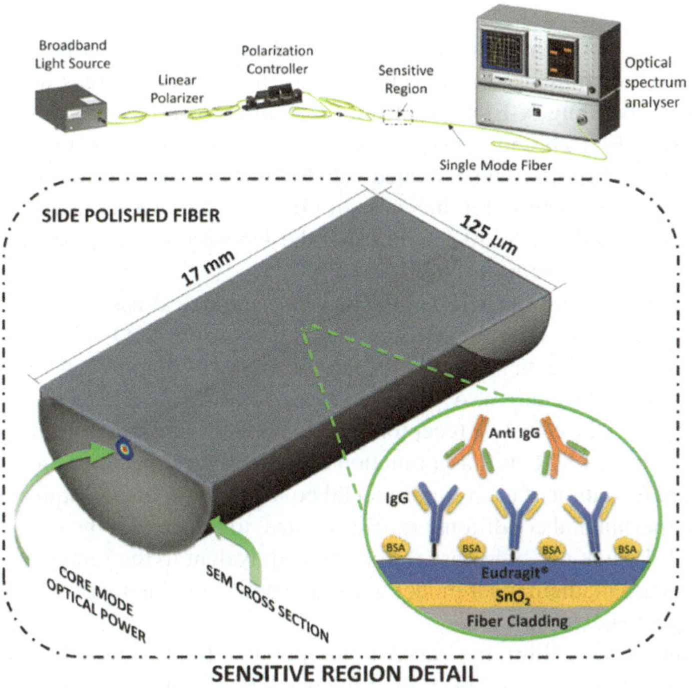


**Fig. 1** Experimental wavelength interrogation system consisting of a broadband light source, a linear polarizer, a polarization controller, an optical spectrum analyser and the sensitive region (D-shaped fibre biosensor). The sensitive region along the D-shaped fibre, the device structure and immunosensing protocol are also detailed

The functionalisation of the sensitive region was achieved by the deposition of a co-polymer (Eudragit L100, Evonik Degussa GmbH) that provided free functionalities, necessary for the antibody immobilization. The fibre was immersed in 2 mM (0.04% w/v) Eudragit L100 in ethanol for 1 min and then let the solvent evaporate in the air for about 15 min. Afterwards, the fibre was inserted inside a thermo-stabilised microfluidics. Reference [13] detailed all the steps performed for the preparation of the biological sensing layer, with the assay completed with the injection of the specific analyte at increasing concentrations from 1 ng $L^{-1}$ up to 10 mg $L^{-1}$. The biosensor specificity was evaluated by spiking the analyte in CRP-free human serum.

A TEM (CM 12 PHILIPS) equipped with an OLYMPUS Megaview G2 camera and a field emission SEM (FESEM, UltraPlus Carl Zeiss Inc.) were used to measure the film thickness. X-ray energy dispersive spectroscopy (X-EDS) tool of FESEM was also used to feature the chemical elements present in the deposited thin film.

## 3 Results and Discussion

The features of $SnO_2$ thin film were firstly studied. A high magnification FESEM image of the functionalised sensor is detailed in Fig. 2a, whereas a SEM image and the corresponding X-EDS spectrum taken in correspondence of the thin film are detailed in Fig. 2b and c, respectively. Figure 2d and e account for the TEM images of another similar sensor coated with $SnO_2$ film and $SnO_2$ + Eudragit layers, respectively. An averaged film thickness of (160 $\pm$ 5) nm was obtained for the first case and it increased up to (220 $\pm$ 5) nm after the deposition of the polymer, deducing a thickness of the polymer of roughly 60 nm.

The implementation of the IgG/anti-IgG immunological assay was then carried out to monitor in real-time the binding interactions using $SnO_2$-coated LMR fibre biosensors. The biochemical steps of the immunoassay were followed by tracking the shift of the LMR wavelength. A washing step in PBS was mandatory to measure the LMR shift due to the effective receptor-analyte bond. Figure 3 details the calibration curve showing the LMR shift as a function of the analyte concentration. The respective standard deviation of each experimental point (15 subsequent acquisitions under the same experimental conditions) is also detailed, together with the sigmoidal fit by using the Hill equation (blue curve), formally equivalent to the Langmuir isotherm, a well-accepted mathematical model used to quantify the interaction level between ligand binding sites.

By means of the calibration curve, it was possible to obtain the biosensor LOD. Considering the blank signal plus $3\sigma$ of the blank (0.18 nm), a LOD of 0.15 ng $L^{-1}$ (1 fM) was obtained; if $3\sigma$ of the maximum standard deviation obtained from all the experimental points (0.27 nm) is considered, a different LOD of 0.6 ng $L^{-1}$ (4 fM) was attained. Moreover, the performances were compared with other highest-performance fibre-based biosensors from the literature. Biosensors based on long period fibre gratings [13, 14] claimed a LOD of the order of $\mu$g $L^{-1}$ (tens of pM) or slightly less.

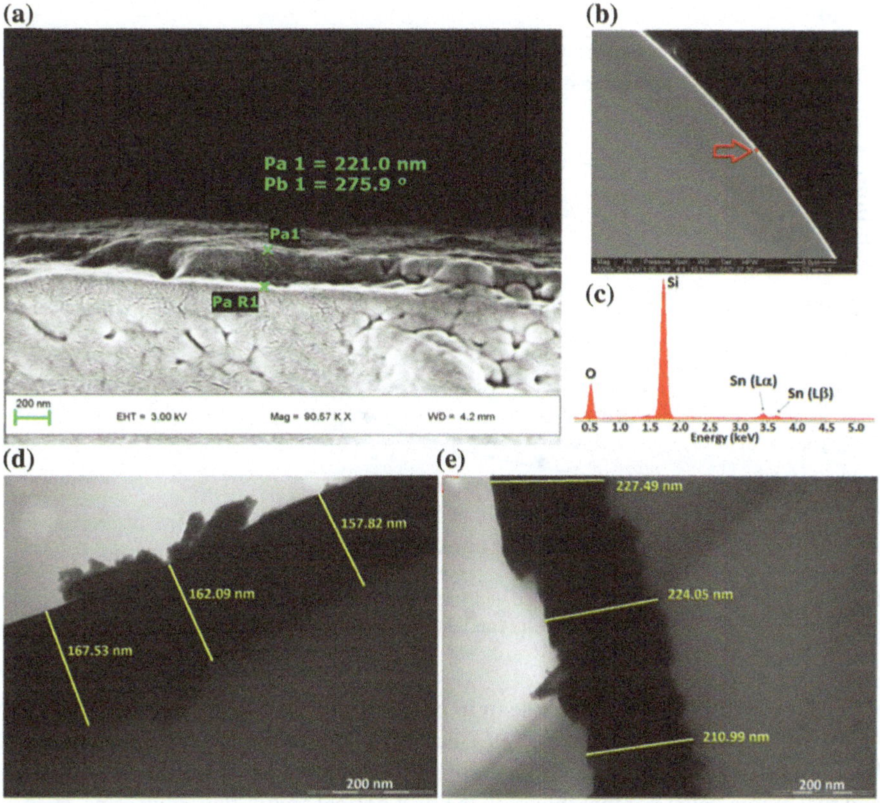


**Fig. 2** **a** FESEM image of the cross-section of a D-shaped fiber coated with both $SnO_2$ thin film and Eudragit layer. **b** SEM image under less magnification taken for X-EDS microanalysis. **c** X-EDS spectrum of the elements present in the thin film region detailing the spectral energy lines of interest. TEM images of the cross-section of a D-shaped fiber coated with $SnO_2$ film **d** and with Eudragit layer too (**e**)

## 4 Conclusion

We proposed an optically absorbing material, i.e. $SnO_2$, that was deposited on D-shaped single-mode fibres for generating lossy mode resonances, which are a recently-explored physical phenomenon that allowed the development of label-free biosensors able to obtain high sensitivity and resolution in measuring refractive index changes of the fibre-surrounding medium.

TEM and SEM imaging allowed tuning the LMRs in the selected wavelength range and, hence, using a standard experimental wavelength interrogation system in NIR. In addition, the proposed biosensor was integrated into a thermo-stabilised microfluidics, thus addressing all the typical requirements that a real biosensor must possess towards the development of an in situ, portable, lightweight and high-sensitivity optical device for biochemical and biomedical applications.

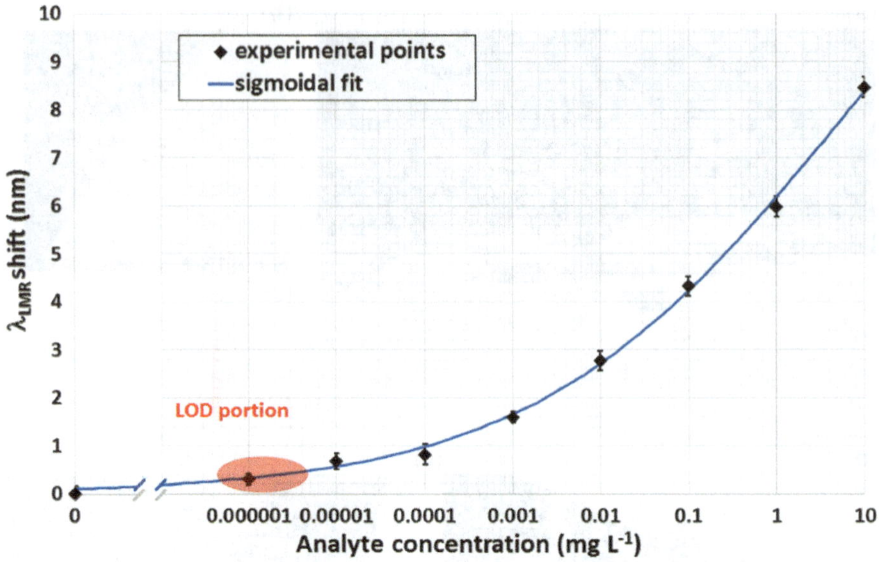


**Fig. 3** Calibration curve of the $SnO_2$-coated LMR-based D-shaped fibre biosensor, together with the sigmoidal fit of the experimental points and their respective standard deviation (black error bars)

The results have proved a big leap in performance thanks to the ability to detect analyte concentrations down to ng $L^{-1}$ (few fM) in human serum, enhancing the LOD by three orders of magnitude when compared with other fibre-based configurations and reaching a value comparable with the best optical technology platforms, such as SPR or localised SPR.

**Acknowledgements** This work was supported by the National Research Council of Italy (CNR) for the Short Term Mobility program 2017, by the Government of Navarra (project no. 72/2015) and by the Spanish Agencia Estatal de Investigaci6n (AEI) and European Regional Development Fund (FEDER) (TEC2016-78047-R,TEC2016-79367-C2-2-R).

## References

1. Rajan G (2015) Optical fiber sensors: advanced techniques and applications. CRC Press, Taylor & Francis Group, Chap. 1
2. Chiavaioli F, Baldini F, Tombelli S, Trono C, Giannetti A (2017) Biosensing with optical fiber gratings. Nanophotonics 6
3. Caucheteur C, Guo T, Liu F, Guan B-O, Albert J (2016) Ultrasensitive plasmonic sensing in air using optical fibre spectral combs. Nature Communications 7:13371
4. Stern L, Desiatov B, Mazurski N, Levy U (2017) Strong coupling and high-contrast all-optical modulation in atomic cladding waveguides. Nat Commun 8:14461
5. Gramotnev DK, Bozhevolnyi SI (2010) Plasmonics beyond the diffraction limit. Nat Photonics 4:83–91

6. Andreev A, Pantchev B, Danesh P, Zafirova B, Karakoleva E, Vlaikova E, Alipieva E (2005) A refractometric sensor using index-sensitive mode resonance between single-mode fiber and thin film amorphous silicon waveguide. Sens Actuators B Chem 106:484–488
7. Del Villar I, Arregui FJ, Zamarreño CR, Corres JM, Bariain C, Goicoechea J, Elosua C, Hernaez M, Rivero PJ, Socorro AB, Urrutia A, Sanchez P, Zubiate P, Lopez D, De Acha N, Ascorbe J, Matias JR (2017) Optical sensors based on lossy-mode resonances. Sens Actuators B Chem 240:174–185
8. Yang F, Sambles JR (1997) Determination of the optical permittivity and thickness of absorbing films using long range modes. J Modern Opt 44:1155–1164
9. Del Villar I, Hernaez M, Zamarreño CR, Sánchez P, Fernández-Valdivielso C, Arregui FJ, Matías IR (2012) Design rules for lossy mode resonance based sensors. Appl Opt 51:4298–4307
10. Homola J (2008) Surface plasmon resonance sensors for detection of chemical and biological species. Chem Rev 108:462–493
11. Chiavaioli F, Zubiate P, Del Villar I, Zamarreño CR, Giannetti A, Tombelli S, Trono C, Arregui FJ, Matias JR, Baldini F (2018) Femtomolar detection by nanocoated fibre label-free biosensors. ACS Sens 3:936–943
12. Chiavaioli F, Gouveia C, Jorge P, Baldini F (2017) Towards a uniform metrological assessment of grating-based optical fiber sensors: from refractometers to biosensors. Biosensors 7:32
13. Chiavaioli F, Biswas P, Trono C, Jana S, Bandyopadhyay S, Basumallick N, Giannetti A, Tombelli S, Bera S, Mallick A, Baldini F (2015) Sol–gel-based titania–silica thin film overlay for long period fiber grating-based biosensors. Anal Chem 87:12024–12031
14. Quero G, Consales M, Severino R, Vaiano P, Borriello A, Diodato L, Zuppolini S, Giordano M, Nettore IC, Mazzarella C, Colao A, Macchia PE, Santorelli F, Cutolo A, Cusano A (2016) Long period fiber grating nano-optrode for cancer biomarker detection. Biosens Bioelectron 80:590–600

# Test Platform for Data Fusion Application in Indoor Positioning

**M. Carratù, S. Dello Iacono, M. Ferro, V. Paciello and A. Pietrosanto**

**Abstract** A low-cost platform for development and testing of attitude and positioning algorithms is presented. In addition to an IMU a barometer is added to the data fusion to further improve the quality. An array of MEMS microphones and laser sensor is the key measurement instrument for the control strategy. The platform is accompanied by the implementation of a 10 DoF AHRS algorithm used to improve the movements in indoor environment.

**Keywords** Inertial Measurement Units (IMU) · Attitude Heading Reference Systems (AHRS) · Data fusion · Micro Electro-Mechanical Systems (MEMS)

## 1 Introduction

Inertial Measurement Units (IMUs) are commonly used in industrial and consumer applications, ranging from automotive, robotics, wearable devices for entertainment or sport, in the monitoring of medical environments [1] and even in autonomous surveillance systems. An IMU is primary composed by accelerometer, gyroscope and magnetometer, they are mainly Micro Electro-Mechanical Systems (MEMS) sensor. This facilitates the integration in IoT application where power consumption represents a problem [2].

The sensors included in the IMU are already vastly adopted in several field like automotive where they can be used as a replacement for mechanical sensors [3, 4], or even as complement to pre-existing sensors to implement fault diagnosis [5, 6]. The fusion of measurements of acceleration data, gyroscope and magnetometer sensor further improves the quality of measurement and can bring information about motion and orientation. It is possible to evaluate the relative position between the platform and the motion direction [7, 8]. Different advanced data fusion techniques [9, 10]

M. Carratù · S. Dello Iacono (✉) · M. Ferro · A. Pietrosanto
Department of Industrial Engineering, University of Salerno, Fisciano, Italy
e-mail: delloiaconos@unisa.it

V. Paciello
Department of Information and Electrical Engineering, University of Cassino and Southern Lazio, Cassino, Italy



G. Di Francia et al. (eds.), *Sensors and Microsystems*, Lecture Notes in Electrical Engineering 629, https://doi.org/10.1007/978-3-030-37558-4_50

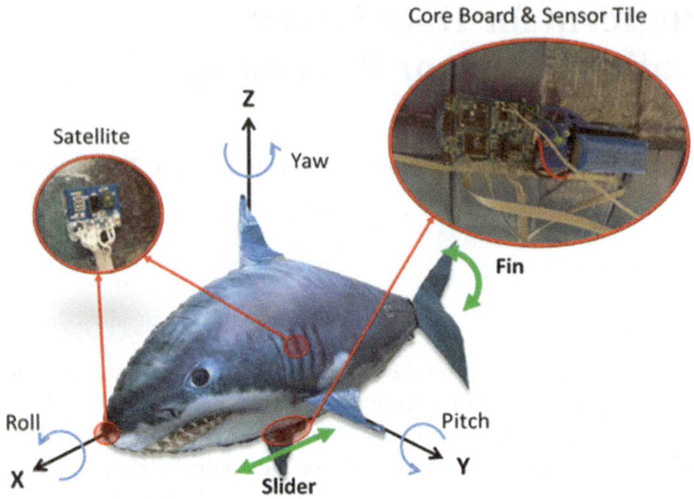


**Fig. 1** Free flying balloon, shaped like a shark, equipped with sensors and custom boards

are used in most of these applications to calculate accurate measurements regarding relative spatial positioning.

In addition to the IMU, a digital barometer could be integrated in an IMU to obtain an absolute measurement of altitude. The common implementation of Attitude Heading Reference System (AHRS) with only gyroscopes, accelerometers and magnetometers is unable to give information about the absolute position of a device in the space, but with the addition of the barometer it is possible to obtain information about the absolute altitude.

The authors propose a low-cost platform for the implementation and validation of Authors propose a low-cost platform for the implementation and validation of algorithms for indoor navigation and target following scenarios of critical devices where a common Global Positioning System (GPS) is not available.

More in details, raw data from digital sensors have been fused to compute the attitude of a free-flying balloon [11] (see Fig. 1) equipped with a custom low-cost board.

## 2 Test Platform

The free-flying balloon adopted as test bed has the appearance of a shark (as depicted in Fig. 1). It is equipped with two actuators: the first one allows the movement of the caudal fin for the forward propulsion and the second one shifts a weight under the trunk for the fixing of the altitude.

The IMU platform used is located on the board attached to the bottom of the balloon. It contains a tri-axial accelerometer and gyroscope LSM6DS, tri-axial

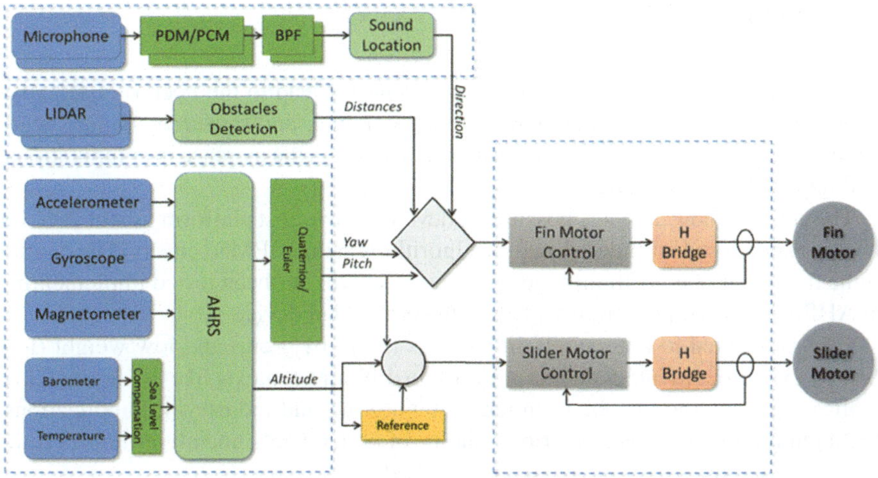


**Fig. 2** Details of the data fusion and target following algorithms

magnetometer LSM303AGR and a LPS22HB pressure sensor, all the sensors are manufactured by STMicroelectronics in an integrated evaluation board.

An AHRS algorithm with 9 Degrees of Freedom (DoF) has been implemented [11] using the data coming from tri-axial accelerometer, tri-axial gyroscope, tri-axial magnetometer. The absolute pressure sensor has been integrated reaching 10 DoF (see Fig. 2).

As reported in Fig. 2, an array of digital MEMS microphone has been integrated in the algorithm to obtain target following capability: a predefined sound source can be localized and followed using an algorithm based on the RMS value of the received signal.

Multiple Time of Flight ranging sensors have been added to the prototype; more in details six VL53L0X sensor by STMicroelectronics are placed in couples of two on the three sides of the balloon and connected to the control board. They are used for obstacle detection in conjunction with the tracking algorithm. For the elaboration of data coming from all the presented sensors and for the management of the both communication and control strategy, a STM32L4 ARM Cortex-M4 based microcontroller has been used, it is mounted on the control board together with the sensors.

When the target following strategy is enabled, then altitude and relative target position are used to control the prototype flight direction. On the other hand, if target following strategy is not enabled, the prototype follows a predefined route using only data coming from the modified AHRS. In both cases, a supervision algorithm avoids possible obstacles like walls. When the shark does not detect any target, it is capable of floating in a stationary position, reacting to light wind and fluctuations thanks to the 9 DoF AHRS algorithm enhanced by the addition of the barometer.

## 3 Conclusion

The shielding of the GPS signal in indoor scenarios lead to the need of alternative strategies for the absolute positioning using complex and resources hungry algorithms. The presented low-cost platform well deals with the increase in computational load required by this application.

The authors show how it is possible to develop a low-cost platform (about 20$) for the test of enhanced attitude estimation algorithms using MEMS sensors. The control of the flying balloon positioning in an indoor scenario has been tested implementing an AHRS proposed in literature [12] on the embedded device.

Care is devoted to both small dimensions (3 cm by 4 cm) and low weight (less than 50 g) in order to integrate this platform in different contexts like drones, medical applications, automotive, environmental monitoring and industry 4.0 applications. The attention to low power in term of choice of both devices and algorithms permits a longer time duration of the presented test bed.

The IMU platform alone gives information about the movement but the inclusion of an absolute reference as the barometer allows a more precise estimation of the position that can be used in the control strategy.

## References


1. Carratù M, Pietrosanto A, Sommella P, Paciello V (2019) A wearable low-cost device for measurement of human exposure to transmitted vibration on motorcycle. In: 2019 IEEE International Workshop on Metrology for Industry 4.0 and IoT, Naples
2. Carratù M, Dello Iacono S, Hoang ML, Pietrosanto A (2019) Energy characterization of attitude algorithms. In: IEEE 17th international conference on industrial informatics INDIN 2019, Helsinki
3. Carratù M, Pietrosanto A, Sommella P, Paciello V (2018) Semi-active suspension system for motorcycles: From the idea to the industrial product. In: 2018 IEEE international instrumentation and measurement technology conference: discovering new horizons in instrumentation and measurement, Houston, May 2018
4. Carratù M, Pietrosanto A, Sommella P, Paciello V (2018) Measuring suspension velocity from acceleration integration. In: IEEE 16th international conference on industrial informatics, INDIN, Porto, July 2018. https://doi.org/10.1109/indin.2018.8472039
5. Capriglione D, Carratù M, Pietrosanto A, Sommella P, Online fault detection of rear stroke suspension sensor in motorcycle. IEEE Trans Instrum Meas 68(5), 1362–1372. https://doi.org/10.1109/tim.2019.2905945
6. Capriglione D, Carratù M, Liguori C, Paciello V, Sommella P, A soft stroke sensor for motorcycle rear suspension. Meas J Int Meas Confed 106:46–52. https://doi.org/10.1016/j.measurement.2017.04.011
7. Carratù M, Dello Iacono S, Paciello V, Pietrosanto A (2019) IMU self-alignment in suspensions control system. In: 2019 international instrumentation and measurement technology conference, I2MTC 2019, pp 861–866, Auckland, May 2019
8. Carratù M, Dello Iacono S, Paciello V, Pietrosanto A (2019) Self-alignment procedure for IMU in automotive context. In: IEEE 17th international conference on industrial informatics INDIN 2019, Helsinki

9. Fortuna L, Graziani S, Rizzo A, Xibilia M (2007) Soft sensors for monitoring and control of industrial processes. Springer Book
10. Diez-Olivan A, Del Ser J, Galar D, Sierra B (2018 Oct) Data fusion and machine learning for industrial prognosis: trends and perspectives towards industry 4.0. Inf Fusion 50. https://doi.org/10.1016/j.inffus.2018.10.005
11. Carratú M, Ciavolino G, Dello Iacono S, Ferro M (2018) Free-flying balloon used as test bed for sensor data fusion application in indoor navigation and target following. In: IEEE international sensors and measurement student contest, Houston, May 2018
12. Madgwick SOH, Andrew J, Harrison L, Vaidyanathan R (2011) Estimation of IMU and MARG orientation using a gradient descent algorithm. In: 2011 IEEE international conference on rehabilitation robotics

# Design and Implementation of a Diagnostic Scheme for Stroke Sensors in Motorcycle Semi-active Suspension Systems

**D. Capriglione, M. Carratù, S. Dello Iacono, A. Pietrosanto, P. Sommella and M. Avoci Ugwiri**

**Abstract** Today the reliability of electronic systems strictly depends on the correct operations of the sensor involved in the normal use. This fact is more evident in applications where the security and safety of end-users are involved, reason why a suitable Instrument Fault Detection Scheme (IFD) including also Isolation feature (IFDI) becomes fundamental. The automotive field is one of the main areas where an IFDI scheme is mandatory. As an example, systems designed to on-line adapt and to electronically control the suspensions of motorcycles are today of great interest for motorcycle and after-market manufacturers. By the way meanwhile semi-active suspension systems can drastically improve the comfort and the traction performance of motorcycle, a promptness detection of faults involving this kind of system is fundamental for the safety and performance of the motorcycle. With this aim, the paper proposes an Instrument Fault Detection and Isolation (IFDI) scheme using analytical redundancy for the fault diagnosis of the front and rear stroke suspension sensors. Both suitable mathematical links and soft sensors based on artificial neural networks are proposed for the residuals generation in order to design and validate the proposed IFDI scheme.

**Keywords** IFDI · Semi-active suspension system · ANN · Soft sensor

## 1 Introduction

Continuously we are concerned by the diagnosis of man-made physical systems characterized by a continuous type operation. Such a system includes industrial production equipment (power station, brewing industry, chemical plants, etc.) transportation vehicles (automobiles, ships, airplanes, etc.). The components in the system susceptible to be in failure or malfunction state are parts of equipment proper as well as sensors and control actuators attached to it. Very often, the diagnosis task consists

D. Capriglione · M. Carratù (✉) · S. Dello Iacono · A. Pietrosanto · P. Sommella · M. Avoci Ugwiri
Department of Industrial Engineering, University of Salerno, Via Giovanni Paolo II, 132 Fisciano, SA, Italy
e-mail: mcarratu@unisa.it

 
G. Di Francia et al. (eds.), *Sensors and Microsystems*, Lecture Notes in Electrical Engineering 629, https://doi.org/10.1007/978-3-030-37558-4_51

to detect and isolate the malfunction [1–4]. Thanks to the increasing of microcontrollers and signal processing devices, smart computing and artificial intelligence it is henceforward possible to improve the efficiency of the detection using the on-line Fault Detection (FD) and Instrument Fault Detection and Isolation (IFDI). However, the challenge still remains in the reliability and the fault tolerance [5–8], since most fault tolerances required redundancy in the hardware components in order to make the systems more reliable [9–11]. In [12, 13] an approach has been undertaken with the aim of improving in a motorcycle the comfort and the traction performance in both racing and original equipment manufacturer applications. This paper proposes the extension of the diagnostic procedure to the rear suspension stroke sensor. A complete scheme based on analytical redundancy and plausibility checks was proposed. A suitable test campaign has been carried out in one hand for detecting different kind of fault which symptom and time dynamic are different. In another hand, the test evaluated the performance of the proposed IFDI scheme.

## 2 Diagnosis Scheme: Methods and Results

Since insured system integrity is a requirement, systems need to be fault tolerant, reliable and safe. To set up a fault tolerant system, it must be able to recognize faults sensible and fast. This is possible by the plausibility checking. In contrast, analytical redundancy provides higher potential in term of precision and detection time. This paper presents both approaches due to their complementarity.

a. *Plausibility checks*

In general, plausibility checks enable the controller to take adequate error handling measure depending on the safety relevance of the system and the severity of the fault. In automotive IFDI, they are typically adopted to verify the electrical consistency of the sensor signal by considering the output range. In this draft, Eqs. (1) and (2) proposed a mathematical model for the detection and isolation of the short circuit ($F_{sc}$) and open circuit fault ($F_{oc}$), for each stroke sensor $S_{\mathrm{i}}$ (with i = 1, 2).

$$F_{SC,Si} = 1\ if: \sum_{n=1}^{k} V_{Si}(T_d - nT)/k > V_{L,Si} - 3\Delta_q \tag{1}$$

$$F_{OC,Si} = 1\ if: \sum_{n=1}^{k} V_{Si}(T_d - nT)/k < V_{H,Si} - 3\Delta_q \tag{2}$$

In the above equations, $V_{L,Si}$ and $V_{H,Si}$ are the extrema of the sensor output range. $\Delta_q$ is the quantization error and $T_d$ the detection time. It is important to notice that the digital block computes the derivative $V_{Si}^{'}$ of the stroke sensor output and the number of the thresholds exceeding during the observation interval $T_{obs}$. Equation (3) establish the necessary condition on $T_{obs}$.

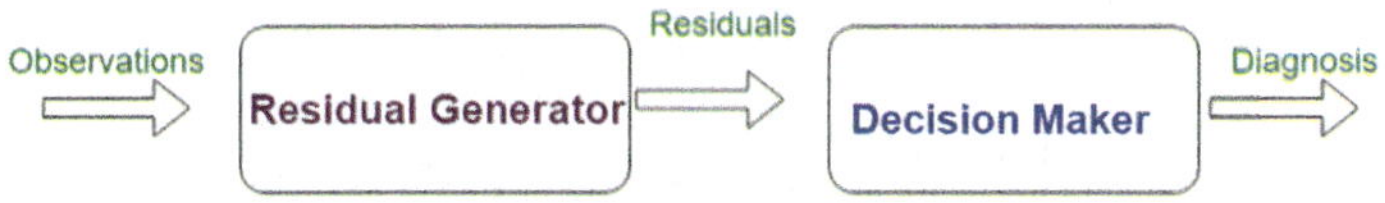


**Fig. 1** Consistence of analytical redundancy

$$F_{Aging,s_i=1} \text{ if } \begin{cases} \left|V'_{s_i}(n_0T)\right| > \mathrm{V}'_{s_i,MAX} \\ \exists n \geq \mathrm{n}: \left|V'_{s_i}[(n_0+k)T]\right| \mathrm{V}'_{s_i,MAXmax} \\ and\ \mathrm{k}_iT < \mathrm{T}_{obs}, \forall i \leq n \end{cases} \quad (3)$$

Regarding the above condition, the faulty sensor is isolated within the detection time $k_{max}$.

b. *Analytical redundancy*

Very often, diagnosis algorithm based on analytical redundancy consist of two blocks: residual generator and decision maker, shown in Fig. 1.

In this paper, the residual generator compares the prediction $S_{1,m}$ and $S_{2,p}$ of the software based on the NARX neural networks with the outputs $S_{1,m}$ and $S_{2,m}$ of the corresponding sensors. A technique based on moving the average is employed for computing more accurate residuals:

$$E_{mean_1,L}(i) = \frac{1}{L_s}\sum_{k=0}^{L_S-1}\left|\frac{S_{1,p}(i-k) - S_{1,m}(i-k)}{S_{1,m(i-k)}}\right| \quad (4)$$

$$E_{mean_2,L}(i) = \frac{1}{L_s}\sum_{k=0}^{L_S-1}\left|\frac{S_{2,p}(i-k) - S_{2,m}(i-k)}{S_{2,m(i-k)}}\right| \quad (5)$$

$L_s$ represents the number of samples including in the moving window of the length L. Actually, residuals are quantities that presents the inconsistency between the actual variables and the mathematical model. They are ideally equal to zero. The plant observables include the measurable values for measured plant variables and command values for the controlled inputs.

Proposed Decision maker detected and isolated the losing calibrated fault when the corresponding block exceed a fixed threshold $Th\%$ longer than an integer multiple $n_{uncal}$ of the sliding window $L$ according to the following relations:

$$F_{uncal,s_1}(iT) = 1 \text{ if } \begin{cases} E_{mean_1,L}(jT) > Th_{\%} \\ \forall j \in [i - (n_{uncal} \times Ls); \mathrm{i}] \\ E_{mean_2,L}(jT) > Th_{\%} \\ \forall j \in [i - (n_{uncal} \times Ls); \mathrm{i}] \end{cases} \quad (6)$$

$$F_{uncal,s_2}(iT) = 1 \text{ if } \begin{cases} E_{mean_2,L}(jT) > Th_{\%} \\ \forall j \in [i - (n_{uncal} \times \text{Ls}); \text{i}] \\ E_{mean_2,L}(jT) > Th_{\%} \\ \exists i \in [i - (n_{uncal} \times \text{Ls}); \text{i}] \end{cases} \quad (7)$$

## 2.1 Proposed Scheme

Based on methods described previously, Fig. 2 presents for automatic detection and isolation of different faults such as: *open circuit*, *short circuit*, *aging* and *losing calibration* a meaningful scheme which easily should act on the operating behavior of the front and rear stroke sensor in terms of highlighted symptoms. Both approaches analytical redundancy and plausibility checks are adopted according to different level of fault danger and desired detection time.

It should be noted that the plausibility checks use also to be adopted for detecting and isolating the aging fault for each stroke sensor by verifying the consistency of the stroke velocity with the expected range.

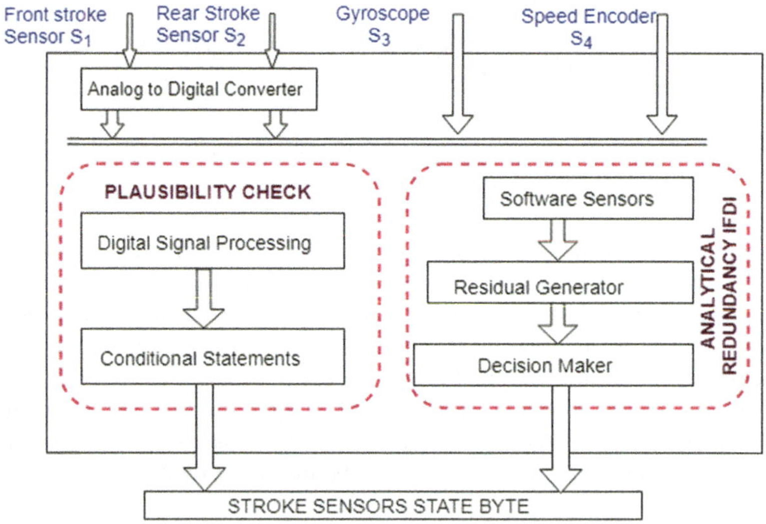


**Fig. 2** The proposed IFDI scheme for the suspension stroke sensors

## 2.2 Overview of the Experimental Results

For the purpose of performance analysis and validation of the proposed scheme, a SUZUKI GSX-1000 equipped with a suitable semi-active suspensions system was used. Linear potentiometers used are the SLS3130 Penny and Giles, one gyroscope of type ST L3GD20 and one velocity encoder of type Dorman 970–711, were used for this study. A Controller Area Network (CAN) operating at 1 Mbps was used for acquiring signals provided by sensors. Signals were recorded on 40 km of travelling with a sampling frequency of 100 Hz. For all faults, in order to statistically analyze the performance of the proposed scheme, a number of 100 trials have been realized for each considered fault, generated each time by randomly varying the instant at which the fault was applied.

In Fig. 3, Correct decision percentage (CD%), False alarm percentage (FA%) and Missed Detection percentage (MD%) are evaluated for each kind of fault.

Some observations can be noted from these results:

- Detecting and isolating *open circuit*, *short circuit* and *aging* faults thanks to the proposed scheme is effectively possible. The reliability has high percentage since it is always 100%, FA % and MD % are also at 0%.
- As for the *losing calibration* faults, as expected, the higher the input/output curve slope variation, the higher the CD % and the lower MD %. Moreover, the performance achieved for $S_1$ and $S_2$ are very similar.

As far as the dynamic performance is concerned, Fig. 4 shows $t_{IFDI}$ value for each considered fault by reporting the mean value ($\mu$), the standard deviation ($\sigma$) and maximum value ($M$) observed for $t_{IFDI}$.

As expected, chosen window length of 50 samples allows the detection of short circuit and open circuit, this means $t_{IFDI} = 0.5$ s. Highest variability of $t_{IFDI}$ tell

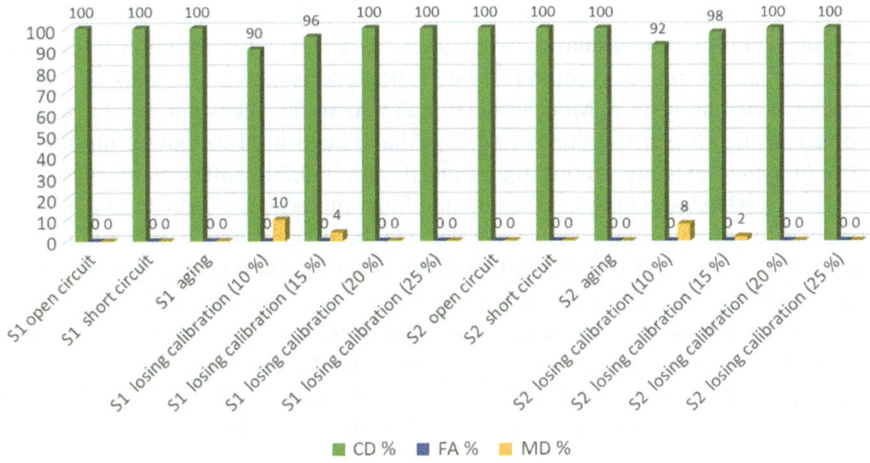


**Fig. 3** Diagnosis performance of the proposed IFDI scheme

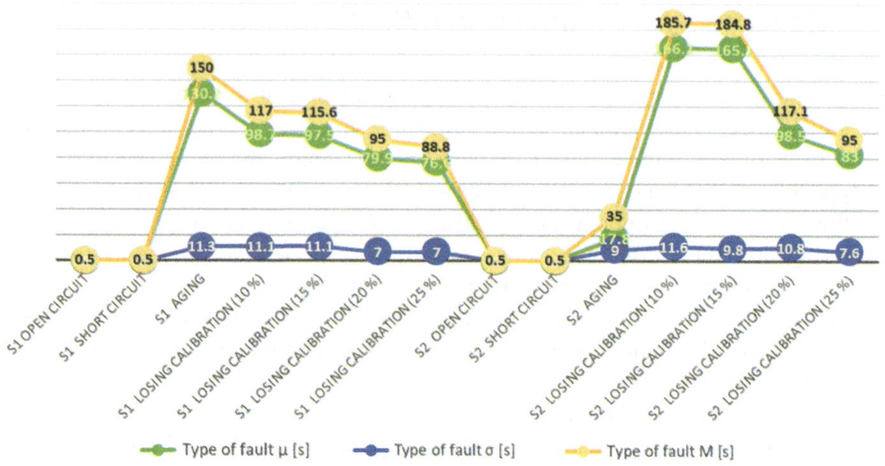


**Fig. 4** Time performance of the proposed scheme

more generally about *aging* and *losing calibration* faults. *Aging* faults are detected and isolated with $t_{IFDI} \leq 150$ s and $t_{IFDI} \leq 35$ s for $S_1$ and $S_2$ respectively. *losing calibration* faults show $t_{IFDI}$ values which decrease (down to maximum values of 88.8 s and 95.0 s for $S_1$ and $S_2$, respectively) as the entity of the fault increases (i.e. 25%) whereas the detection and isolation require more time (up to maximum values of 117.0 s and 185.7 s for $S_1$ and $S_2$, respectively) when smaller losing calibrations (i.e. 10%) are considered.

## 3 Conclusion

The proposed IFDI scheme shows two main paths designed for accomplish different sensitivity and promptness required for detecting and isolating the faults in a semi-active suspensions system. More in details, the quantities measured by the sensors involved in the suspension control strategy have been used to extract the analytical redundancy existing among them and design suitable soft sensors (based on NARX networks) for generating the residual functions adopted in the fault diagnosis scheme. The experimental results have verified the diagnostic and dynamic performance against different typology of faults proving the very good behavior of the proposed IFDI procedure. The scheme has shown the correct detection and isolation of the faults in all considered situations (with Correct Decision percentage approaching 100%) in an acceptable delay for the kind of faults considered. As an example, open and short circuit faults were always detected and isolated up to 0.5 s, whereas aging and un-calibration faults were always detected up to 200 s.

## References

1. Catelani M, Ciani L, Luongo V, Singuaroli R (2010) Evaluation of the safe failure fraction for an electromechanical complex system: remarks about the standard IEC61508. In: 2010 IEEE instrumentation and measurement technology conference proceedings, Austin, TX, pp 949–953. doi:10.1109/IMTC.2010.5488034
2. Catelani M, Ciani L, Luongo V (2012) A new proposal for the analysis of safety instrumented systems. In: 2012 IEEE international instrumentation and measurement technology conference proceedings, Graz, pp 1612–1616. doi:10.1109/I2MTC.2012.6229556
3. Carratù M, Pietrosanto A, Sommella P, Paciello V (2017) Velocity prediction from acceleration measurements in motorcycle suspensions. In: 2017 IEEE international instrumentation and measurement technology conference (I2MTC), Turin, pp 1–6. doi:10.1109/I2MTC.2017.7969943
4. Betta G, Pietrosanto A (2000) Instrument fault detection and isolation: state of the art and new research trends. IEEE Trans Instrum Meas 49(1):100–107. https://doi.org/10.1109/19.836318
5. Capriglione D, Carratù M, Pietrosanto A, Sommella P (2018) Real-time implementation of an IFD scheme for motorcycle sensors. In: 2018 IEEE international instrumentation and measurement technology conference (I2MTC), Houston, TX, pp 1–6. doi:10.1109/I2MTC.2018.8409833
6. Capriglione D, Carratù M, Pietrosanto A, Sommella P (2018) NARX ANN-based instrument fault detection in motorcycle. Meas J 117:304–311. https://doi.org/10.1016/j.measurement.2017.12.026
7. Capriglione D, Carratù M, Liguori C, Paciello V, Sommella P, A soft stroke sensor for motorcycle rear suspension. Meas J 106:46–52. https://doi.org/10.1016/j.measurement.2017.04.011
8. Carratù M, Pietrosanto A, Sommella P, Paciello V (2018) Semi-active suspension system for motorcycles: From the idea to the industrial product. In: 2018 IEEE international instrumentation and measurement technology conference (I2MTC), Houston, TX, pp 1–6. doi:10.1109/I2MTC.2018.8409829
9. Gertler J (1991) Analytical redundancy methods in fault detection and isolation. School of Information Technology and Engineering, George Mason University. Germany
10. Dai X, Gao Z (2013) From model signal to knowledge: a data-driven perspective of fault detection and diagnosis. IEEE Trans Ind Informat 9(4):2226–2238
11. Venkatasubramanian V, Rengaswamy R, Kavuri SN (2003) A review of process fault detection and diagnosis: Part II: Qualitative models and search strategies. Comput Chem Eng 27(3):313–326
12. Capriglione D, Carratù M, Pietrosanto A, Sommella P (2019) Online fault detection of rear stroke suspension sensor in motorcycle. IEEE Trans Instrum Meas 68(5):1362–1372. https://doi.org/10.1109/TIM.2019.2905945
13. Capriglione D, Carratù M, Pietrosanto A, Sommella P (2018) Analytically redundancy based IFDI scheme for semi-active suspension systems in motorcycle. J Phys Conf Ser

# A New Conditioning Method for PZT-Based Microsensors

A. Fort, C. Trigona, E. Panzardi and V. Vignoli

**Abstract** A conditioning circuit able to simultaneously actuate and measure the output of an integrated MEMS cantilever mass-sensor fabricated in PiezoMUMPs technology is presented. The MEMS is a resonator with an embedded aluminum nitride (AlN) layer. The basic idea is to use the active layer properties to excite the device and simultaneously read the mechanical response by detecting the resonant frequency. In order to do this a suitable conditioning system has been developed, and a hybrid solution based on microelectronic devices has been realized. The proposed measurement system allows to detect the main parameters characterizing the electro-mechanical device response in real time, allowing the monitoring of the sensed quantity.

**Keywords** Mass sensing · AlN based MEMS · Cantilever resonator

## 1 Introduction

Resonant sensors and oscillating sensing elements have been applied to many and different application fields in recent years [1]. In particular, resonant mass sensors for chemical and bio-sensing have proved to be one of the most successful solutions, especially exploiting micro-scale systems [2] and MEMS technology. It should be noted that this family of sensors find applicability in various fields including industrial applications, e.g. chemical, biomedical, automotive, etc. As known MEMS resonant mass sensors are based on a mechanical oscillator, such as out-of-plane beam, bridge, and more generally second order mass-spring-damper systems [2]. The method of measurement, in order to estimate the measurand (i.e. mass, viscosity, temperature, etc.), is to evaluate the natural mechanical frequency of the resonator and how it

A. Fort · E. Panzardi (✉) · V. Vignoli
D.I.I.S.M., Dipartimento di Ingegneria dell'Informazione e Scienze Matematiche, University of Siena, Via Roma 56, 53100 Siena, Italy
e-mail: panzardi@diism.unisi.it

C. Trigona
D.I.E.E.I., Dipartimento di Ingegneria Elettrica Elettronica e Informatica, University of Catania, Viale Andrea Doria 6, 95125 Catania, Italy



G. Di Francia et al. (eds.), *Sensors and Microsystems*, Lecture Notes in Electrical Engineering 629, https://doi.org/10.1007/978-3-030-37558-4_52

changes as function of the physical quantity to be measured [2, 3]. It is worth noting that several mechanisms can be used to perform the conversion of the mechanical vibration into an electric signal, and/or vice versa to excite the mechanical structure through an electrical signal, including electromagnetic, capacitive, piezoresistive, piezoelectric, etc. [4–7]. In particular, in the perspective to implement a low-power sensor, simple, and able to be electrically excited and, at the same time, to measure the target quantity, a self-sustained approach based on active materials (i.e. piezoelectric) is particularly attractive. In this context, some recent process has been proposed in MEMS technology in order to realize microsensors based on aluminum nitride (AlN) layers, as piezoelectric active material. The literature presents several works on such sensors but, very often, the active layer is used as sensing embedded element to transduce the vibration of the oscillator into an output voltage, whereas the actuation is realized by using external transducers, such as piezoceramic solutions or shaker-based platforms [4–7]. It should be noted that typically the signals, during the sensing procedures, are very weak and the idea to excite the system and to read its output (with the same layer) during its mechanical transient is something interesting.

In this work the feasibility of this method is shown and, on the basis of the results obtained by authors in [8], an autonomous mass measurement system based on a MEMS resonance structure and microelectronic device is proposed.

## 2 Measurement System Setup and Working Principle

The basic idea of the proposed measurement system is to exploit the piezoelectric characteristics of the resonant structure embedded in a MEMS device in order to implement a "self-generating" system. Indeed, an AlN layer embedded in the resonant structure allows for the simultaneous actuation and sensing of the electromechanical response of the device [8, 9]. The mass sensing is obtained exploiting a cantilever beam structure embedded in the device which presents a natural frequency of about 200 Hz. A change in the resonant frequency $\Delta f$ indicates a change in the cantilever system mass according to the well-known relation of the resonant structures as:

$$\Delta f = \frac{1}{2\pi}\sqrt{\frac{k}{m + \Delta m}}. \tag{1}$$

where $k$ and $m$ are the spring constant and the effective mass of the cantilever, respectively and $\Delta m$ indicate the occurring mass change.

The proposed measurement system is summarized in Fig. 1 and it consists of three main parts such as a microcontroller unit (MCU), the device under test (DUT), which is the MEMS structure, and a properly designed conditioning electronics.

The microcontroller device is the STM32L476 board, it contains a 32-bit microprocessor with a maximal clock frequency of 80 MHz. In particular, the microprocessor has been used to generate a voltage pulse signal $V_p$ via the DAC (12-bit) to excite

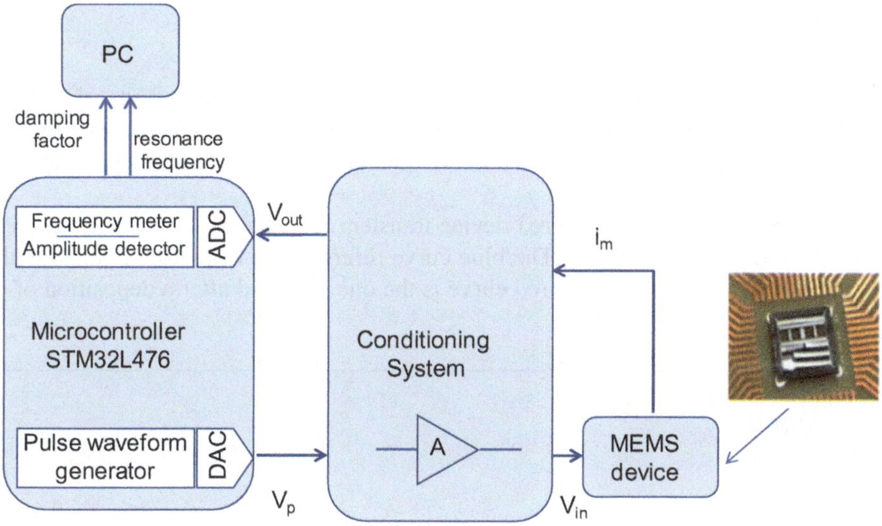


**Fig. 1** Schematic of the proposed measurement system architecture

the device resonant transient response and to acquire the MEMS output via a precise 12-bit ADC. The sampling frequency is set to 100 kHz and frames with length of 0.3 s are acquired and processed. The data processing includes the frequency measurement of the device transient response and its damping factor. The processing method is based on the enhanced features of the synchronization mechanism of the MCU timers and on the MCU implemented software algorithms for the waveform peak detection.

The conditioning system contains a first stage (A), used to properly amplify the voltage signal $V_p$ to be applied to the DUT ($V_{in}= AV_p$), and a dedicate front-end electronics which transduces the generated piezoelectric current $i_m$ into a voltage signal $V_{out}$.

The mass sensing features of the device has been deeply investigated by authors in [8], the device has a mass sensitivity of about 1 Hz/10 μg and a very stable and repeatable response with a resonant frequency standard deviation of about 0.004 Hz and 0.038 Hz for short and long term, respectively.

Accordingly, by setting the MCU timer clock (*TIM2CLK*) to 8 MHz it is enough to guarantee a minimum measured frequency $f_m$ without a timer counter overflow given by

$$f_m = \frac{TIM2CLK}{N_c} \approx 0.002\,\text{Hz}; \tag{2}$$

where $N_c= 2^{32}$ and indicate the length of the timer counter.

## 3 Measurement Results

The system provides the possibility to perform the real time monitoring (1 s) of the device resonance frequency and the transient response damping factor. In Fig. 2 is reported the device resonance frequency during approximately 6 min of measurement.

Whereas in Fig. 3 the acquired device transient response is reported in the frequency domain for two cases. The blue curve refers to the response obtained with the pristine device whereas the red curve is the one obtained after a deposition of a

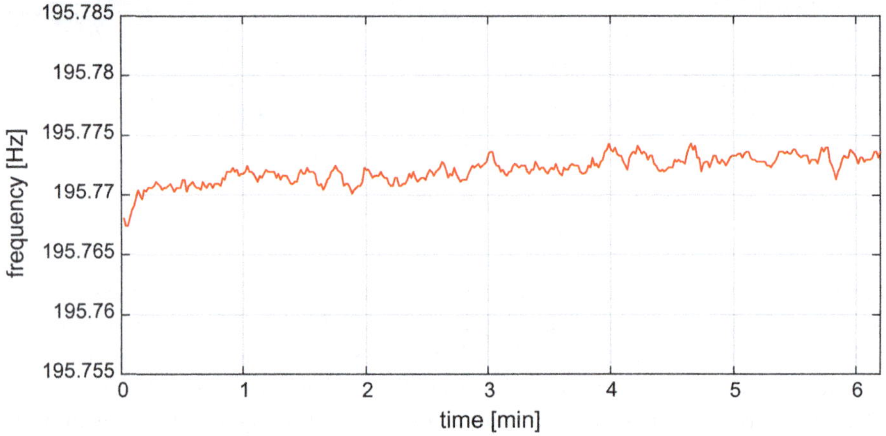


**Fig. 2** Measured resonance frequency as a function of time

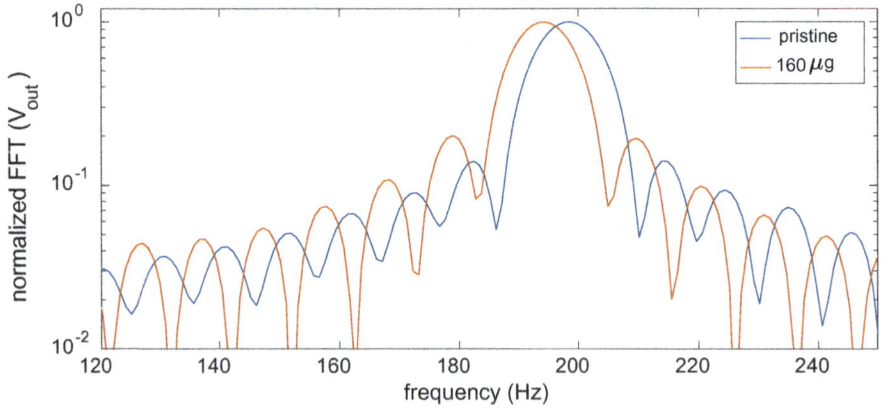


**Fig. 3** The normalized FFT of the pristine device response compared to the one with a deposited mass of 160 μg at the end of the beam

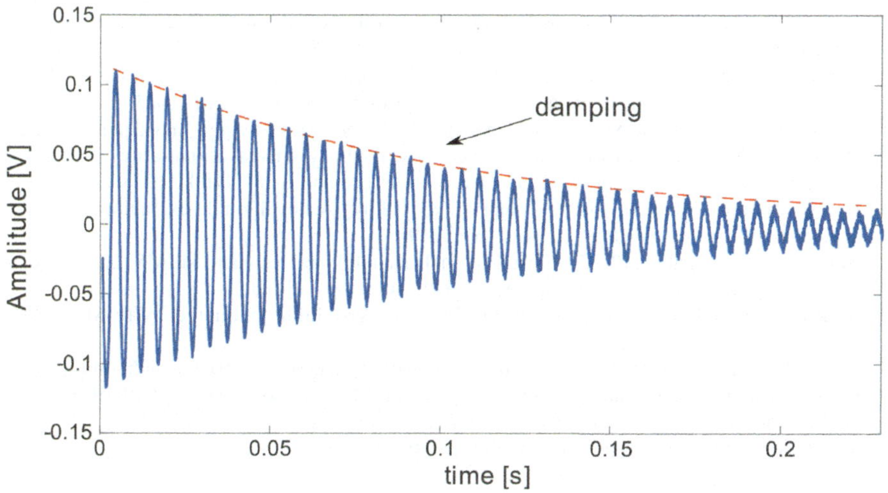


**Fig. 4** A time window of the acquired device transient response

rigid mass of 160 μg on the free end of the cantilever beam. The obtained results shown an appreciable oscillation frequency shift due to the mass deposition.

Figure 4 shown an example of an acquired time window of the device transient response. The related data are used by the MCU implemented firmware to perform the oscillation frequency measurements and the damping factor assessment.

## 4 Conclusion

In this work the authors propose a new method for mass measurements based on PZT-microsensors and a microcontroller unit. The feasibility of a "self-generating" measurement system exploiting the piezoelectric characteristics of the MEMS device and an ad hoc designed low-cost measurement system is presented. The measurement system exploits the capability of a microcontroller unit for the data processing. The working frequency range of the MEMS device and the features of the selected microcontroller guarantee the implementation of a robust, compact and low power solution, suitable for in field measurement.

## References

1. Zhao C, Montaseri MH, Wood GS, Pu SH, Seshia AA, Kraft M et al (2016) A review on coupled MEMS resonators for sensing applications utilizing mode localization. Sens Actuators A 249:93–111

2. Andò B, Baglio S, Baù M, Ferrari V, Sardini E, Savalli N, Trigona C (2010) Numerical and experimental investigation on contactless resonant sensors. Sens Actuators A 162(2):329–335
3. Davis ZJ, Svendsen W, Boisen A (2007) Design, fabrication and testing of a novel MEMS resonator for mass sensing applications. Microelectron Eng 84(5–8):1601–1605
4. Trigona C, Algozino A, Maiorca F, Andò B, Baglio S (2014) Design and characterization of PiezoMUMPs microsensors with applications to environmental monitoring of aromatic compounds via selective supramolecular receptors. Procedia Eng 87:1190–1193
5. Hou B, Zhou B, Song M, Lin Z, Zhang R (2016 July) A novel single-excitation capacitive angular position sensor design. Sensors 16(8):1196
6. Brand O, Pourkamali S (2015) Electrothermal excitation of resonant MEMS. In: Resonant MEMS: fundamentals, implementation and application, pp 173–201
7. Ando B, Ascia A, Baglio S, Trigona C (2007) Resonant ferrofluidic inclinometers. In: SENSORS, 2007 IEEE, Atlanta, GA, pp 776–779
8. Fort A, Panzardi E, Addabbo T, Mugnaini M, Vignoli V, Trigona C (2019) Conditioning circuit for simultaneous sensing and actuation in piezoelectric MEMS resonators. In: 2019 IEEE sensors applications symposium (SAS), Sophia Antipolis, France, pp 1–6
9. http://www.memscap.com/products/mumps/piezomumps

# Outstanding $NO_2$ Sensing Performance of Sensors Based on $TiO_2$/Graphene Hybrid

**M. L. Miglietta, B. Alfano, J. P. Santos, I. Sayago, T. Polichetti, E. Massera, Paola Delli Veneri, G. Di Francia, Carlos Sanchez and Jesús Lozano**

**Abstract** Titanium oxide nanoparticle decorated graphene sensors have been studied for the detection of low concentrations of nitrogen dioxide at several temperatures. A remarkable response, in terms of sensitivity and response time, has been obtained at room temperature.

**Keywords** Graphene · Sensor · Nanoparticle · Nitrogen dioxide

## 1 Introduction

Nitrogen dioxide ($NO_2$) is one of the major urban pollutants. The last report of the European Environment Agency shows that $NO_2$ concentrations above limit values were widely distributed across Europe [1]; consequently the demand for detecting this toxic gas at environmentally relevant concentrations is pressing. Graphene is able to detect gas analytes even at room temperature (RT) and has shown a particular sensitivity towards $NO_2$ but its sensing performances are somehow jeopardized by its kinetic behavior: slow response, recovery rate and poor reversibility are in fact major drawbacks [2–4]. To overcome these issues numerous functionalization approaches have been proposed. In particular, it was shown that the addition of metal oxides to graphene is able to tune the selectivity of the material as well as to improve the overall sensing performances toward $NO_2$ [5], achieving in some cases a sensitivity gain of around two order of magnitude [6]. Herein, we show the outstanding effect of $TiO_2$ decoration on graphene based sensors for $NO_2$ detection at environmental conditions. Devices based on this hybrid material operate at room temperature and show a more than hundredfold increase of the sensing response as well as a significant

M. L. Miglietta (✉) · B. Alfano · T. Polichetti · E. Massera · P. D. Veneri · G. Di Francia
ENEA, CR-Portici, P.le E. Fermi 1, 80055 Naples, Italy
e-mail: mara.miglietta@enea.it

J. P. Santos · I. Sayago · C. Sanchez
NOySI, Instituto de Tecnologías Físicas y de la Información ITEFI-CSIC, Madrid, Spain

J. Lozano
Departamento de Electricidad, Electrónica y Automática, Universidad de Extremadura, Badajoz, Spain

 
G. Di Francia et al. (eds.), *Sensors and Microsystems*, Lecture Notes in Electrical Engineering 629, https://doi.org/10.1007/978-3-030-37558-4_53

reduction of the sensing times. These sensing features are ascribed mainly to the enhanced reactivity of $TiO_2$ nanoparticles promoted by the interaction with graphene nanosheets. The sensitivity achieved in environmental conditions opens the way to the $NO_2$ monitoring in extremely low concentrations.

# 2 Materials and Methods

## 2.1 Sensing Layer Fabrication

Graphite flakes were dispersed into a mixture of ultrapure water and i-propanol and sonicated. Un-exfoliated graphite crystallites were removed by centrifuging. The concentration of the graphene suspension was $0.1 \pm 0.01$ mg ml$^{-1}$. Graphene powder was obtained by freeze drying of graphene suspension. 2.5 mg of graphene powder was mixed with 4 mg of $TiO_2$ nanoparticles (25 nm) and microwave irradiated for 5 min at 1000 W. The resulting powder was dissolved in isopropyl alcohol/water. The process is illustrated in Fig. 1.

The sensing material was deposited onto silicon microhotplates by drop casting and onto on silicon and oxidized silicon substrates for morphological and structural characterizations.

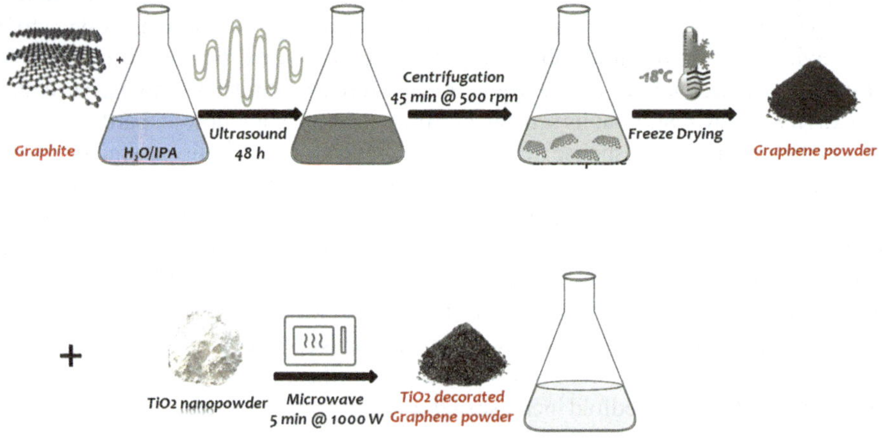


**Fig. 1** Sensing layer fabrication process

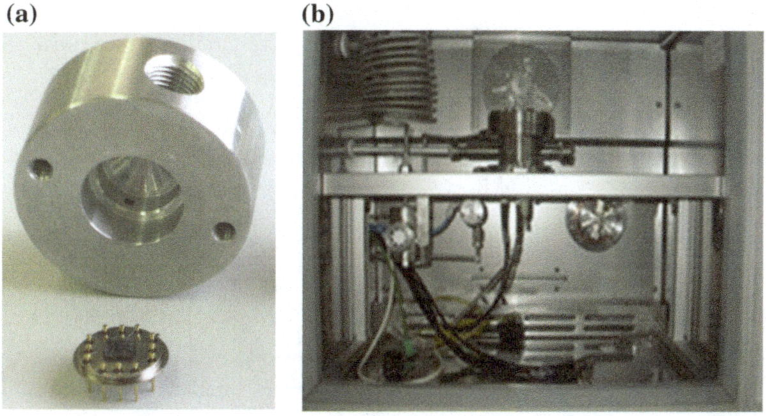


**Fig. 2** **a** Sensor chamber and microhotplate, **b** thermostatic box

## *2.2 Sensing Layer Characterization*

The sensing layers were characterized by SEM (Quanta 3D FEG, FEI Company), TEM (FEI TECNAI G12 Spirit-Twin) and Raman (Renishaw InViaReflex spectrometer).

## *2.3 Measurement Setup*

The sensors have been characterized in an automated gas line (GSCS, Kenosistec equipment) with the test chamber of 1 $cm^3$ of volume shown in Fig. 2a placed in a thermostatic box (Fig. 2b).

The measurement parameters were: Flow rate, 500 mL/min; RH, 50%; operation temperature 25–300 °C; adsorption time, 10 min; desorption time 20–90 min. $NO_2$ concentration were varied from 0.1 to 1 ppm.

# 3 Results and Discussion

## *3.1 Sensing Layer Characterization*

Raman spectra of pristine graphene (black line) and $GTiO_2$ (red line), shown in Fig. 3, exhibit the typical profile of carbon-based material with three prominent features: the D band (~1350 $cm^{-1}$), ascribed to breathing mode of the carbon rings; the G

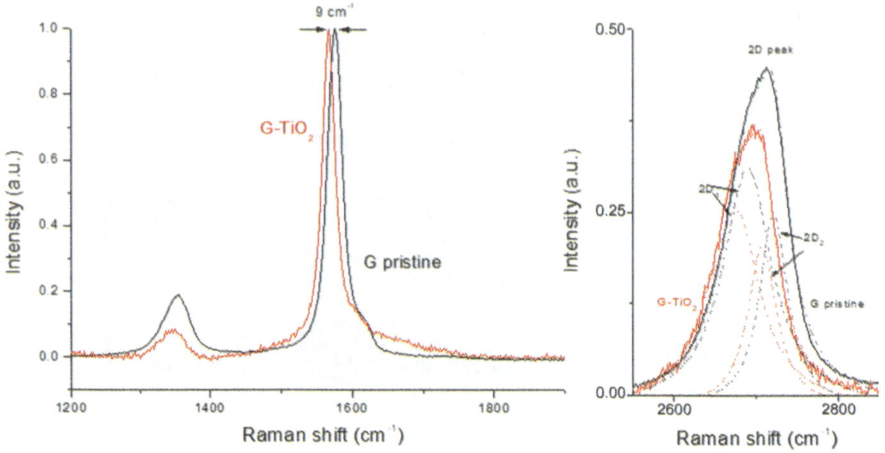


**Fig. 3** Raman spectra of Gr (black line) and $GTiO_2$ (red line)

peak (~1580 $cm^{-1}$) which is related to the in-plane bond stretching of sp2 pairs; the 2D band (~2700 $cm^{-1}$) that represents the overtone of the D band.

A red shift of both G and 2D peaks can be noted: this may be indicative of the fact that the p-doped starting graphene, in the interaction with an electron donor material, such as $TiO_2$, is partially neutralized, causing a redshift of the peaks.

SEM and TEM images are shown in Fig. 4. $TiO_2$ nanoparticles are distributed around the flakes showing some agglomeration. In SEM image a continuous nanoparticle film on top of the graphene flakes is appreciated.

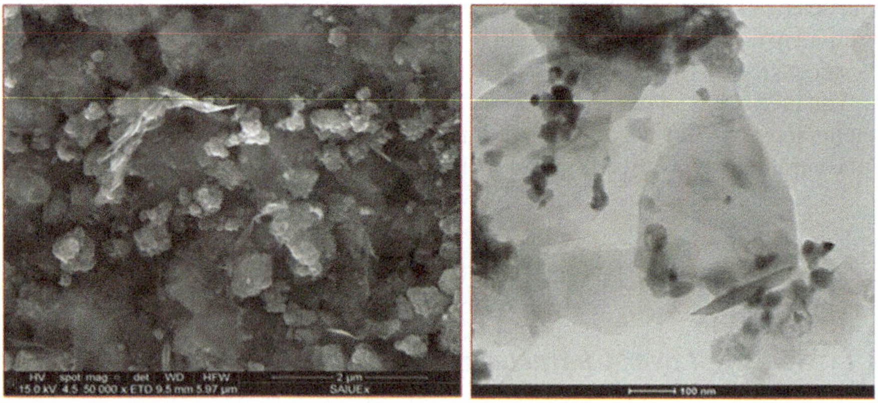


**Fig. 4** SEM (left) and TEM (right) images of the $TiO_2$ loaded graphene layer

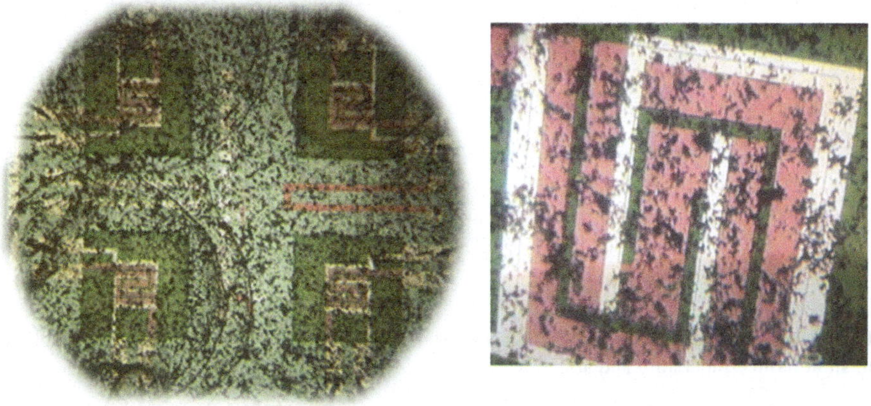

**Fig. 5** Optical image of the sensing layer onto the whole chip (left) and detail of one of the membranes (right)

## *3.2 Device Characterization*

Optical images of the deposited layer onto the silicon microhotplate can be seen in Fig. 5.

Increasing the operating temperature results in a decrease of the chemiresistor resistance. $NO_2$ exposition induces an increase of the sensor resistance so the behaviour is that of an n-type semiconductor. Dynamic response curves show the resistance as a function of time. Static response curves (or simply response curves) show the percentage variation of the resistance with respect to its initial value: r(%) = 100 * R − $R_0/R_0$ where R is the sensor conductance upon its exposure to $NO_2$ and $R_0$ is the sensor baseline conductance. Responses were very high at room temperature (over 900% to 250 ppb) and experimented an abrupt decrease at the other operating temperatures. In the range 100–300 °C the response increased from 13 to 27%. Response times were very fast, in the order of seconds for the whole temperature range, although desorption times were longer. These responses are two orders of magnitude greater than pristine graphene sensors [7].

In order to check the influence of the graphene we also fabricated a $TiO_2$ nanoparticle sensor. Its temperature behaviour was quite different. It only began to detect $NO_2$ at 150 °C and had a máximum response of 145% at 250 °C, much lower than that of the $GTiO_2$ sensor (Fig. 6).

The effect of the carrier gas on the sensor response was also investigated. Figure 7 shows the response at room temperature to several $NO_2$ concentrations in air and nitrogen. Higher responses in nitrogen were obtained.

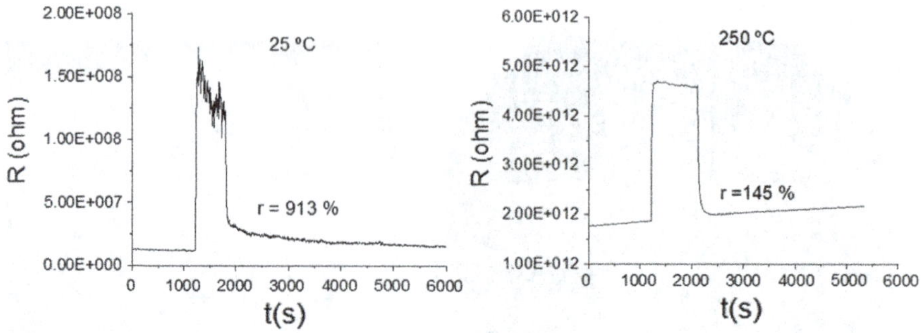


**Fig. 6** Dynamic response of the $GTiO_2$ sensor (left) and the $TiO_2$ sensor (right) to 250 ppb of $NO_2$ in air at room temperature and at 250 °C

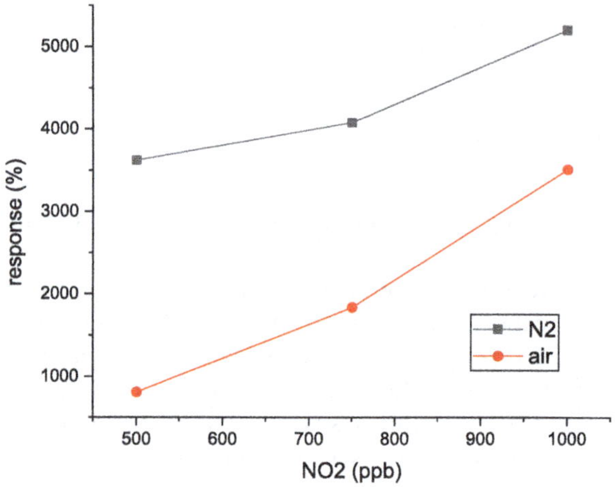


**Fig. 7** Response of $GTiO_2$ sensor at room temperature in air and nitrogen

## 4 Conclusion

Herein, we have shown the outstanding effect of $TiO_2$ decoration on graphene based sensors for $NO_2$ detection at environmental conditions. The sensitivity achieved in environmental conditions opens the way to the $NO_2$ monitoring in extremely low concentrations. Detection mechanism deserves further investigation.

## References

1. Guerreiro C, Gonzalez Ortiz A, de Leeuw F, Viana M, Horalek J (2016) Air quality in Europe—2016 report. https://doi.org/10.2800/413142

2. Schedin F, Geim AK, Morozov SV, Hill EW, Blake P, Katsnelson MI, Novoselov KS (2007) Detection of individual gas molecules adsorbed on graphene. Nat Mater. https://doi.org/10.1038/nmat1967
3. Miglietta ML, Massera E, Romano S, Polichetti T, Nasti I, Ricciardella F, Fattoruso G, Di Francia G (2011) Chemically exfoliated graphene detects $NO_2$ at the ppb level. Procedia Eng. https://doi.org/10.1016/j.proeng.2011.12.282
4. Lu G, Park S, Yu K, Ruoff RS, Ocola LE, Rosenmann D, Chen J (2011) Oxidation resistance of graphene-coated Cu and Cu/Ni alloy. ACS Nano 5(2):1321–1327
5. Chatterjee SG, Chatterjee S, Ray AK, Chakraborty AK (2015) Graphene-metal oxide nanohybrids for toxic gas sensor: a review. Sens Actuators B 221:1170–1181
6. Neri G, Leonardi SG, Latino M, Donato N, Baek S, Conte DE, Russo PA, Pinna N (2013) Sensing behavior of $SnO_2$/reduced graphene oxide nanocomposites toward $NO_2$. Sens Actuators B 179:61–68
7. Santos JP, Polichetti T, Hontañón E, Sayago I, Aleixandre M, Alfano B, Miglietta M, Di Francia G, Lozano J (2018) Study of graphene based nanosensors for the detection of nitrogen dioxide. Paper presented at the Proceedings of the 2018 12th Spanish conference on electron devices, CDE. https://doi.org/10.1109/cde.2018.8597167

# Graphene Patterning via Photolithography

**B. Alfano, E. Massera, M. L. Miglietta, T. Polichetti, Eugenia Bobeico, Paola Delli Veneri and G. Di Francia**

**Abstract** We present a proof of concept to scale graphene based device fabrication combining Liquid Phase Exfoliated (LPE) method for the production of graphene flakes and standard photolithography for patterning high-resolution areas of graphene films (film widths down to 0.5 μm). The subsequent silver electrodes are created to test the patterned graphene films towards low $NO_2$ concentrations at room temperature. The results prove that the combination of LPE method and standard photolithography could be a powerful strategy for building high-performance graphene based device with low cost and in large-scale.

**Keywords** Photolithography · Graphene patterning · $NO_2$ detection

## 1 Introduction

In the last decade, various areas of research have focused the attention on possible exploitation of graphene as functional material. Specifically, the sensor sector intends to take advantage of its outstanding electrical properties in order to obtain more innovative and increasingly better performing devices [1]. Recent studies have been carried out in order to use this exceptional material for technological applications [2, 3].

There are a lot of new manufacturing techniques to produce large scale and high-quality graphene films [4–6]. Nevertheless, the only way to get a graphene large-volume manufacturability is to succeed in a large scale transfer to the final substrate used for the specific application.

The current technology for transferring is slow, consequently this step could represent a bottleneck in the industrial-scale production [7–9]. Among transfer techniques, standard photolithography could be a concrete helpful to realize graphene based micro-nano/device on a massive scale.

B. Alfano (✉) · E. Massera · M. L. Miglietta · T. Polichetti · E. Bobeico · P. D. Veneri · G. Di Francia
ENEA, CR-Portici, P.le E. Fermi 1, 80055 Naples, Italy
e-mail: brigida.alfano@enea.it

 
G. Di Francia et al. (eds.), *Sensors and Microsystems*, Lecture Notes in Electrical Engineering 629, https://doi.org/10.1007/978-3-030-37558-4_54

In this work, we present a proof of concept to combine Liquid Phase Exfoliated (LPE) method for the massively production of graphene flakes with standard photolithography to transfer efficiently graphene, patterning high-resolution areas onto appropriate substrate for devices fabrication.

The method is accomplished by photolithography printing polymer pattern, LPE graphene solution drop-casting and removing the polymer pattern via lift-off process to create the patterned graphene film (PGF).

Furthermore, we demonstrate the sensing behavior of the PGF by creating a device, operating at room temperature towards low concentration of $NO_2$ (200–800 ppb).

## 2 Experimental

### *2.1 Material Fabrication*

Pristine graphene was obtained by a liquid phase exfoliation (LPE) of graphite flakes in a hydro-alcoholic solution as described in detail in our previous publication [6]. Briefly, graphite powder were dispersed into a water/iso-propyl alcohol mixture (7:1 v/v) and sonicated for several hours. An subsequent step of centrifugation is useful to separate not exfoliated material obtaining a final suspension of few-layer graphene.

After an initial cleaning of the $SiO_2$/Si substrate, a photoresist layer (Clarion, AZ 9260) was spun coated onto the silicon substrate and heated (pre-bake) in an oven (115 °C 20 min). Then, an ultra-violet photolithography defines windows in the photoresist, exposing portions of the photoresist that are to be removed. An AZ 400 K developer was used to remove the exposed areas, revealing the underlying $SiO_2$/Si substrate. A cleaning process removed the remaining photoresist and yields a photoresist layer with $SiO_2$/Si patterned areas on its surface. Next, graphene solution was drop casted over the polymer patterned surface. A post baking process was subsequently conducted in a convection oven (115 °C 5 min) to remove solvents and simultaneously improve the physical bond between the substrate and the graphene. Finally, the photoresist was completely removed and lifted-off by dipping the substrate into acetone bath. The final result is the formation of well-defined graphene areas with widths down to 0.5 μm.

Figure 1 shows the flowchart for the scalable fabrication of graphene area based on standard photolithography technology.

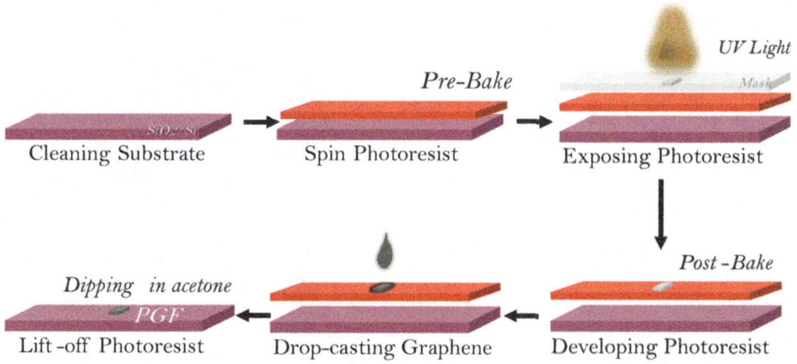


**Fig. 1** Schematic of fabricating patterned graphene film (PGF) by photolithography

## 2.2 *Experimental Apparatus*

Transmission Electron Microscopy (TEM) images were acquired on FEI TECNAI G12 Spirit-Twin. The Raman spectra were acquired by means of Renishaw InViaReflex spectrometer for 514 nm wavelength incident laser light, in backscattering configuration.

Surface images of the patterned graphene film and their roughness were analysed using an optical profilometer (KCL Tencor). The profiles were indicative of complex shape of graphene surface. The roughness parameter Ra (average roughness) is inadequate for this irregular material, hence Rq (root–mean–square roughness) parameter is evaluated.

Electrical contacts were deposited onto PGF for device testing. The sensing measurements were performed in a Gas Sensor Characterization System (Kenosistec) that has electrical feed-through connections to an HP parameter analyzer. The current change of the sensor was measured upon exposure to various gases while a constant bias was applied.

## 2.3 *Experimental Results*

*Graphene Ink characterization*

Figure 2 displays a TEM image of pristine graphene where overlapped flakes are clearly observable.

*Patterned Graphene Film characterization*

A big problem for fabricating graphene devices through photolithography is originated from effect due to residual photoresist, which would damage the graphene

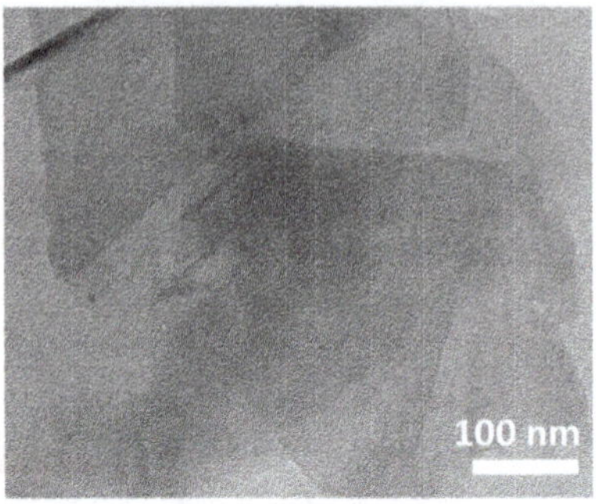


**Fig. 2** TEM image of the pristine graphene

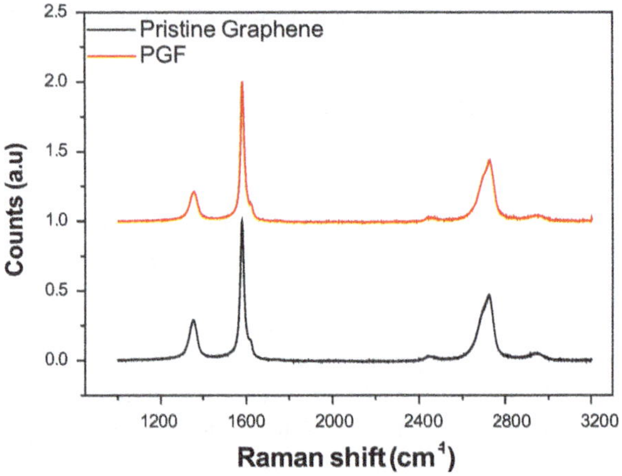


**Fig. 3** Typical Raman spectra of graphene before and after photolithography steps

properties. The Raman spectroscopy performed on PGF showed that the process did not modify the peculiar characteristics of graphene material (Fig. 3).

The patterned graphene was characterized with both optical microscopy (OM) and profilometer in order to analyze PGF surface morphology.

OM was used (Fig. 4a) to observe graphene deposition. It displays relatively smooth morphology across the surface but with high surface area due to randomly oriented graphene flakes.

As can be seen in Fig. 4b, the profile on the surface displays some deep gaps. These gaps are distributed regularly and the whole material can be considered almost homogeneous. Even if it is impossible to define exactly the roughness of bottom and top graphene layers, the root-mean-square roughness (Rq) is around 16 Å and the difference between the highest peak and the deepest valley is 150 Å

Electrical property is a significant parameter before to characterize the performance and potential of sensing device based on PGF. Metal contacts composed of silver paste were deposited onto PGF surface. Current-voltage (I-V) characteristic measured for a typical device is plotted in Fig. 5.

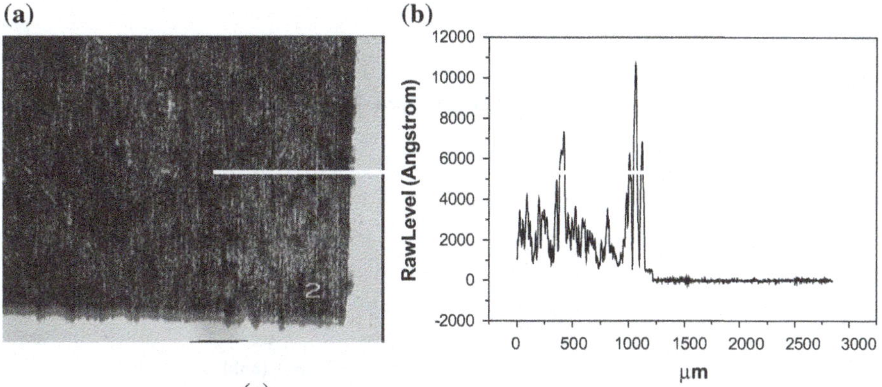


(c)

| Roughness parameter | (Å) |
|---|---|
| $R_a$ | 12 |
| $R_{RMS}$ | 16 |
| $R_t$ | 146 |

**Fig. 4** **a** Optical microscopy image of PGF. **b** Profile along the line seen in the (**a**)

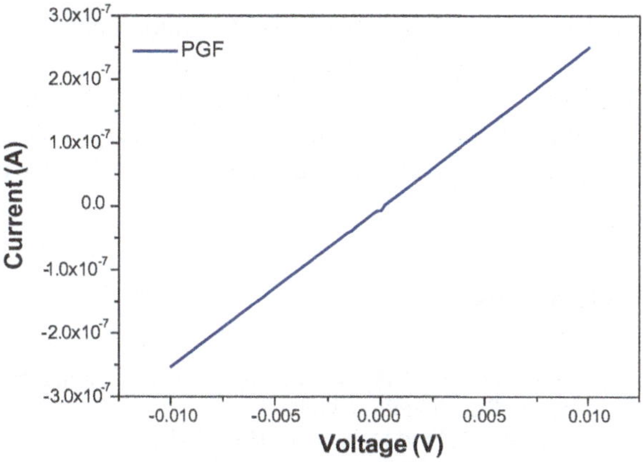


**Fig. 5** I-V characteristics of PGF

By photolithography, the graphene patterned film is sufficiently adhered to the substrate as evidenced by Ohmic behavior.

The sensing capability of the patterned graphene using photolithographic method was first characterized upon $NO_2$ analyte, setting temperature and relative humidity at 22 °C and 50%, respectively.

Figure 6 shows the calibration curve for PGF based device.

This sensing characteristic shows how the PGF is linearly sensitive to variations in the $NO_2$ in the sub-ppm range at RT.

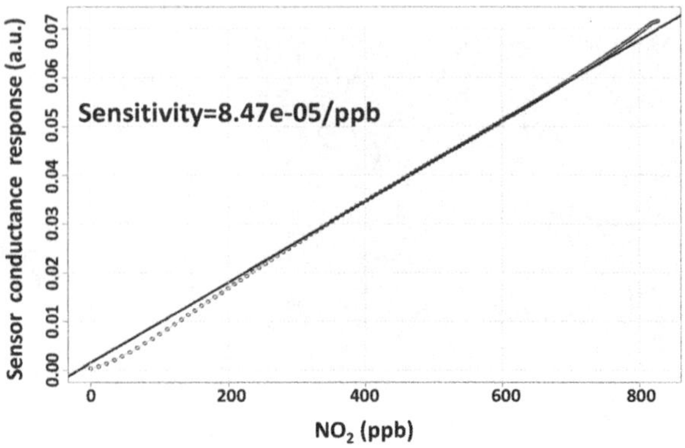


**Fig. 6** Calibration curve for PGF based device

# 3 Conclusion

Results indicated that the Liquid Phase Exfoliation (LPE) method for the production of graphene flakes and standard photolithography for patterning high-resolution areas of graphene films could be a well-suited for a rapid solution in order to prototyping graphene film on substrates for numerous applications including chemical sensing.

In conclusion, this work presents a proof of concept to scale graphene based device fabrication by combining Liquid Phase Exfoliation (LPE) method for the production of graphene flakes and standard photolithography for patterning high-resolution areas of graphene films. The device was completed with silver electrodes and tested towards low $NO_2$ concentrations at room temperature. The results prove that the combination of LPE method and standard photolithography could be a powerful strategy for building high-performance graphene based device with low cost and in large-scale.

# References

1. Neto AC, Guinea F, Peres NM, Novoselov KS, Geim AK (2009) The electronic properties of graphene. Rev Mod Phys 81(1):109
2. Bae S, Kim SJ, Shin D, Ahn JH, Hong BH (2012) Towards industrial applications of graphene electrodes. Phys Scr 2012(T146):014024
3. Zurutuza A, Marinelli C (2014) Challenges and opportunities in graphene commercialization. Nat Nanotechnol 9(10):730
4. Lee Y, Bae S, Jang H, Jang S, Zhu SE, Sim SH, … Ahn JH (2010) Wafer-scale synthesis and transfer of graphene films. Nano Lett 10(2):490–493
5. Yamada T, Kim J, Ishihara M, Hasegawa M (2013) Low-temperature graphene synthesis using microwave plasma CVD. J Phys D Appl Phys 46(6):063001
6. Fedi F, Miglietta ML, Polichetti T, Ricciardella F, Massera E, Ninno D, Di Francia G (2015) Mater. A study on the physicochemical properties of hydroalcoholic solutions to improve the direct exfoliation of natural graphite down to few-layers graphene. Res. Express 2:035601–035608
7. Suk JW, Kitt A, Magnuson W, Hao Y, Ahmed S, An J, Swan A, Goldberg BB, Ruoff RS (2011) Transfer of CVD-grown monolayer graphene onto arbitrary substrates. ACS Nano 5:6916

8. Liang BA, Sperling I, Calizo G, Cheng CA, Hacker Q, Zhang Y, Obeng K, Yan H, Peng Q, Li et al (2011) Toward clean and crackless transfer of graphene. ACS Nano 5:9144
9. Lee S, Bae H, Jang S, Jang S-E, Zhu SH, Sim YI, Song BH, Hong, Ahn J-H (2010) Wafer-scale synthesis and transfer of graphene films. Nano Lett 10:490

# Conducted Disturbance Robustness Evaluation of a Magnetic Short-Range Localization System

**G. Betta, D. Capriglione, G. Cerro, L. Ferrigno and F. Milano**

**Abstract** In this article, a metrological characterization of the performance of a wireless localization system is carried out, considering the typical disturbances affecting the medical environment. Such system will be used in specific biomedical applications, including a glove for accurate tracking of unconstrained movements. To this purpose, the authors participate to a national project "Six DOF scalable finger tracking system", whose goal is to develop a short-range wireless localization system with six degrees of freedom. The realized system will guarantee the real-time tracking of the fingers of the hand to provide support for the rehabilitation of patients with reduced motor skills. Obtained results, which exhibit a worsening in localization error due to disturbance effects, paths the way to a more robust transmission method able to reject or minimize single ton disturbance effect.

**Keywords** Wireless localization system · Magnetic measurement · Electromagnetic compatibility

## 1 Introduction

In the biomedical field, wireless monitoring systems can offer an important solution [1] whenever the condition of the line of sight (LoS), relative to instruments or organs, is not guaranteed. Interesting applications involve open surgery, traumatology and remote virtual sonography [2]. Recently, the use of wireless tracking has also been proposed in radiotherapy [3] to compensate for movement, in those patients whose age–related decline leads to a reduction in motor functioning. Therefore, there is a need for systems that provide a low-cost objective measurement of the functional capabilities of free subjects in the home environment. In detail, a localization system based on the alternating magnetic fields generation and their measurement through

G. Betta · G. Cerro · L. Ferrigno · F. Milano (✉)
Department of Electrical and Information Engineering, University of Cassino and Southern Lazio, Cassino, Italy
e-mail: filippo.milano@unicas.it

D. Capriglione
Department of Industrial Engineering, University of Salerno, Fisciano, Italy

 
G. Di Francia et al. (eds.), *Sensors and Microsystems*, Lecture Notes in Electrical Engineering 629, https://doi.org/10.1007/978-3-030-37558-4_55

circular and planar coils has been developed in [4]. This system is robust with respect to multipath effect, it can be realized with low–cost components, and enables the position and orientation of the mobile node to be measured simultaneously. In this work the authors, stemming from their experience in magnetic sensors [5, 6], monitoring systems [7, 8], RF measurements [9–12] and electromagnetic compatibility [13, 14], perform a characterization of system performance when it is subjected to conducted disturbances, whose features are accurately chosen among those deriving from medical environment.

## 2 The Localization System and the Algorithm Used

The designed system requires the mobile node (Tx coil) to be active, generating a magnetic field that is measured by a series of passive beacons (Rx coils). To improve the usable capacity of the system and the consumption of the active mobile node, the coils have been transformed into resonant circuits by connecting appropriately sized capacitors in parallel. In addition, an instrumentation amplifier was added to each beacon. The measurement model used is based on magnetic dipole moments, according to which the magnetic field vector $\vec{B}$ generated by the active node in a given position is described by Eq. (1).

$$\vec{B} = \frac{\mu_0}{4\pi}\frac{m_{tx}}{d^3}\left[3\left(\hat{n}_{tx}\cdot\hat{n}_d\right)\hat{n}_d - \hat{n}_{tx}\right] \tag{1}$$

where $\mu_0$ is the magnetic is the permeability of the vacuum, $d$ and $\hat{n}_d$ describe the module and the direction, respectively, of the distance vector $\vec{d}$ between the Tx coil and the considered position, $m_{tx}$ e $\hat{n}_{tx}$ describe the module and the direction of the magnetic dipole moment vector $\vec{m}_{tx}$ that characterizes the active node. The term $m_{tx}$ is in turn equal to the product $N_{tx}S_{tx}$I, in which $N_{tx}$ and $N_{tx}$ are respectively the number of turns and the area of the Tx coil, crossed by a known current $I$. The output voltage to the i-th receiving circuit ($V_i$) is given by Eq. (2).

$$V_i = 2\pi f_0 N_i S_i G_i \vec{B}_i \cdot \hat{n}_i \tag{2}$$

where $f_0$ is the operating frequency, $N_i$ and $S_i$ are respectively the number of turns and the area of the i-th Rx coil, $\vec{B}_i$ is the induced magnetic field in the i-th beacon position, $\hat{n}_i$ describes the direction of the i-th beacon, and $G_i$ is a gain term, which takes into account both the merit factor of the resonator, and the gain of the amplifier. Once the beacons output voltages have been measured, the position of the mobile node can be estimated using a numerical technique. Assuming that the position and orientation of the Tx coil are described by the variable $\theta$, the position estimate has been realized minimizing with respect to $\theta$, through the Nelder–Mead algorithm, the cost function reported in Eq. (3).

$$F(\theta) = \sum_{i=1}^{17} \left[\widetilde{V}_{rms,i} - V_{rms,i}(\theta)\right]^2 \tag{3}$$

where $\widetilde{V}_{rms,i}$ indicates the voltage measured at the i-th beacon, while $V_{rms,i}(\theta)$ indicates the voltage that should be measured if the position and orientation were equal to $\theta$.

## 3 Disturbances in the Medical Environment

Like any electrical or electronic device, electromedical equipment can be subject to phenomena related to electromagnetism. Each electromedical device can represent a source of disturbance and, at the same time, it can suffer the influence of nearby devices capable of generating electromagnetic waves. The possible effects of EM interference on the functioning of an electromedical device are varied and can lead to serious consequences for the health of patients. In the medical field it is not difficult to find high frequency signal sources in the vicinity of particularly sensitive receptor devices, increasing the risk of possible discomfort. In this work, electromagnetic immunity tests were carried out for conducted disturbances introduced in the localization system. All the tests were conducted in the semi–anechoic chamber of the University of Cassino. Among the different types of disturbances present in the medical environment, conducted disturbances induced by radiofrequency (RF) fields were chosen. The reference standard for this type of disturbance is EN 61000–4–6. In particular, the analyzed frequency range goes from 150 kHz to 80 MHz, with disturbance signals modulated in amplitude at 80% of the depth and with a modulation frequency of 1 kHz. The voltage levels used for the disturbances are 1 and 3 Vrms.

## 4 Set-up Explanation

The localization system adopted is composed of 17 receiving devices, positioned in fixed and known positions, and by a mobile transmitter free to move in a space of 30 × 30 × 30 cm. The mobile node powered by a sinusoidal signal having 10 V peak-to-peak voltage and a frequency of 200 kHz. The receiving devices are connected through a data acquisition (DAQ) to process the induced voltages as explained in the previous section. In a first phase, a test was carried out in ideal conditions, i.e. without the introduction of the disturbances. In particular, the Tx coil was placed in a known position and an average location error ($\varepsilon$) was obtained, in accordance with 4, of 7.2 mm and a standard deviation ($\sigma$) of 0.18 mm. The number of measures repeated during the test was 100.

$$\varepsilon = \sqrt{\varepsilon_x^2 + \varepsilon_y^2 + \varepsilon_z^2} \tag{4}$$

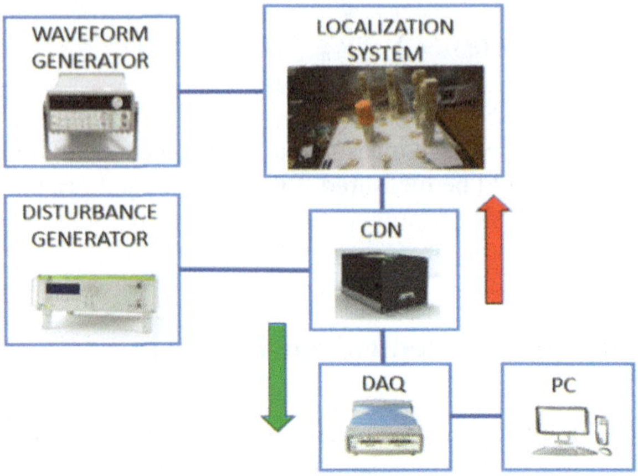


**Fig. 1** Schematic set-up used for conducted immunity tests

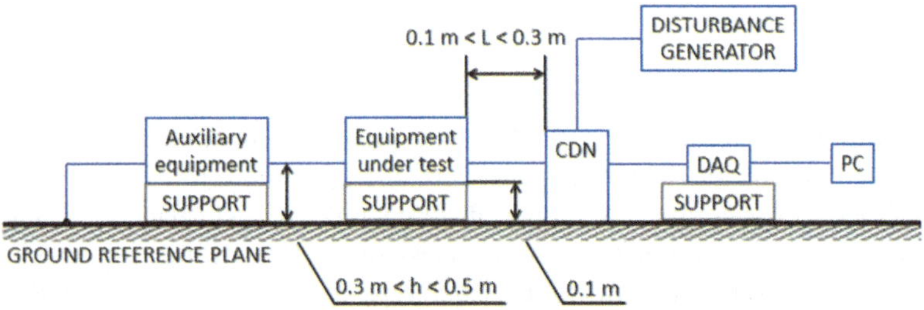


**Fig. 2** Immunity test to RF conducted disturbances

Subsequently, the EM TEST CWS 500 certified noise generator and the CDN–S25 coupling/decoupling network were introduced for immunity tests in the presence of disturbances. The CDN allows to introduce the disturbances both towards the dreceiving devices and towards the data acquisition. Both directions were considered for testing. Figure 1 shows the block diagram of the set-up adopted for the injection of disturbances, while Fig. 2 shows the distances to be respected, according to the standard, for the connection of all the devices.

## 5 Obtained Performance

In order to carry out a metrological characterization of the performance of the system, against the introduction of disturbances, the Tx coil was placed in the same position assumed during the ideal test. In Figs. 3, 4, 5 and 6, the obtained results are reported in terms of the mean error in the three spatial directions and standard deviation,

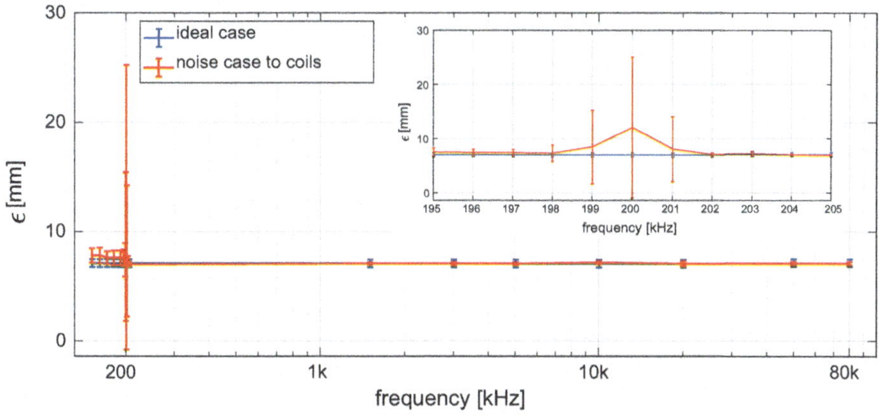


**Fig. 3** Performances obtained with disturbance towards Rx coils and voltage equal to 1 V

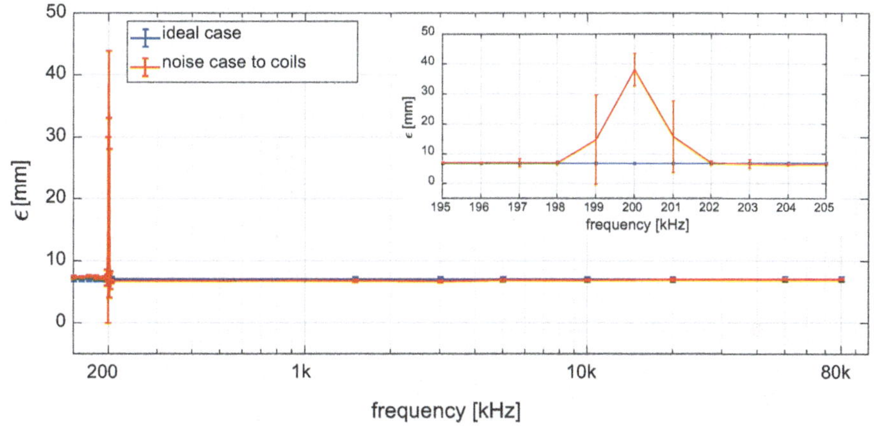


**Fig. 4** Performances obtained with disturbance towards Rx coils and voltage equal to 3 V

compared with the performances obtained in the case of absence of disturbances. To check the compatibility between the various measurements, a coverage factor equal to 2 has been adopted.

# 6 Conclusions

The results obtained show a performance degradation when the frequency of the disturbance is close to the power supply frequency of the transmitting coil. In this situation, the repeatability of the measurement process also deteriorates compared to the ideal case without disturbance. In all other cases, in which the frequency of the

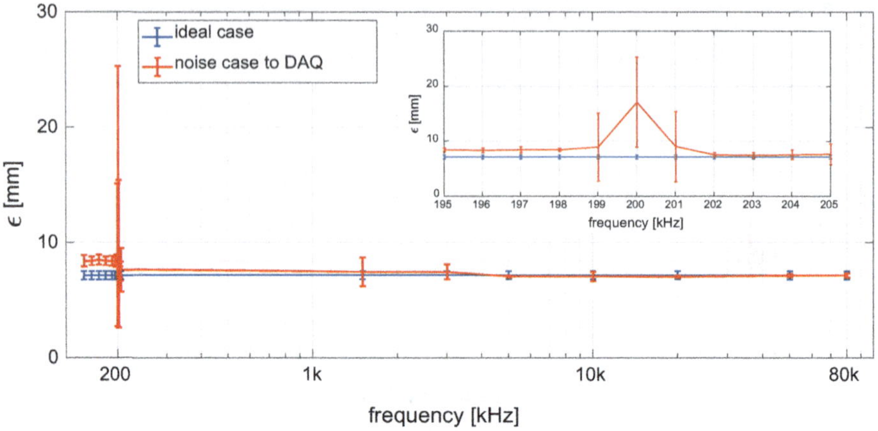


**Fig. 5** Performances obtained with disturbance towards DAQ and voltage equal to 1 V

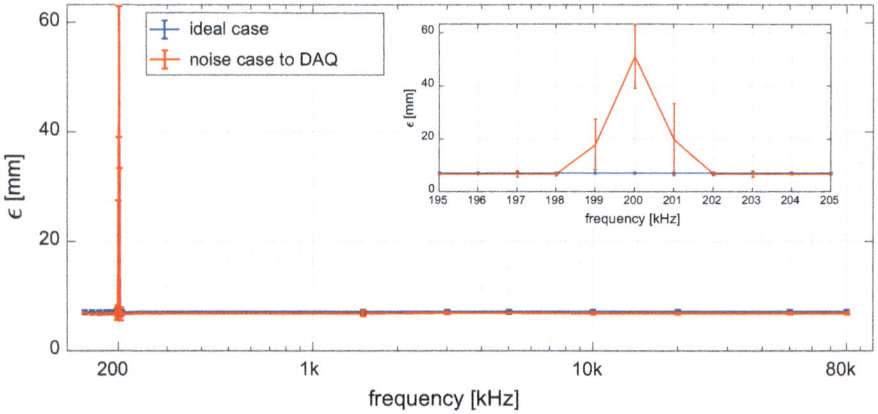


**Fig. 6** Performances obtained with disturbance towards DAQ and voltage equal to 3 V

disturbance is far from that of the transmitting coil, there is compatibility between the performance obtained in the ideal case and that achieved by the injection of the disturbances.

## References


1. Clifton DA, Wong D, Clifton L, Wilson S, Way R, Pullinger R, TarassenkoL (2013) A large-scale clinical validation of an integrated monitoring system in the emergency department. IEEE J Biomed Health Inf 17(4):835–842. https://doi.org/10.1109/jbhi.2012.2234130
2. Ferrer-Roca O et al (2006) Tele-virtual sonography. J Perinat Med 34(2):123–129

3. Choi T et al (2009) Wireless magnetic tracking system for radiation therapy. In: 2009 IEEE/NIH life science systems and applications workshop, pp 148–151. https://doi.org/10.1109/LISSA.2009.4906731
4. Moschitta A, De Angelis A, Dionigi M, Carbone P (2017) Analysis of simultaneous 3d positioning and attitude estimation of a planar coil using inductive coupling. In: 2017 IEEE international instrumentation and measurement technology conference (I2MTC), pp 1–6. https://doi.org/10.1109/i2mtc.2017.7969848
5. Bernieri A, Betta G, Ferrigno L, Laracca M (2013) Improving performance of gmr sensors. IEEE Sens J 13(11):4513–4521. https://doi.org/10.1109/JSEN.2013.2271275
6. Bernieri A, Betta G, Ferrigno L, Laracca M (2012) Multi-frequency eddy current testing using a gmr based instrument. Int J Appl Electromagn Mech 39(1–4):355–362
7. Sorrentino L, Bellini C, Capriglione D, Ferrigno L (2015) Local monitoring of polymerization trend by an interdigital dielectric sensor. Int J Adv Manuf Technol 79(5–8):1007–1016
8. Amicone D, Bernieri A, Ferrigno L, Laracca M (2009) A smart add-on device for the remote calibration of electrical energy meters. In: 2009 IEEE instrumentation and measurement technology conference, pp 1599–1604. https://doi.org/10.1109/IMTC.2009.5168710
9. Angrisani L, Capriglione D, Cerro G, Ferrigno L, Miele G (2016) Optimization and experimental characterization of novel measurement methods for wide-band spectrum sensing in cognitive radio applications. Measurement 94:585–601. https://doi.org/10.1016/j.measurement.2016.08.036
10. Capriglione D, Cerro G, Ferrigno L, Miele G (2016) Analysis and implementation of a wavelet based spectrum sensing method for low snr scenarios. In: 2016 IEEE 17th international symposium on a world of wireless, mobile and multimedia networks (WoWMoM), pp. 1–6. https://doi.org/10.1109/WoWMoM.2016.7523585
11. Betta G, Capriglione D, Cerro G, Ferrigno L, Miele G (2015) The effectiveness of savitzky-golay smoothing method for spectrum sensing in cognitive radios. In: 2015 XVIII AISEM annual conference, pp 1–4. https://doi.org/10.1109/AISEM.2015.7066819
12. Betta G, Capriglione D, Ferrigno L, Miele G (2008) Experimental investigation of the electromagnetic interference of zigbee transmitters on measurement instruments. IEEE Trans Instrum Meas 57(10):2118–2127. https://doi.org/10.1109/TIM.2008.922105
13. Capriglione D, Chiariello A, Maffucci A, Accurate models for evaluating the direct conducted and radiated emissions from integrated circuits. Appl Sci 8(4). https://doi.org/10.3390/app8040477
14. Betta G, Capriglione D, Carobbi CF, Migliore MD (2011) The accurate calibration of emc antennas in compact chambers–measurements and uncertainty evaluations. Comput Stand Interfaces 33(2):201–205. https://doi.org/10.1016/j.csi.2010.06.012

# Fabrication and Characterizations of Pristine and Metal Oxide Nanoparticles Decorated Graphene Sheets

**T. Polichetti, M. L. Miglietta, B. Alfano, E. Massera, F. Villani, G. Di Francia and P. Delli Veneri**

**Abstract** Poor selectivity, slow response and long recovery times are the main bottlenecks that have emerged since the fabrication of the first graphene-based sensor, consequently limiting its application in the sensing field. To mitigate these drawbacks, the following research has turned to the decoration of the material with metal and metal oxide nanoparticles, thus demonstrating their ability to modulate the selectivity of the material and to improve the overall performance. Starting from these premises, we present different materials obtained by functionalizing graphene with ZnO, $SnO_2$ and $TiO_2$ metal oxide nanoparticles in comparison with pristine graphene.

**Keywords** Graphene functionalization · MOX nanoparticles · ZnO · $SnO_2$ · $TiO_2$

## 1 Introduction

Since the fabrication of the first graphene-based sensor device, the huge potential of this material in sensing field has highlighted several limits, including poor selectivity, slow response and long recovery times [1]. Different methods have been employed in order to overcome the above mentioned drawbacks: for instance thermal heating [2, 3] and the exposure to UV radiation [4] have been proved to speed up the desorption process, since they can provide the energy needed to separate the analyte molecules from the sensing material. Nevertheless, the aforesaid approaches require a complicate design of the sensing device with the integration on board of a micro-joule heater or a UV-light emitting diode.

In order to surmount these downsides, successive research has focused on changing the properties of the material through its functionalization. A technique to achieve this goal is to decorate the graphene surface with metal and metal oxide nanoparticles (NPs), and actually several works have demonstrated the possibility of modulating the selectivity of the material, improving its overall detection performances [5–9].

T. Polichetti (✉) · M. L. Miglietta · B. Alfano · E. Massera · F. Villani · G. Di Francia · P. Delli Veneri
ENEA, P.le E. Fermi 1, 80055 Portici (Naples), Italy
e-mail: tiziana.polichetti@enea.it



G. Di Francia et al. (eds.), *Sensors and Microsystems*, Lecture Notes in Electrical Engineering 629, https://doi.org/10.1007/978-3-030-37558-4_56

This paper aims to illustrate a simple and effective technique for functionalizing graphene (Gr) with of metal oxides NPs capable of producing different types of hybrid materials, in the present case graphene functionalized with ZnO (GZnO), graphene functionalized with $SnO_2$ ($GSnO_2$) and graphene functionalized with $TiO_2$ ($GTiO_2$).

## 2 Experimental

The preparation of the materials was carried out by a sonication-assisted exfoliation of graphite flakes in a hydro-alcoholic solution as described elsewhere [10]. Briefly, 80 ml of water/IPA mixture (7:1 v/v) was employed to disperse 80 mg graphite flakes. A black, homogeneous suspension of few-layer graphene was obtained after a sonication treatment in a low power bath for 48 h and a following centrifugation at 500 rpm for 45 min. Such suspension was successively freeze-dried and the resulting powders were mixed with NPs metal oxide (3:1 mol/mol) and finally microwave irradiated for 5 min at 1000 W. The samples were imaged by FEI TECNAI G12 Spirit-Twin TEM operating at 120 kV and by FESEM LEO 1530-2 coupled with an Energy Dispersive Analysis System (EDS) working at an acceleration voltage of 5 kV. The morphological characterizations were supplemented by AFM imaging, performed by Digital Instruments Dimension Nanoscope IV in tapping mode. Structural characterization was performed by using a Raman spectrometer (Renishaw inVia Reflex) @ 514 nm excitation, in backscattering configuration.

## 3 Results and Discussion

Pristine graphene, GZnO, $GSnO_2$ and $GTiO_2$ were initially characterized by means of TEM imaging techniques. The samples were prepared by dipping standard holey carbon grids (400mesh) into the colloidal suspensions and drying at room conditions.

The images of graphene and its hybrids are displayed in Fig. 1; in particular, Fig. 1a depicts pristine graphene, which appears consisting of overlapped flakes with lateral size ranging between 300 and 500 nm. Figure 1b–d show the morphology of the three hybrid materials where the GR flakes appear covered by metal oxide NPs, re-arranged into scattered clusters, mostly deposited on the edges of the GR sheets.

SEM analyses on GZnO, $GSnO_2$ and $GTiO_2$ were carried out on samples prepared by drop-casting few microliters of the solutions directly onto Si substrates and diplayed in Fig. 2a, 2c and 2e, respectively. The micrographs confirm the morphology identified through TEM. The related EDS analysis provides evidence that the NPs visible on the surface and at the flakes edges are made up of metal oxides, thus proving that the applied synthesis technique truly produces the adhesion of the NPs to the GR surface (Fig. 2b, 2d and 2f).

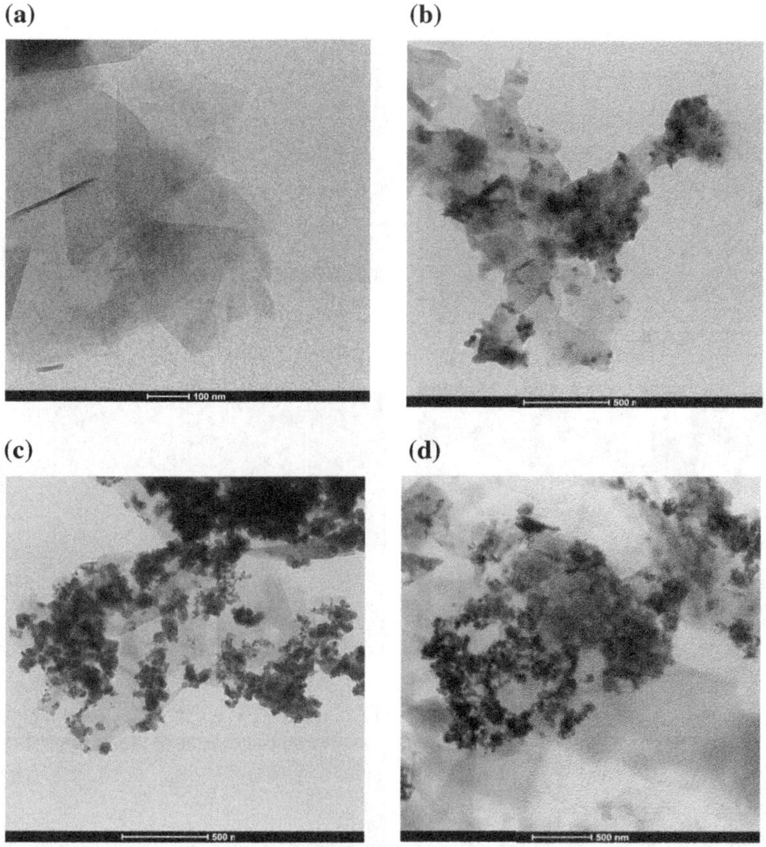


**Fig. 1** TEM images of **a** pristine Gr, **b** GZnO, **c** $GSnO_2$ and **d** $GTiO_2$

The samples drop-casted on Si substrate were also characterized by Atomic Force Microscopy. In Fig. 3 both the topography and the phase image of a graphene flake detail, decorated with $TiO_2$ NPs, are reported as an example; the picture highlights the material stratified structure with a number of layers ranging between 3 and 5. The NPs, whose mean size varies between 15 and 25 nm, aggregate in clusters having an average diameter of 250–300 nm, and mostly stick to the edges of the material.

To verify the presence of a real interaction between the nanoparticles and the graphene surface, Raman investigations have been carried out; in this case spectra were acquired on films drop-casted onto oxidized silicon wafers ($SiO_2$ 300 nm). Raman spectra of the three hybrid were compared to that of pristine graphene (Fig. 4).

All the analyzed materials exhibit a typical spectrum of carbon-based materials, with the presence of the three main bands, the D, the G and the 2D band. The comparison of these profiles with that of the pristine material highlights an appreciable redshift of both G and 2D peaks, systematically observed for all graphene materials

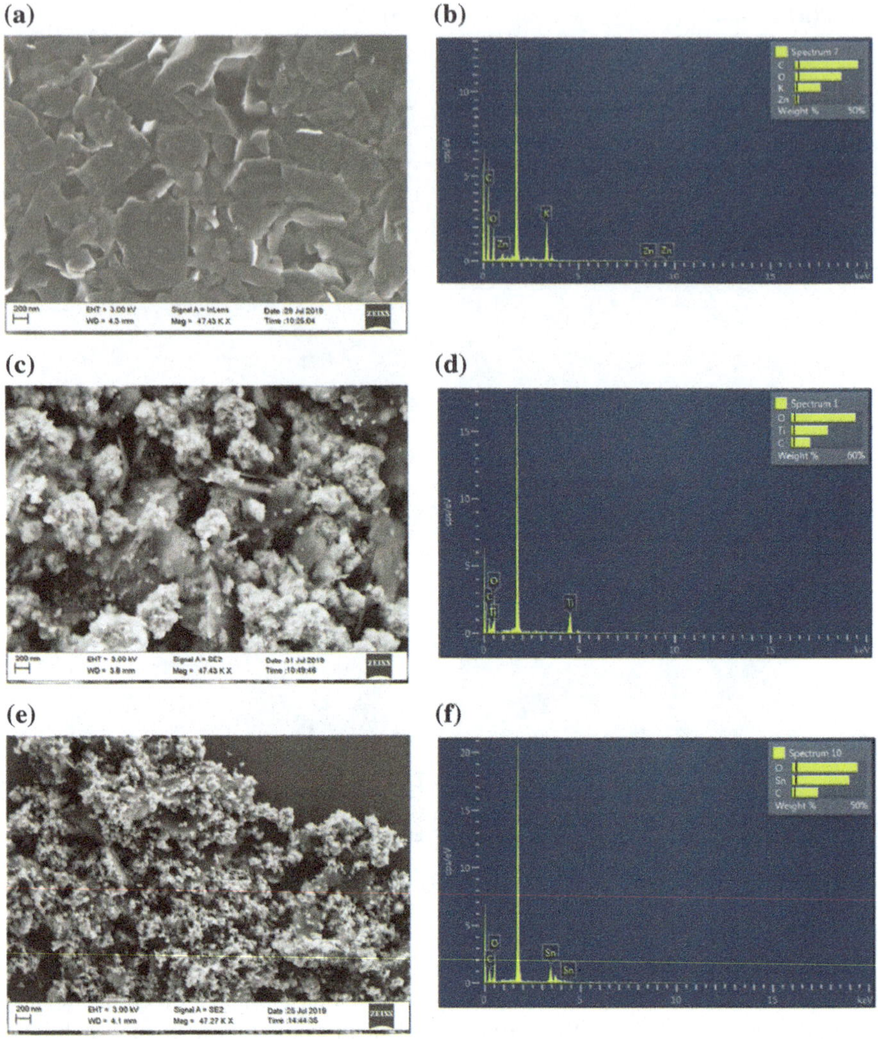


**Fig. 2** SEM images of functionalized materials: **a** and **b** GZnO and related compositional analysis; **c** and **d** $GSnO_2$ and related compositional analysis; **d** and **f** $GTiO_2$ and related compositional analysis

decorated with metal oxide herein presented. It is plausible to ascribe this displacement to the circumstance that as prepared graphene nanosheets are naturally p-type doped, and consequently the G peak and 2D band are upshifted with respect to intrinsic material [11]. Decoration with electron donor materials, such as metal oxides, lead to a partial doping neutralization, expressed in a redshift of the peaks. This

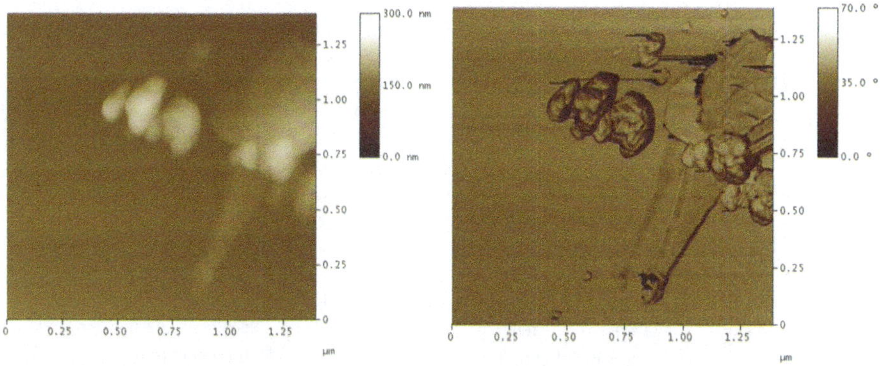


**Fig. 3** Detail of a graphene flake decorated with $TiO_2$ NPs, observed under the AFM. On the left the topography, on the right the phase image

suggests that the functionalization technique here adopted determines a real interaction between metal oxide nanoparticles and graphene, thus modifying the physical properties of the material which certainly reflect on its sensing behavior.

## 4 Conclusion

In this contribution a simple approach to functionalize the graphene surface was illustrated. Morpho-structural analyzes indicate that the technique is able to grow metal oxide nanoparticles on the material flakes, also modifying their physical properties. These results provide good indications on the potential to modulate the selectivity and sensitivity of the graphene towards the analytes. On this regard, measurements in the test chamber towards the main environmental pollutants are ongoing and part of the results are the subject of another publication in this volume.

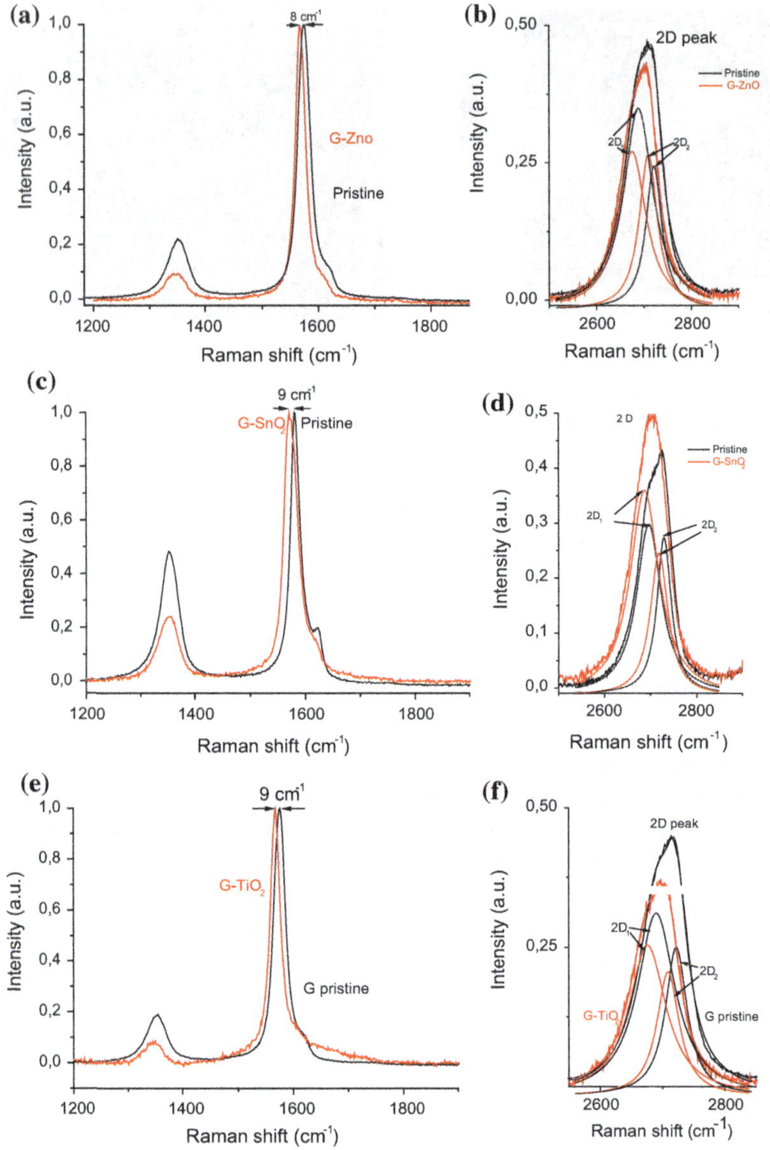


**Fig. 4** Raman spectra of functionalized materials (red solid line) compared to that of pristine graphene (black solid line): **a** G band and **b** 2D band of GZnO; **c** G band and **d** 2D band of $GSnO_2$; **e** G band and **f** 2D band of $GTiO_2$

## References

1. Schedin F, Geim AK, Morozov SV, Hill EW, Blake P, Katsnelson MI, Novoselov KS (2007) Detection of individual gas molecules adsorbed on graphene. Nat Mater 6:652
2. Yavari F, Chen Z, Thomas AV, Ren W, Cheng HM, Koratkar N (2011) High sensitivity gas detection using a macroscopic three-dimensional graphene foam network. SciRep 1:166
3. Iezhokin I, Offermans P, Brongersma SH, Giesbers AJM, Flipse CFJ (2013) High sensitive quasi freestanding epitaxial graphene gas sensor on 6H-SiC. Appl Phys Lett 103:053514
4. Chen G, Paronyan TM, Harutyunyan AR (2012) Sub-ppt gas detection with pristine graphene. Appl Phys Lett 101:053119
5. Alfano B, Massera E, Miglietta ML, Polichetti T, Schiattarella C, Di Francia G (2018) Graphene decoration for gas detection, lecture notes in electrical. Engineering 457:35–40
6. Kang IS, So HM, Bang GS, Kwak JH, Lee JO, Won Ahn C (2012) Recovery improvement of graphene-based gas sensors functionalized with nanoscale heterojunctions. Appl Phys Lett 101:123504
7. Mao S, Cui S, Lu G, Yu K, Wen Z, Chen J (2012) Tuning gas-sensing properties of reduced graphene oxide using tin oxide nanocrystals. J Mater Chem 22(22):11009–11013
8. Latif U, Dickert FL (2015 Dec 4) Graphene hybrid materials in gas sensing applications. Sensors (Basel) 15(12):30504-24. Epub 2015 Dec 4
9. Wang LL, Ng WB, Jackman JA, Cho N-J (2016) Graphene functionalized natural microcapsules: modular building blocks for ultrahigh sensitivity bio electronic platforms. Adv Funct Mater 26:2097–2103
10. Fedi F, Miglietta ML, Polichetti T, Ricciardella F, Massera E, Ninno D, Di Francia G (2015) A study on the physicochemical properties of hydroalcoholic solutions to improve the direct exfoliation of natural graphite down to few-layers graphene. Mater Res Express 2:035601
11. Das A, Pisana S, Chakraborty B, Piscanec S, Saha SK, Waghmare UV, Novoselov KS, Krishnamurthy HR, Geim AK, Ferrari AC, Sood AK (2008) Monitoring dopants by Raman scattering in an electrochemically top-gated graphene transistor. Nat Nanotechnol 3:210–215

# Chemicals Detection in Water by SENSIPLUS Platform: Current State and Ongoing Progress

**Carmine Bourelly, M. Ferdinandi, M. Molinara, L. Ferrigno and Roberto Simmarano**

**Abstract** The challenge to detect contaminants inside water solutions is addressed in this paper, through the use of an integrated, low-cost, smart and IoT platform, namely SENSIPLUS. In particular, the complete process from the sensing phase to classification and results analysis is provided with further investigations about the limitations of the current proposal and the description of a further processing technique that promises to improve classification accuracy. The classification is performed by adopting machine learning techniques, particularly Artificial Neural Network, that well fits the implementation on a low-cost microcontroller, as the one SENSIPLUS platform uses.

**Keywords** Water quality · Contaminant detection · Machine-learning · Sensors · IoT

## 1 Introduction

Life quality is a crucial aspect that employs several scientific sectors, trying to monitor every single pieces of data that could be related to the improvement or worsening of human and other species' lives. Water [1, 2] purity is one of the most monitored quantities, since such element is pervasive on the earth and its level influences the way people live and the planet survival. Many reasons affect water contamination, as the wide-spreading of the industrial sector and the usage of polluting substances in people's daily routine, whose disposal often conveys to water. Therefore, two different operations are necessary: human behavioural change and reliable monitoring of water state all over the planet. Literature offers important solutions [3, 4] in this sense that mainly focus their attention on two distinct aspects: the sensing phase, i.e.

C. Bourelly · R. Simmarano
Sensichips s.r.l., Via delle Valli 46, 04011 Aprilia, Italy

M. Ferdinandi (✉) · M. Molinara · L. Ferrigno
Department of Electrical and Information Engineering, Università degli studi di Cassino e del Lazio Meridionale, Cassino, Italy
e-mail: m.ferdinandi@unicas.it

 
G. Di Francia et al. (eds.), *Sensors and Microsystems*, Lecture Notes in Electrical Engineering 629, https://doi.org/10.1007/978-3-030-37558-4_57

the usage of suitable sensors and sensing materials to be sensitive to water contaminants, and data processing techniques that, starting from sensors' data could achieve classification purposes. In terms of sensing, the common knowledge leads any novel research effort to adopt sensors array, since a single sensor sensitivity appears limited when several contaminant typologies are to be detected. Large number of sensors, needed to compose sensors array, implies high volumes of data and this aspect pushes the processing techniques to be focused on Artificial Intelligence (AI). On this aspect, many scientific references, which mostly treat the problem as separate issues, can be found. In this work the authors, stemming from their past experience in data processing [5–7], impedance measurement [8, 9], AI applications [10], communication networks [11, 12], propose a fully-integrated system allowing both the sensing and classification phases, using an ad hoc platform, namely SENSIPLUS [13], having low-cost, low-power and smart capabilities. To the best of our knowledge, this is one of the few examples in literature showing the full system working from raw data to classification output. The paper is organized in the following sections: the procedure and the set-up are reported in Sect. 2, obtained results are described in Sect. 3, further raw data processing to improve classification accuracy is treated in Sect. 4. Conclusions follow in Sect. 5.

## 2 The Adopted Set-up and Procedure

The set-up is focused on the adoption of the SENSIPLUS platform, owned by Sensichips S.r.l., already adopted for application regarding people's health in industry environment. It is endowed with an ad hoc chip, able to perform electrical impedance measurements, and other sensors, as temperature, adopted to corroborate contaminant classification. As for contaminant-detection sensors, two interdigitated electrodes (IDEs), built in gold and copper, are used. They are completely immersed in water and communicate with the platform as in Fig. 1.

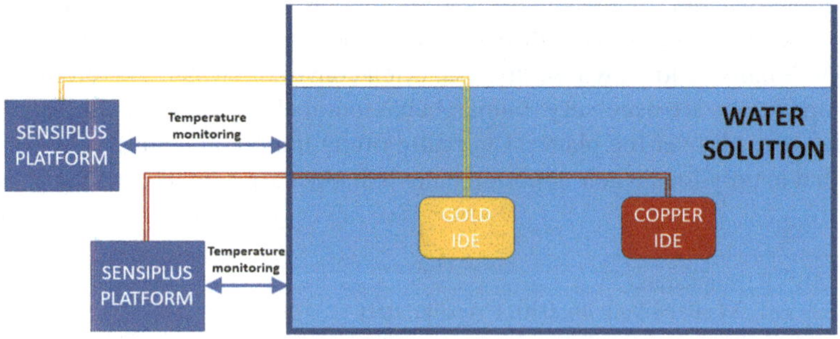


**Fig. 1** Block diagram of the developed set-up

Gold and copper material differently react to the contaminants and their responses can be merged to obtain more selectivity for classification purposes. In terms of adopted procedure, a three-step method has been employed:

- clean water impedance measurement for 350 s, in order to monitor environmental background;
- contaminant injection in water;
- measurement of electrical parameter (namely R, C) for another time period equal to 1300 s.

The impedance measurements are carried out at two different frequency stimuli: 2 and 78 kHz, in order to exploit diversity in frequency response by the used sensors.

In terms of used contaminants, typologies are: alcohol, bleach, detergent, sodium chloride and sulfuric acid. Obtained measurements consist of 8 values for each time-step, i.e. resistance and capacitance at the chosen frequencies for each sensor.

The so-big amount of data is pre-processed, according to a technique known as Principal Component Analysis (PCA), and results are then used to train, in the first phase, and test, in the second one, an artificial neural network (ANN) that outputs the classified items.

## 3 Results

Since the amount of data is quite high and they are not expressed in the same measurement unit, a normalization process is required before applying neural network algorithms. In this sense, the PCA technique is used. It extracts the most important components of the raw data and this process reduces the amount of data being input of the ANN. In particular, for each time-step, the useful data are reduced from 8 to 3 and, according to the metric used to evaluate data loss in PCA application, more than 95% of data are kept with such a transformation.

In Fig. 2, the results of PCA is proposed. Here, different substances are reported in a 3D representation with dimensions reporting the value of the specific substance projected on that component. The more the substances diverge each other, the best the classifier is able to differentiate them.

After having trained the ANN, classification results are reported in Fig. 3. Each line represents the percentage detection rate for a specific substance. On x-axis, truly injected contaminants are reported. Ideally, each line should have its peak (equal to 100%) in correspondence of the correct substance and be constantly equal to zero at all other cases. Some contaminants (alcohol and bleach) got almost 100% recognition rate, while others are slightly confused with substances having similar chemical composition. Worst situation is represented by Sulfuric Acid, confused with Detergent.

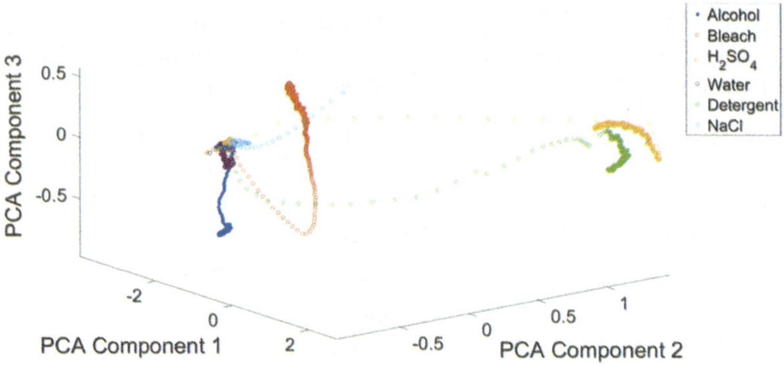


**Fig. 2** Principal component analysis decomposition

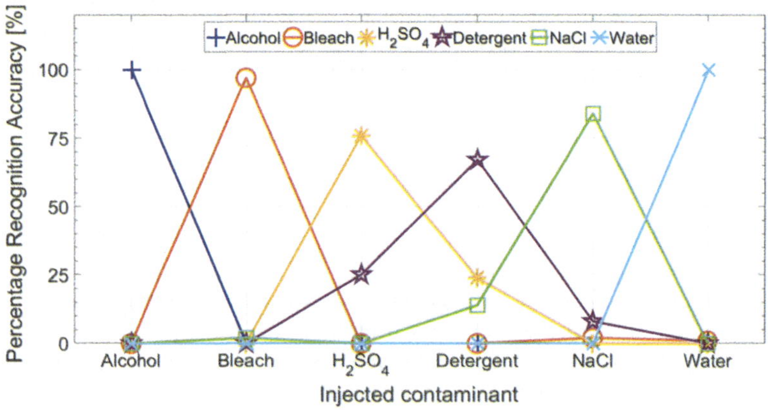


**Fig. 3** Classification results obtained with ANN technique

## 4 Further Development in Data Pre-processing

In order to improve the capability to detect contaminants, another technique is currently being used to acquire raw data. It is known as cyclic voltammetry. In this case, the electrodes in water have a potential that ramps up with time to come back to its initial value in a cyclic manner. The underlying idea is that there is a special potential difference that induces redox reactions in water, thus provoking electron release and, consequently, increase in current flow. This process can be useful for classification purposes because different substances should have different "breakthrough" potentials, provoking large current increase. Estimating such threshold value, it could be possible to understand which substance has induced the phenomenon.

## 5 Conclusions and Future Developments

In this work, first classification results for contaminant detection in water have been proposed. In detail, data have been acquired from sensors according to a parallel R-C equivalent circuit, processed through PCA, finally tested with an Artificial Neural Network. Obtained results show very good performance for some substances and some confusion for chemically similar contaminants. For that, a different processing technique, namely cyclic voltammetry is actually being adopted to improve data pre-treatment and allow the neural network to have higher classification rates. In future, classification results obtained with the novel techniques are envisioned to be proposed and compared with the actual ones.

## References

1. Goel P (2006) Water pollution: causes, effects and control, new age international
2. Brack W et al (2017) Towards the review of the european union water framework directive: recommendations for more efficient assessment and management of chemical contamination in european surface water resources. Sci Total Environ 576:720–737. https://doi.org/10.1016/j.scitotenv.2016.10.104
3. Li J, Cao S (2015) A low-cost wireless water quality auto-monitoring system. Int J Online Eng (iJOE) 11(3):37–41
4. Schmidt W, Raymond D, Parish D, Ashton IG, Miller PI, Campos CJ, Shutler JD (2018) Design and operation of a low-cost and compact autonomous buoy system for use in coastal aquaculture and water quality monitoring. Aquacult Eng 80:28–36
5. Betta G, Capriglione D, Cerro G, Ferrigno L, Miele G (2015) The effectiveness of savitzky-golay smoothing method for spectrum sensing in cognitive radios. In: 2015 XVIII AISEM annual conference, pp 1–4. https://doi.org/10.1109/aisem.2015.7066819
6. Capriglione D, Cerro G, Ferrigno L, Miele G (2016) Analysis and implementation of a wavelet-based spectrum sensing method for low snr scenarios. In: 2016 IEEE 17th international symposium on a world of wireless, mobile and multimedia networks (WoWMoM), pp 1–6. https://doi.org/10.1109/wowmom.2016.7523585
7. Angrisani L, Capriglione D, Cerro G, Ferrigno L, Miele G (2016) Optimization and experimental characterization of novel measurement methods for wide-band spectrum sensing in cognitive radio applications. Measurement 94:585–601. https://doi.org/10.1016/j.measurement.2016.08.036
8. Cerro G, Ferdinandi M, Ferrigno L, Laracca M, Molinara M (2018) Metrological characterization of a novel microsensor platform for activated carbon filters monitoring. IEEE Trans Instrum Meas 67 (10) (2018) 2504–2515. https://doi.org/10.1109/tim.2018.2843218
9. Cerro G, Ferdinandi M, Ferrigno L, Molinara M (2017) Preliminary realization of a monitoring system of activated carbon filter RLI based on the SENSIPLUS microsensor platform. In: 2017 IEEE international workshop on measurement and networking (M N), pp 1–5. https://doi.org/10.1109/iwmn.2017.8078361
10. Bruschi P, Cerro G, Colace L, De Iacovo A, Del Cesta S, Ferdinandi M, Ferrigno L, Molinara M, Ria A, Simmarano R, Tortorella F, Venettacci C (2018) A novel integrated smart system for indoor air monitoring and gas recognition. In: 2018 IEEE international conference on smart computing (SMARTCOMP), pp 470–475. https://doi.org/10.1109/smartcomp.2018.00048
11. Di Cara D, Luiso M, Miele G, Sommella P (2013) A smart measurement network for optimization of electrical grid operation. In: Proceedings of 19th IMEKO TC 4 symposium and 17th IWADC workshop, advances in instrumentation and sensors interoperability, pp 649–654

12. Angrisani L, Capriglione D, Ferrigno L, Miele G (2011) Packet jitter measurement in communication networks: a sensitivity analysis. In: 2011 IEEE international workshop on measurements and networking proceedings (M&N), pp 146–151. https://doi.org/10.1109/iwmn.2011.6088488
13. Sensichips. URL sensichips.com

# Thickness Measurements with Eddy Current and Ultrasonic Techniques

**G. Betta, L. Ferrigno, M. Laracca, A. Rasile and A. Sardellitti**

**Abstract** In the industrial applications, innovative solutions for thickness measurements are investigated in this last years. In particular, technical solutions characterized by proper metrological performance and at the same time small size, low cost, and simple integrability are followed with the aim of operating within the modern industrial context. In these scenarios, the paper describes a two thickness measurement techniques directly derived from Non-Destructive Testing (NDT) are analyzed and compared. In particular, the first technique is based on the induction of eddy currents (EC) in conductive materials, while the second technique is based on ultrasonic (UT) propagation within materials. A deep experimental analysis is carried out to verify the metrological performance of the two proposed techniques, and the obtained results have shown that is allows reaching metrological performance suitable for the typical industrial applications.

**Keywords** Eddy current · Ultrasonic · Thickness measurement · Non destructive testing

---

G. Betta · L. Ferrigno · M. Laracca · A. Rasile · A. Sardellitti (✉)
Department of Electrical and Information Engineering, University of Cassino and Southern Lazio, Via G. Di Biasio, 43, 03043 Cassino, Italy
e-mail: alessandro.sardellitti@d-solutions.it

G. Betta
e-mail: betta@unicas.it

L. Ferrigno
e-mail: ferrigno@unicas.it

M. Laracca
e-mail: m.laracca@unicas.it

A. Rasile
e-mail: a.rasile@unicas.it



G. Di Francia et al. (eds.), *Sensors and Microsystems*, Lecture Notes in Electrical Engineering 629, https://doi.org/10.1007/978-3-030-37558-4_58

# 1 Introduction

In recent years, the concept of safety in the industrial sector, especially in the automotive sector, is increasingly at the center of attention. It is necessary that the structural components of a vehicle fall within the safety intervals established by increasingly restrictive regulations.

In this context, thickness measurements play a crucial role, because they define the quality of these components in terms of strength and elasticity. The methods used to measure thickness in the industrial sector are many and fall into the category of non-destructive tests (NDT).

Contrary to destructive tests, in which the component to be tested can be damaged or destroyed during the examination process, non-destructive tests allow the inspection of materials without destroying them and guaranteeing their use once the test is complete. The main objective of NDT is to guarantee the integrity and reliability of the product, to maintain a uniform quality level [1].

The NDT can be applied during the production phase, to guarantee the quality of the materials produced and improve the production processes. These aspects are very important in all those application fields such as the automotive sector, aerospace and so on, where the integrity of the components has a direct impact on the safety of human beings. The presence of a wide variety of materials and components has led to the development of different investigation techniques based on different physical principles: ultrasound, thermography, radiography, electromagnetism, etc.

In this paper, the authors, starting from their experience on NDT and magnetic sensors [2–5], ultrasound sensors [6], impedance measurement [7, 8], electromagnetic measurement [9] and monitoring systems [10–12], propose the analysis and the comparison of two thickness measurement techniques directly derived from non-destructive tests (NDT). In particular, the first technique is based on the induction of parasitic currents (EC) in conductive materials, while the second technique is based on ultrasonic propagation (UT) inside the materials.

# 2 Thickness Measurements Techniques

## *2.1 Thickness Measurements with Eddy Current Technique*

The typical basic operating principle of eddy currents is very simple. An excitation coil induces eddy currents in the test sample while magnetic sensors or sensing coils measure the reaction magnetic fields. The analysis of the reaction magnetic field allows defining the characteristics of the investigated material.

In particular, from previous studies and through the mathematical model proposed by Dodd and Deeds [13–16], it is possible to associate the thickness of the analyzed sample to a specific frequency (f). In detail, using a system that involves the use of two coils, one for the induction of the magnetic field in the material and one for detecting the reaction magnetic field, it is possible to identify the thickness of the material by observing the variation of the imaginary part of the overall inductance of the two coils (Im(ΔL)) as a function of frequency. It is possible to calculate this variation by evaluating the imaginary part of the difference between the impedance measured on the material (Z(f)) and the impedance measured in air ($Z_{air}(f)$), normalized in relation to the frequency value, Eq. (1).

$$\mathrm{Im}(\Delta L) = \mathrm{Im}((Z(f) - Z_{air}(f))/(j \cdot 2 \cdot \pi \cdot f)) \tag{1}$$

The thickness of the sample is related to the frequency (f) identified by the minimum point of the curve obtained, by Eq. 2, where $\alpha_0$ is the spatial frequency characteristic of the two coils used as probe, which depends on their geometry, σ is the electrical conductivity of the material and $\mu_0$ is the magnetic permeability of the vacuum.

$$\text{Thickness} = (2 \cdot \alpha_0)/(\sigma \cdot \mu_0 \cdot 2 \cdot \pi \cdot f) \tag{2}$$

## 2.2 *Thickness Measurements with Ultrasonic Technique*

As for the Ultrasonic Non-Destructive Test (US-NDT), they can be performed by several techniques [17, 18]. Techniques based on the pulse-echo approach, which represents the amplitude of the detected echoes versus the signal covered distance or the elapsed time, are the most widely adopted. In such applications, the ultrasonic beam is injected into the object tested by a transmitting probe, reflected by its edges and finally detected by a receiving probe. In some cases, the same probe acts as a transmitting and receiving probe. In this way, the evaluation of the number of echoes received, their amplitudes and their shape characteristics can reveal the geometric characteristics of the sample. In detail, knowing the ultrasonic propagation speed in the considered material (v), the thickness is identified by Eq. (3), where t is the time measured between the transmitted signal and the received one.

$$\text{Thickness} = t \cdot v \tag{3}$$

In particular, using the pulse-echo technique, Eq. (3) is divided by a factor 2, Eq. (4), since the measured time is relative to the round-trip path of the ultrasonic signal in the material.

$$\text{Thickness}_{\text{pulse-echo}} = (t \cdot v)/2 \tag{4}$$

## 3 The Measurement Set Up and the First Experimental Results

A schematic block diagram of the measurement setup is shown in Fig. 1. The measurement station realized, common for both measurement techniques used, consists of a hardware section, including the instrumentation for the execution of the test, and a software section dedicated to the generation of excitation signals for the probes, conditioning and digital processing of the acquired data.

Referring to Fig. 1, the hardware section is based on the use of a TIE PIE Engineering Handyscope HS5-540XMS-W5™ digital scope card. It is a high-speed USB oscilloscope, in which the arbitrary analog output channel is used to provide excitation signals to the probes, while the two analog input channels acquire the received signals. Finally, a Labview™ based software running on a Personal Computer guides the TIE PIE system through a suitable graphical environment.

In the application of the induced current method, a probe consisting of two coaxial coils, one as excitation coil and the other as a receiver coil, were used [3], both having

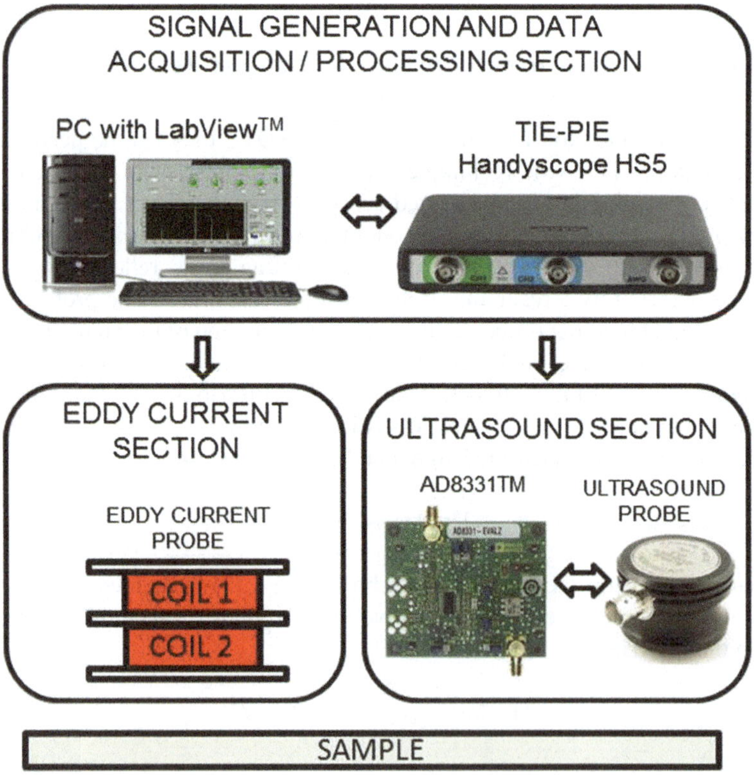


**Fig. 1** A schematic block diagram of the measurement setup for eddy current and ultrasound test

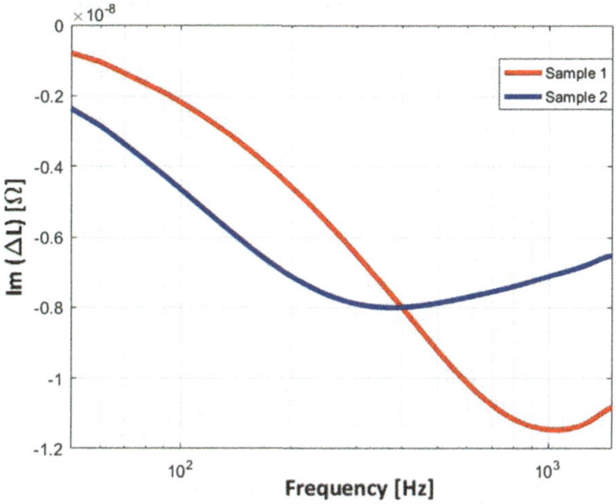


**Fig. 2** The trend of Im(ΔL) when the frequency of the excitation signal changes on the examined plates

the same. The tests were carried out going to feed the upper coil of the probe with a constant sinusoidal current while the voltage on the lower coil was measured. Several tests have been carried out going to vary the frequency of the excitation signal from 200 Hz to 20 kHz, measuring for each frequency value the total impedance of the probe. Figure 2 shows the trends of the quantities of interest obtained in the tests performed on the plates described in Fig. 5.

As regards the application of the ultrasonic method, the probe used is a piezo-ceramic contact probe, that is Olympus Videoscan V109RB™ with a frequency bandwidth of 5 MHz, used both as an excitation transmission probe and as an echo reception probe respectively.

For signal conditioning for the probe, an Analog Device AD8331TM amplifier is used, with a gain up to 44 dB and a maximum frequency of 120 MHz.

The tests were carried out by feeding the transmission probe with different excitation signals (chirp signals, sinusoidal signals and impulsive signals), with frequency ranges from 3 to 7 MHz.

Several tests have been carried out by varying the duration of each excitation signal and measuring the time delay of the reflection echoes, using a special processing software based on the matched and filter technique. Figure 3 shows an example of an ultrasonic signal, transmitted and received in the time domain, by the measurement system described above. While Fig. 4 shows the same signal after digital processing, in which it is possible to observe that the thickness of the analyzed sample is identified by the peak of the signal.

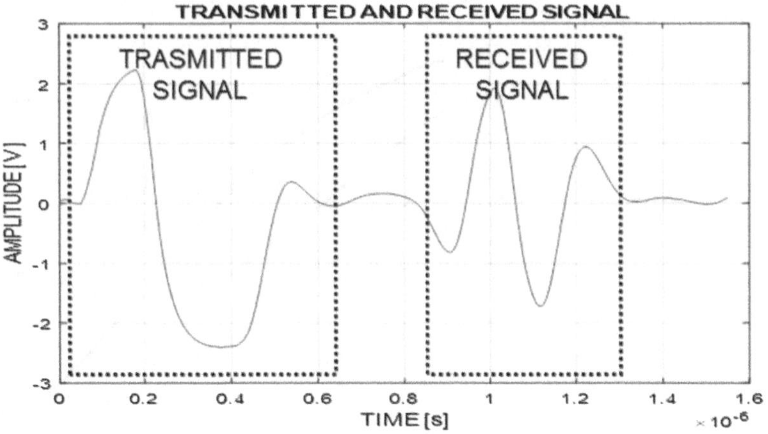


**Fig. 3** Example of the ultrasonic signal transmitted and received for thickness measurement

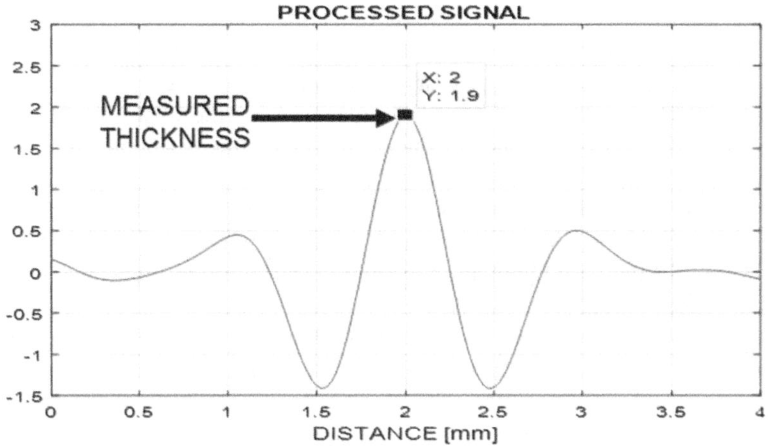


**Fig. 4** The ultrasonic signal after digital processing for thickness measurement

## 3.1 *First Experimental Results*

All tests were performed on two 2024T3 aluminum plates, Fig. 5. Both plates have a square shape, with a side of 20 cm. The thickness of the aluminum plates is 2 mm and 1 cm respectively.

The results obtained are shown in Table 1. It can be observed that with both measurement methods the maximum error committed is less than 0.8%.

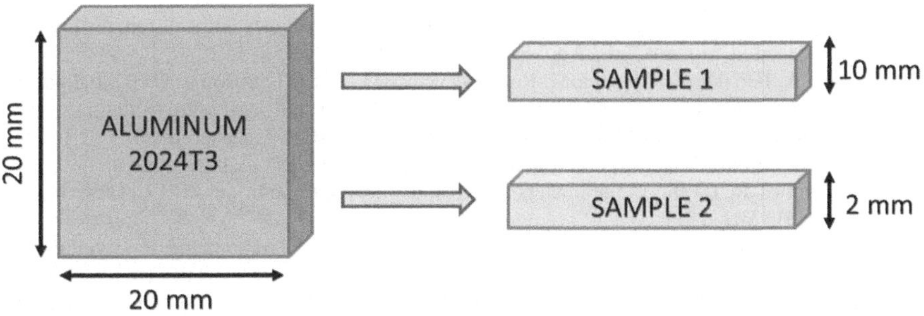


**Fig. 5** The samples tested for thickness measurements

**Table 1** First experimental results

| Sample thickness [mm] | Measured thickness [mm] | | Error [%] | |
|---|---|---|---|---|
| | EC | UT | EC | UT |
| 2 | 2.011 | 2.003 | 0.55 | 0.15 |
| 10 | 9.921 | 10.047 | −0.79 | 0.47 |

## 4 Conclusion

The document proposes an experimental comparison of two non-destructive techniques for measuring thickness. The results obtained show, for both measurement techniques, that the metrological performances are acceptable for the context of industrial applications and are comparable to those of the solutions currently used. Current efforts are focused on the development of methods for increasing the immunity of the system to external disturbances.

Future work will concern the improvement of the analyzed techniques and the development of experimental set-ups able to improve performance and guarantee its applicability in industrial environments.

## References

1. Directive 85/374/EEC, amended by Directive 99/34/EC, "EU Directive on the responsibility for defective products," 1999
2. Bernieri A, Betta G, Ferrigno L, Laracca M (2013) Improving performance of GMR sensors. IEEE Sens J 13(11):4513–4521
3. Bernieri A, Ferrigno L, Laracca M, Tamburrino A (2007) Improving GMR magnetometer sensor uncertainty by implementing an automatic procedure for calibration and adjustment. In 2007 IEEE instrumentation and measurement technology conference
4. Betta G, Ferrigno L, Laracca M (2002) Calibration and adjustment of an eddy current based multi-sensor probe for non-destructive testing. In: 2002 proceedings of the ISA/IEEE sensors for industry conference, Sicon/02, pp 120–124

5. Bernieri A, Betta G, Ferrigno L, Laracca, M (2012) Multi-frequency eddy current testing using a GMR based instrument. Int J Appl Electromagnet Mech 39:355–362
6. Bernieri A, Ferrigno L, Laracca M, Rasile A, Ricci M (2018) Ultrasonic NDT on aluminum bars: An experimental performance comparison of excitation and processing techniques. Meas: J Int Meas Confederation 128:393–402. https://doi.org/10.1016/j.measurement.2017.10.040
7. Ferrigno L, Laracca M, Liguori C, Pietrosanto A (2012) An FPGA-based instrument for the estimation of R, L, and C parameters under nonsinusoidal conditions. IEEE Trans Instrum Meas 61(5):1503–1511
8. Sorrentino L, Bellini C, Capriglione D, Ferrigno L (2015) Local monitoring of polymerization trend by an interdigital dielectric sensor. Int J Adv Manuf Technol 79:1007–1016
9. Betta G, Capriglione D, Ferrigno L, Miele G (2008) Experimental investigation of the electromagnetic interference of ZigBee transmitters on measurement instruments. IEEE Trans Instrum Meas 57(10):2118–2127
10. Amicone D, Bernieri A, Ferrigno L, Laracca M (2009) A smart add-on device for the remote calibration of electrical energy meters. In: Proceedings of IEEE instrumentation and measurement technology conference, I2MTC 2009, pp 1599–1604
11. Ferrigno L, Paciello V, Pietrosanto A (2011) Visual sensors for remote metering in public networks. In: 2011, proceedings of IEEE instrumentation and measurement technology conference, pp 1006–1011
12. Bernieri A, Betta G, Ferrigno L, Laracca M (2012) Electrical energy metering in compliance with recent european standards. In: 2012, proceedings of international instrumentation and measurement technology conference, pp 1541–1545
13. Dodd CV, Deeds WE (1968) Analytical solutions to eddy-current probe-coil problems. J Appl Phys 39:2829
14. Lu M, Yin L, Peyton AJ, Yin W (2016) A novel compensation algorithm for thickness measurement immune to lift-off variations using eddy current method. IEEE Trans Instrum Meas 65(12)
15. Wang H, Li W, Feng Z (2015) Noncontact thickness measurement of metal films using eddy-current sensors immune to distance variations. IEEE Trans Instrum Meas 64(9):8
16. Pallav P, Gan TH, Hutchins DA (2007) Elliptical-Tukey chirp signal for high-resolution, aircoupled ultrasonic imaging. IEEE Trans Ultr, Ferr Freq Contr 54(8):1530–40
17. Ricci M, Senni L, Burrascano P (2012) Exploiting pseudorandom sequences to enhance noise immunity for air-coupled ultrasonic nondestructive testing. IEEE Trans Instrum Meas 61(11):2905–2915
18. Ykhlef F, Arezki M, Guessoum A, Berkani D (2004) A wavelet denoising method to improve detection with ultrasonic signal. In: Proocedings of 2004 IEEE international conference on industrial technology, pp 1–352 6

# Long Period Fiber Grating Sensors Fabricated by Electric Arc Discharge Technique

**Anubhav Srivastava, Flavio Esposito, Agostino Iadicicco and Stefania Campopiano**

**Abstract** This work reports about the fabrication and characterization of Long Period Gratings (LPGs) sensors in pure silica core optical fibers having significant differences in physical and geometrical design, by means of Electric Arc Discharge (EAD) technique. EAD leads to a point-by-point LPG inscription, due to localized tapering of the transversal size of the core and cladding regions along the fiber, and to changes of the silica refractive index due to the stress relaxation induced by local hot spots. LPG in standard fiber is well known for its physical, chemical and biological sensing while specialty fibers permit to widen the horizon of application of fiber optic technology towards unconventional field of research. For instance, pure silica fibers are extensively appealing in high energy and space applications. This work aimed at identifying an appropriate "recipe" for each fiber, to fabricate LPGs with strong and narrow attenuation bands, trivial power loss supported by smaller grating length. Hence, a suitable combination of arc power, arc time, fiber tension and electrodes gap, is a must for the appropriate core and cladding modulation and in turn for the desired LPG spectral features. Finally, the surrounding refractive index (SRI) and temperature characterization of these LPGs were performed to investigate the sensitivity features.

**Keywords** Electric arc discharge technique · Pure silica fiber · Long period grating sensor

## 1 Introduction

A long period grating (LPG) whose period typically lies in the range of 100 μm–1 mm, is a wavelength selective in-fiber device in which periodic longitudinal modulation can be created by acting on the refractive index of silica and/or on the waveguide geometry along an optical fiber, that couples the fundamental core mode and forward propagating cladding modes at specific resonant wavelengths. As a result,

A. Srivastava · F. Esposito · A. Iadicicco (✉) · S. Campopiano
Department of Engineering, University of Naples Parthenope, Centro Direzionale di Napoli Isola C4, 80143 Naples, Italy
e-mail: iadicicco@uniparthenope.it



G. Di Francia et al. (eds.), *Sensors and Microsystems*, Lecture Notes in Electrical Engineering 629, https://doi.org/10.1007/978-3-030-37558-4_59

the transmission spectrum is characterized by discrete attenuation bands which obey the phase-matching condition given by:

$$\lambda_{res,i} = \left(n_{eff,co} - n_{cl,i}\right) \cdot \Lambda \quad (1)$$

where $n_{eff,co}$ and $n_{cl,i}$ are the effective refractive index of the core and of the ith order cladding mode respectively and $\Lambda$ is the grating spatial period. Vengsarkar et al. for the first time in 1996 highlighted about LPGs devices for gain equalization in optical communication systems along with its sensitivity towards the effects induced by strain and temperature, which in turn established the foundation for their use in different domain of sensing applications [1, 2]. LPG being one of the most demanding devices nowadays, is getting employed in various biological and chemical applications due to its highly intrinsic property of sensitivity to surrounding refractive index (SRI) changes [3, 4].

There are different techniques available to fabricate LPGs, the most important of being namely UV radiation [3], $CO_2$ lasers [5], IR femtosecond lasers [6], and electric arc discharge (EAD) [7].

The EAD technique, which creates the perturbation by applying an arc discharge to the fiber with specific periodicity, has gained significant attention in the recent years due to having an intrinsic property of inscribing the gratings in all kinds of fiber. In addition, it is flexible, economical and requires simple instrumentation since it eliminates the need for expensive laser systems, as well as avoiding the need for pre-hydrogenation of the fiber and consequent post thermal annealing to stabilize the gratings. Since recent years, EAD technique has been applied to fabricate the LPG in pure-silica core and radiation resistant fibers [8–11], Er-doped [12], P-doped fibers [13], polarization-maintaining [14] and micro-structured fibers [15–17]. The most important limitation of EAD technique lies in the difficulty in fabricating the LPGs with smaller spatial periods (i.e. shifting to higher order cladding modes), for which many efforts are still in progress.

Nowadays, specialty optical fibers (i.e. pure silica core fibers) is getting highly demanded for the areas ranging from low radiation levels of medical applications to space-borne equipments and high energy applications, assessed in terms of their resistance under various radiation environments. The predominant aspect concerning radiation (i.e. gamma, neutron…) environments is the radiation-induced attenuation (RIA), for which pure silica core fibers show to have more resistance to it, which in turn can be applied as an environmental sensor in radiation conditions or in radiation dosimetry. Recently we fabricated LPGs in F-doped fiber through EAD and investigated them under different radiation environments [9, 11].

In this work, the fabrication of LPGs is reported with our recent experimental results, by employing EAD technique in different pure silica core optical fibers i.e. endlessly single mode PCF from NKT Photonics and Nufern solid F-doped cladding single mode fiber. For the fiber stated above, a proper parameter set has been identified to fabricate LPGs with deep and narrow attenuation bands (>20 dB), trivial power loss along with grating length (25–30 mm). The sensitivity features towards SRI and temperature changes has also been reported here.

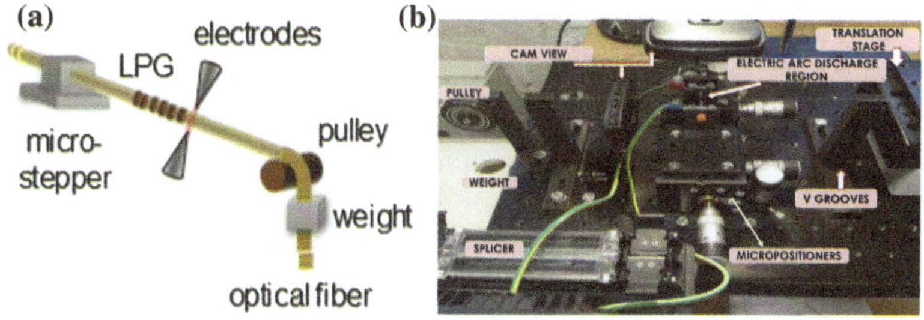


**Fig. 1** **a** Schematic of EAD fabrication set up and **b** Real time experimental set up

## 2 Fabrication Mechanism of LPGs via EAD Technique

The basic working principle for fabrication of LPGs via EAD technique involves a certain stripped section of an optical fiber, firmly fixed on a remotely controlled precision translation stage at one side and through constant axial tension at other side, positioned between the two electrodes in order to create perturbation with even power distribution via arc discharge.

The state-of-the-art schematic and computer assisted step-by-step fabrication platform for LPGs is reported in Fig. 1a and b. Here, the main effect of the arc discharge is of two-fold: (i) a modification in the geometry of the optical fiber, and (ii) a change in the silica refractive index, due to the stress relaxation induced by local hot spots. Overall setup was designed to permit the fabrication of gratings without creating any bending in the fiber during the process. The selection of the power and duration of the electrodes supply current, as well as the axial tension in the fiber, allows the control of the modulation strength and thus LPG spectral features. Moreover, we acted on the gap that is reduced up to 0.9 mm between the two electrodes to permit the fabrication of LPGs with lower periods to achieve higher order cladding modes. Finally, the LPG transmission spectra were recorded by using optical spectrum analyzer (OSA) Yokogawa AQ6370B and a broadband source (SLED in range 1100–1700 nm).

Here, we fabricated LPGs in two different types of pure silica core optical fibers as summarized in Table 1, for which the period and arc discharge parameters were chosen based on the type of fiber used (Fig. 2).

## 3 SRI and Temperature Sensitivity Characterization

The sensitivity characteristics towards SRI and temperature changes for the LPGs fabricated in above stated fibers are discussed in this section and the experimental set up for performing these characterizations are shown in Fig. 3a and b.

The SRI sensitivity arises from the dependence of the effective refractive index of the cladding modes on the external medium refractive index, hence from Eq. (1): we have,

**Table 1** Experimental aspect for selected pure silica core fibers

| Fiber model | $D_{core}$ (μm) | $D_{clad}$ (μm) | MFD (μm) @ 1.55 μm | NA | Core doping | Grating period (μm) | Electrodes gap | Arc step | Arc time (ms) | $\lambda_{res}$ (nm) | Depth (dB) |
|---|---|---|---|---|---|---|---|---|---|---|---|
| NKT ESM 12 | 12.0 | 125 | $10.5 \pm 1.0$ | 0.11 | Pure silica | 550 | 0.9 | 5 | 650 | 1420 | 24.77 |
| Nufern S1310 | 8.0 | 125 | $10.4 \pm 0.8$ | 0.12 | Pure silica | 445 | 0.9 | 2 | 650 | 1577 | 31.83 |

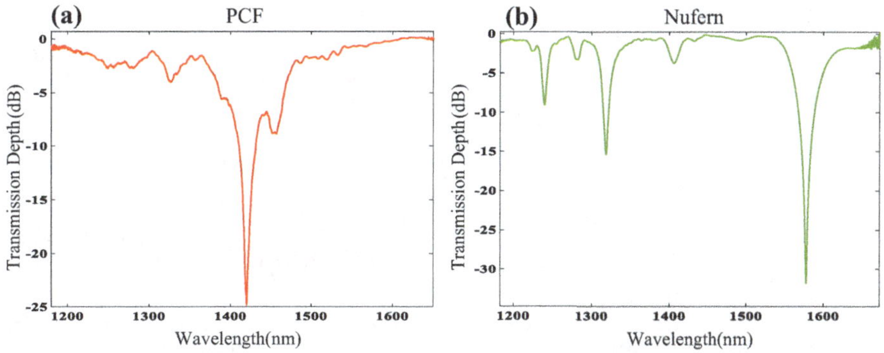


**Fig. 2** Transmission spectrum of LPGs in: **a** PCF ($\Lambda = 550\ \mu$m), **b** Nufern fiber ($\Lambda = 445\ \mu$m)

**Fig. 3** LPG characterization set up for **a** SRI and **b** Temperature

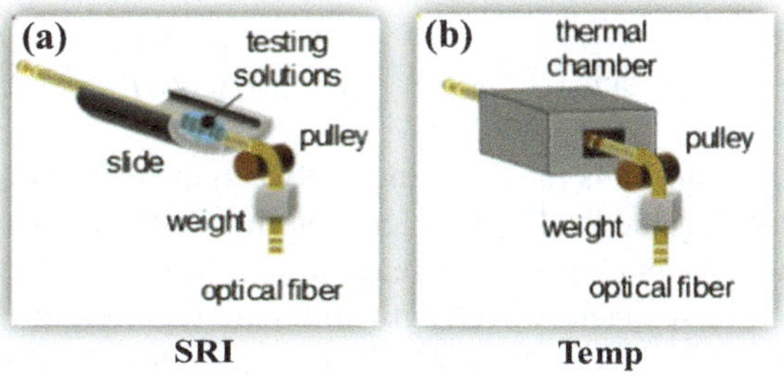


$$\frac{\partial \lambda_{res,i}}{\partial n_{ext}} = \frac{\partial \lambda_{res,i}}{\partial n_{cl,i}} \cdot \frac{\partial n_{cl,i}}{\partial n_{ext}} \tag{2}$$

where $n_{cl,i}$ is the effective refractive index of the cladding mode and $n_{ext}$ is the surrounding medium refractive index.

The LPGs were characterized towards SRI changes, by placing them inside the liquid environment with known refractive index ranges from 1.33 to 1.44. The wavelength shifts of an attenuation band for the different LPGs are reported in Fig. 4a, with respect to the value in air. Generally, for fibers with solid core and cladding regions made of different materials, the attenuation bands experience shifts towards lower wavelengths when SRI increases, with maximum sensitivity occurring at refractive index values approaching to that of the cladding, while in PCF, it experienced red shifts. In particular, the sensitivity ($S_{1.33-1.35}$) for the PCF and Nufern fiber is found to be at +37.94 nm/RIU and −98.47 nm/RIU respectively.

The temperature characterization of LPGs for which the set up as shown in Fig. 3b, has been done by placing the grating region into an aluminium chamber housed on a precision heater having length of 10 cm and diameter of 2 mm, whose real time monitoring was done using a commercial FBG-based sensor as a reference and the range investigated was 20–100 °C. The temperature sensitivity of LPG can be derived from Eq. (1), yielding for a given cladding mode:

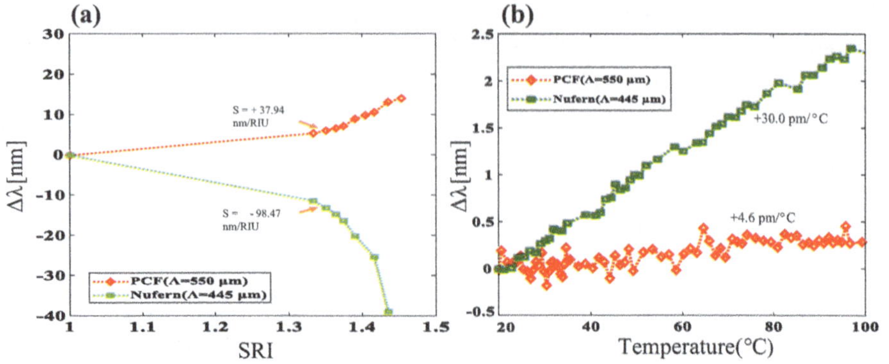


**Fig. 4** Sensitivity analysis of PCF and Nufern LPG in **a** SRI characterization and **b** Temperature characterization

$$\frac{\partial \lambda_{res,i}}{\partial T} = \lambda_{res,i} \cdot \left[ \frac{1}{\delta n_{eff,i}} \cdot \left( \frac{\partial n_{eff,co}}{\partial T} - \frac{\partial n_{cl,i}}{\partial T} \right) + \left( \frac{1}{\Lambda} \cdot \frac{\partial \Lambda}{\partial T} \right) \right] \quad (3)$$

where $\delta n_{eff,i}$, is the difference between the effective refractive index of the core and cladding mode. The first term on the RHS side is related to the core and cladding thermo-optic effects whereas the second term is due to silica thermal expansion.

The resonant wavelength shifts are reported in Fig. 4b with respect to the value at a room temperature of 25 °C. Depending on the type of fiber and its structure, the attenuation bands could experience shifts towards both shorter and longer wavelengths with temperature increasing, following a linear dependence in the range under investigation. Here, the temperature sensitivity is found to be positive for PCF and Nufern fiber with sensitivities noted as 4.6 pm/°C and 30.0 pm/°C, respectively. The lower temperature sensitivity for PCF could be attributed to the fact that the pure silica fibers has lower thermo-optic and thermal expansion co-efficient. Also, since light propagates mainly through the silica, and hence the variations in the effective indices of the core and cladding modes with the refractive index of silica nearly cancels each other. The overall effect is a red shift of only a few picometers which signals about weak sensitivity of PCF LPG to temperature and hence can act as a temperature insensitive device. While Nufern LPG showed higher temperature sensitivity than PCF but lower than standard Ge-doped core fiber due to the fact that Fluorine depresses the cladding thermo-optic coefficient.

## 4 Conclusions

The fabrication of LPGs by means of EAD technique with our recent results, in different pure silica core fibers as stated above has been analyzed and discussed. The above LPGs is presented with strong attenuation bands (20–30 dB), trivial power loss (<0.5 dB), and shorter length and characterized towards SRI and temperature changes. The resonance wavelength was found to be red shifted in PCF while it was blue shifted in Nufern fiber during SRI tests. Temperature sensitivity shows that the different pure silica fiber resulted in the same positive polarity of their thermal response where PCF sensitivity was found to be 4.6 pm/°C, which is much lower than Nufern fiber which was noted as 30.0 pm/°C. These results are useful for the design and fabrication of LPG-based sensors to be employed in bio-chemical sensing and sensitive radiation environments.

**Acknowledgements** This work was supported by the University of Naples "Parthenope" through the "Bando di sostegno alla ricerca individuale per triennio 2015–2017, Annualità 2017".

## References

1. Vengsarkar AM, Lamaire PJ, Judkins JB, Bhatia V, Erdogan T, Sipe JE (1996) Long-period fiber gratings as band rejection filters. J Light Technol 14:58–65
2. Bhatia V (1999) Applications of long-period gratings to single and multi-parameter sensing. Opt Express 4:457–466
3. Esposito F, Ranjan R, Campopiano S, Iadicicco A (2017) Experimental study of the refractive index sensitivity in arc-induced long period gratings. IEEE Photonics J 9:1–10
4. Esposito F, Sansone L, Taddei C, Campopiano S, Giordano M, Iadicicco A (2018) Ultrasensitive biosensor based on long period grating coated with polycarbonate-graphene oxide multi-layer. Sens Actuators: B Chem 274:517–526
5. Wang Y (2010) Review of long period fiber gratings written by $CO_2$ laser. J Appl Phys 108:81101
6. Kondo Y, Nouchi K, Mitsuyu T, Watanabe M, Kazansky PG, Hirao K (1999) Fabrication of long-period fiber gratings by focused irradiation of infrared femtosecond laser pulses. Opt Lett 24:646–648
7. Rego G (2016) Arc-induced long period fiber gratings. J Sens (2016) 3598634
8. Rego G, Fernandez Fernandez A, Gusarov A, Brichard B, Berghmans F, Santos JL, Salgado HM (2005) Effect of ionizing radiation on the properties of arc-induced long-period fiber gratings. Appl Opt 44:6258–6263
9. Esposito F, Ranjan R, Stăncălie A, Sporea D, Neguţ D, Becherescu N, Campopiano S, Iadicicco A (2017) Real-time analysis of arc-induced long period gratings under gamma irradiation. Sci Rep 7:43389
10. Esposito F, Ranjan R, Campopiano S, Iadicicco A (2018) Arc-induced long period gratings from standard to polarization-maintaining and photonic crystal fibers. Sensors 18:918
11. Esposito F, Stancalie A, Negut D, Campopiano S, Sporea D, Iadicicco A (2019) Comparative investigation of gamma radiation effects on long period gratings and optical power in different optical fibers. J Lightwave Technol 37:4560–4566
12. Esposito F, Campopiano S, Iadicicco A (2019) Arc-induced long period gratings in erbium-doped fiber. IEEE Photonics J 11:1–8
13. Ranjan R, Esposito F, Iadicicco A, Campopiano S (2017) Arc-induced long period gratings in phosphorus-doped fiber. IEEE Photonics Technol Lett 29:611–614
14. Esposito F, Srivastava A, Iadicicco A, Campopiano S (2019) Multi-parameter sensor based on single long period grating in panda fiber for the simultaneous measurement of SRI, temperature and strain. Opt Laser Technol 113:198–203
15. Hu H, Du C, Wang Q, Wang X, Zhao Y (2017) High sensitivity internal refractive index sensor based on a photonic crystal fiber long period grating. Instrum Sci Technol 45:181–189

16. Petrovic JS, Dobb H, Mezentsev VK, Kalli K, Webb DJ, Bennion I (2007) Sensitivity of LPGs in photonic crystal fibers fabricated by an electric arc to temperature, strain, and external refractive index. J Lightwave Technol 25:1306–1312
17. Iadicicco A, Ranjan R, Campopiano S (2015) Fabrication and characterization of long period gratings in hollow core fibers by electric arc discharge. IEEE Sens J 15:3014–3020

# Relative Humidity Sensor Based on Tip of Multimode Optical Fiber Integrated with Photonic Crystal of Hydrogel Coated Polystyrene Nanoparticles

**Pasquale Di Palma, Lucia Sansone, Chiara Taddei, Stefania Campopiano, Michele Giordano and Agostino Iadicicco**

**Abstract** A relative humidity (RH) sensing device based on photonic crystal colloids auto-assembled directly on the tip of a multi-mode optical fiber is here described. The core-shell nanospheres are obtained by Polystyrene (PS) nanoporous structures (core) incorporated into a poly(N-isopropylacrylamide) (PNIPAM) hydrogel (shell) and are subsequently deposited via a vertical dip-coating technique on a multimode optical fiber tip. The sensing capability is due to the hydrogel shell that swells depending on the relative humidity that induce wavelength shift and amplitude changes in the photonic bandgap peak. Spectral characterization and sensing analysis are performed, and the obtained results show that spectral characteristics are highly non-linear and strongly manufacturing dependent. A resolution of 1% RH is measured at 30% RH that significantly decreases at higher RH. Hereby, numerical simulations based on the finite element method (FEM) have been performed in order to investigate the sensing mechanism.

**Keywords** Optical fiber sensors · Deflection monitoring · Structural health monitoring · Strain measurement

## 1 Introduction

The measurement of relative humidity (RH) is of a great significance in several industrial and research applications and thus intensive effort has been devoted to the development of reliable sensors. Even if a large number of electrical and optoelectronics humidity sensors have been proposed in last decades [1, 2], there is still a need to search for other low cost, highly stable and low power consumption devices, which can be used in a wide range of areas. On the other side, in-fiber optic sensor

P. Di Palma · S. Campopiano · A. Iadicicco (✉)
Department of Engineering, University of Naples Parthenope, Centro Direzionale Isola C4, Naples 80143, Italy
e-mail: iadicicco@uniparthenope.it

L. Sansone · C. Taddei · M. Giordano
Institute for Polymers, Composites and Biomaterials, IPCB-CNR, 80055, Portici, Italy
e-mail: gmichele@unina.it



G. Di Francia et al. (eds.), *Sensors and Microsystems*, Lecture Notes in Electrical Engineering 629, https://doi.org/10.1007/978-3-030-37558-4_60

technology has gained a lot of popularity offering [3–6] numerous advantages over conventional electrical sensors, such as small size, intrinsic multiplexing capability, chemical inertness, immunity to electromagnetic interference, chemical inertness and remote sensing ability [7–10]. In this scenario, the research of integration between photonic crystal (PhC) lattice structures and optical fiber technology is increased. PhCs can offer unique characteristic to manipulate the light or permitting light matter interaction.

Recently, we demonstrated an efficient and cost effective approach to integrate a PhC made of polymeric nanospheres with optical fiber technology [11]. The PhC was made of commercial polystyrene (PS) nanospheres with 200 nm diameter deposited by self-assembly vertical deposition technique directly on fiber optic tip. The resonances band in the reflected signal shown significant sensitivity to surrounding refractive index. In this paper, a similar methodology is explored with core-shell nanospheres.

## 2 Experimental

With the aim of obtain RH sensing devices, PS-PNIPAM core/shell nanoparticles were chemically prepared as in [12] and then deposited on the tip of an optical fiber. An IR-VIS multimode silica fiber optic, 125/105 μm cladding/core diameter, was selected and then prepared by using a high precision cleaver.

PS-PNIPAM nanospheres are deposited directly on the optical fiber tip by means of an automated dip coater system. An ordered 3D structure of core-shell nanospheres has been formed on the fiber tip after the evaporation of solvent, as shown by the scanning electron microscope (SEM) image in Fig. 1.

In order to test the fabricated device as function of the environmental relative humidity the setup, schematically plotted in Fig. 1, has been used. The optical

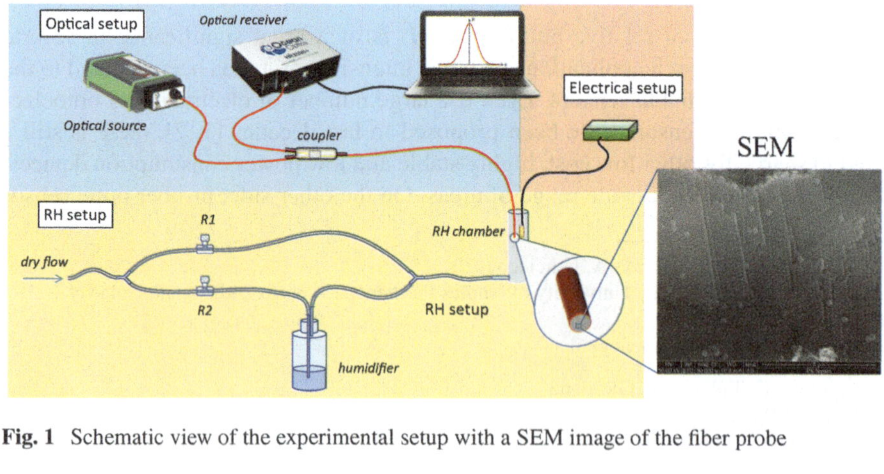


**Fig. 1** Schematic view of the experimental setup with a SEM image of the fiber probe

part of the setup, for the reflectance spectral monitoring, consists of a spectrometer (OceanOptics HR2000 + 190–1100 nm) for reading the optical spectra, a 1 × 2 50/50 coupler that enable the reflection interrogation and a VIS optical source (Avantes AvaLight-HAL-(S)-Mini). Moreover, to change the RH around the probe a dry air source (~20% RH) is switched in two lines equipped with regulators: the first line is connected directly to the chamber, the second one is connected via a humidifier. Hence, by varying the flow regulators R1 and R2, this setup permits to change the relative humidity from 20 to 95% in a chamber where are placed the optical probe and an electrical humidity sensor used as reference (model HIH-4000).

Optical spectra exhibit resonant peaks starting from 470 nm wavelength attributable to the photonic bandgap of the ordered nanospheres structure assembled on the fiber tip.

The spectral characterization of the sensing probe as function of the environmental RH reveals a red shift of the resonant wavelength and a decreasing of the peak amplitude occur increasing the relative humidity, as you can see from the Fig. 2a where all spectra are normalized to the reflectance of the bare fiber signal. This behaviour is attributable to the PhC core-shell nanoparticles and, in particular, to the fact that hydrogel-shells can reversibly swell (or shrink) as function of water absorption amounts.

Concerning the peak wavelength trend vs RH, we can observe that the sensitivity increases with RH itself, Fig. 2b. Indeed, for RH around 30% sensitivity of about 30 pm/RH% is estimated, whereas, when the RH moves to the 80 and to 95% the sensitivity of the sensor moves to 300 and 6000 pm/RH%. Moreover, a good reversibility is obtained. Finally, for comparison we also investigated the bare PS based sample versus RU: as expected, the absence of the hydrogel shell does not induce wavelength shift as function of the RH.

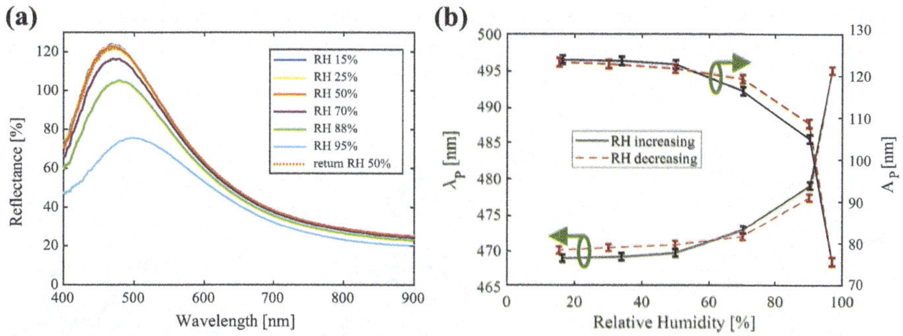


**Fig. 2** **a** Experimental spectra as function of relative humidity; and **b** resonant peak wavelength and peak amplitude during increasing and decreasing of the relative humidity

## 3 Numerical Modelling

In order to investigate the sensing mechanism, numerical simulation based on the finite element method (FEM) has been performed by means of the commercial software COMSOL Multiphysics (RF module). Basing on the swelling capability of the PNIPAM shell, two limit behaviors (schematically depicted in Fig. 3a–c) have been hypothesized and a mathematical model for the spatial period has been built. In the first case it is supposed that the hydrogel expands into the empty spaces between the nanospheres without modifying the spatial period of the photonic crystal when the RH increases. Instead in the second the conservation of the spherical shape in the nanoparticles is assumed, therefore the filling factor is constant, and the spatial period increases with the increase of the RH.

In practice, we hypothesized that an intermediate behavior occurs, then the lattice period $\Lambda$ can be modelled as:

$$\Lambda = \Lambda_0 + 2 \cdot \alpha \cdot \Delta t_{shell}$$

where $\Lambda_0$ is the lattice period for 0% RH, $\Delta t_{shell}$ is the shell thickness variation and the parameter $\alpha$ describes the different possible swelling behaviors of the photonic crystals.

Numerical spectra as function of the shell thickness $t_{shell}$, ranging in 5–15 nm, in the two limit behaviors, $\alpha = 0$ and $\alpha = 1$, are reported in Fig. 4a, b respectively. The simulation parameters are set after the morphological characterizations done with SEM, TEM (transmission electron microscopy) and DLS (dynamic light scattering) techniques as in [13].

In both cases, the increase of $t_{shell}$ (swelling of the hydrogel) induces red shift of the reflected resonant peak. The resonant wavelength shifts to 496.9 and 500.8 nm for $\alpha = 0$ and $\alpha = 1$, respectively, when the $t_{shell}$ moves to 15 nm. Instead, the peak amplitude moves in different direction in dependence on the $\alpha$ value. The former case

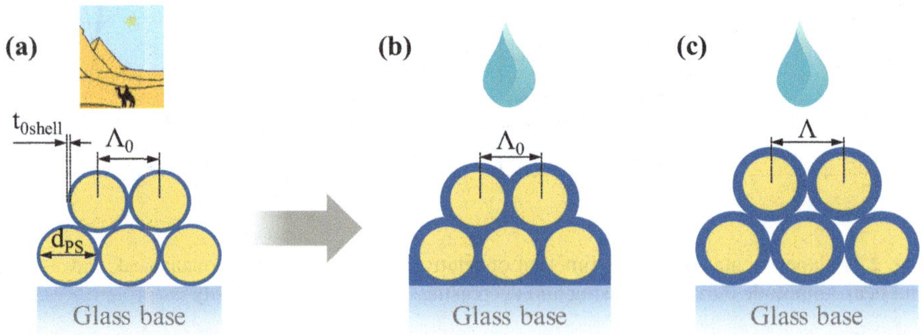


**Fig. 3** Schematic view of the core-shell lattice structure in **a** dry environmental and in wet environmental with **b** $\alpha = 0$ e and **c** $\alpha = 1$

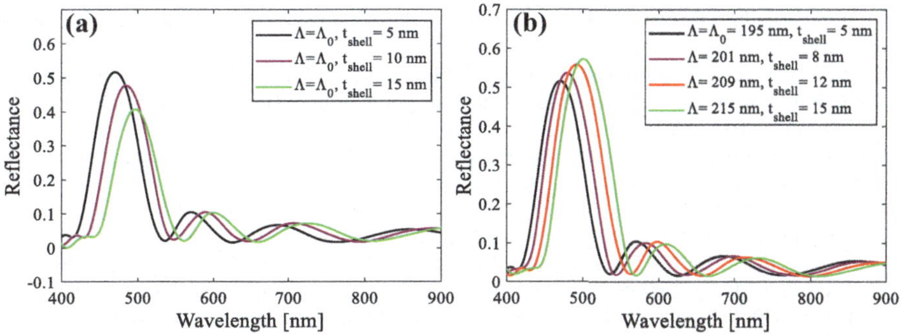


**Fig. 4** Numerical spectra as function of the shell thickness increase when **a** $\alpha = 0$ and **b** $\alpha = 1$

exhibits a decrease of the peak amplitude of 21% behavior whereas in the second case the peak amplitude slightly increases of 11%.

## 4 Conclusion

In this work, a core-shell PS-PNIPAM nanospheres suspension is prepared and deposited on a multimode optical fiber by means of a vertical dip-coating method combining the intrinsic humidity sensitivity of the hydrogels with the intrinsic periodicity dependence of the photonic crystal template.

Under the action of the relative humidity, the PNIPAM shell swells and a red shift in the reflectance spectrum is induced. The fabricated device exhibits a very interesting sensibility when the RH exceeds the 50%.

The simulations done for the two limit behaviors have highlighted that in both cases a resonant red shift occurs, instead the amplitude decrease is principally attributable to the gap filling due to the refractive index contrast decreasing and/or increasing of the filling factor.

The comparison between the numerical and experimental results showed that the swelling of the hydrogel-shells should mainly fill the air gap between the spheres. We think that this is due to the very small PNIPAM Young's modulus.

Further study, experimentation and characterization on this topic are devoted to investigate the reproducibility of the sensor fabrication and sensing characteristics.

## References

1. Yamazoe N, Shimizu Y (1986) Humidity sensors: principles and applications. Sens Actuators 10:379–398
2. Alwis L, Sun T, Grattan KTV (2013) Optical fibre-based sensor technology for humidity and moisture measurement: review of recent progress. Meas J Int Meas Confed 46:4052–4074

3. Ascorbe J, Corres J, Arregui F, Matias I (2017) Recent developments in fiber optics humidity sensors. Sensors 17:893
4. Rajan G, Noor YM, Liu B, Ambikairaja E, Webb DJ, Peng GD (2013) A fast response intrinsic humidity sensor based on an etched singlemode polymer fiber Bragg grating. Sens Actuators, A Phys 203:107–111
5. Su D et al (2013) A fiber Fabry-Perot interferometer based on a PVA coating for humidity measurement. Opt Commun 311:107–110
6. An J, Zhao Y, Jin Y, Shen C (2013) Relative humidity sensor based on SMS fiber structure with polyvinyl alcohol coating. Opt Int J Light Electron Opt 124:6178–6181
7. Esposito F, Ranjan R, Campopiano S, Iadicicco A (2018) Arc-induced long period gratings from standard to polarization-maintaining and photonic crystal fibers. Sensors 18:918
8. Di Palma P, Palumbo G, Della Pietra M, Canale V, Alviggi M, Iadicicco A, Campopiano S (2019) Deflection monitoring of bi-dimensional structures by fiber Bragg gratings strain sensors. IEEE Sens J 99:1–1
9. Di Palma P, Sansone L, Pannico M, Musto P, Campopiano S, Iadicicco A, Giordano M (2019) Metallic-dielectric colloidal photonic crystal on the multimode optical fiber tip: preliminary results as optical fiber SERS probe. In: Proc. SPIE 11028, optical sensors 2019, p 110282F
10. Li H-N, Li D-S, Song G-B (2004) Recent applications of fiber optic sensors to health monitoring in civil engineering. Eng Struct 26:1647–1657
11. Di Palma P, Taddei C, Borriello A, De Luca G, Giordano M, Iadicicco A, Campopiano S, Sansone L (2017) Self-assembled colloidal photonic crystal on the fiber optic tip as a sensing probe. IEEE Photonics J 9:1–11
12. Di Palma P, Taddei C, Campopiano S, Giordano M, Iadicicco A, Sansone L (2018) Miniaturized fiber optic probe based on colloid crystals of hydrogel coated nanoparticles for relative humidity measurements. In: 26th international conference on optical fiber sensors, pp 1–4
13. Di Palma P, Sansone L, Taddei C, Campopiano S, Iadicicco A, Giordano M (2019) Fiber optic probe based on self-assembled photonic crystal for relative humidity sensing. J Lightwave Technol 37:1–9

# LSTM Networks for Particulate Matter Concentration Forecasting

**S. Ferlito, F. Bosso, S. De Vito, E. Esposito and G. Di Francia**

**Abstract** Last years have witnessed an increasing interest in particulate matter present in the air and a greater awareness for its toxic effects. Fine particulate matter ($PM_{2.5}$) concentration, i.e. particulate with diameter less than 2.5 μm, pose a severe risk to public health. This type of particulate matter due to its tiny dimensions can get deep into lungs and eventually even reach the bloodstream. Recent studies relate high levels of $PM_{2.5}$ to lungs and heart disease. It can be very beneficial to be able to provide reliable forecasts for $PM_{2.5}$ in the short-term horizon (24 h ahead). In this paper, a deep learning model based on a Long Short-Term Neural (LSTM) network has been evaluated. The dataset taken into consideration to validate experimentally the proposed model is publicly available at UC Irvine Machine Learning Repository and consists of hourly $PM_{2.5}$ levels, from Jan 1st, 2010 to Dec 31st, 2015, for Beijing city, as well as atmospheric variables. The developed LSTM network is able to find useful patterns in previous values of $PM_{2.5}$ that in conjunction to exogenous atmospheric variable provide reliable forecasting of $PM_{2.5}$ in the short period 1–6 h, showing a progressive reduction in accuracy as the forecasting horizon increase. The accuracy of the proposed LSTM, expressed by the root mean square error (RMSE), has been evaluated using a rolling window validation methodology. Performance results, within the forecasting horizon of three hours, are interesting and qualify the proposed model as a "soft" indicative measurement method in the framework of current EU AQ Monitoring.

**Keywords** Deep learning · Long short-term memory · $PM_{2.5}$ forecasting

## 1 Introduction

Air pollution is a serious problem that afflicts the urban population and to a greater extent densely populated cities. Among different air pollutant, $PM_{2.5}$ is a particulate matter with an aerodynamic diameter less or equal to 2.5 μm. As stated by last researches in the field [1], this type of particulate matter poses a severe risk for

S. Ferlito (✉) · F. Bosso · S. De Vito · E. Esposito · G. Di Francia
ENEA Research Centre Portici, Portici, Italy
e-mail: sergio.ferlito@enea.it

 
G. Di Francia et al. (eds.), *Sensors and Microsystems*, Lecture Notes in Electrical Engineering 629, https://doi.org/10.1007/978-3-030-37558-4_61

human health causing respiratory and cardiovascular disease. China is one of the regions that is greatly affected by problems caused to $PM_{2.5}$ exposure [2]. As reported in some recent papers [3], the data collected from air quality monitoring stations in 338 big cities in China show that the range of annual average concentrations of $PM_{2.5}$ in 2015 was 11–125 $\mu g/m^3$ with an average value of 50 $\mu g/m^3$. In year 2016 has been registered an increase of $PM_{2.5}$ levels, that varies in the range 12–158 $\mu g/m^3$, although with an average value of 47 $\mu g \times m^{-3}$ with $PM_{2.5}$ being the main pollutant for more than 80.3% of the days with severe pollution [4] (in the year 2015 has registered the 66.8%). $PM_{2.5}$ can consequently be considered as the main cause of atmospheric particulate pollution in China. It can be very beneficial, in this scenario, to be able to provide reliable forecasts for $PM_{2.5}$, especially in the short-term horizon (24 h ahead) so as have at our disposal a useful tool to support decision making for environmental agencies [5]. During the last years, efforts by researchers all around the world have produced several papers focused on methods useful to provide forecasts for $PM_{2.5}$ levels. Techniques usually employed to predict $PM_{2.5}$ concentration levels can be grouped into three main categories: physical-model-based, data-driven models, and hybrid. Belongs to the first category, physical-model-based, models such as Community Multiscale Air Quality (CMAQ) [6]. In the current era of Big Data and Internet of Things (IoT), there is a huge mass of data related to air quality, i.e. measured $PM_{2.5}$ concentration levels as well as atmospheric data. This gave rise to many data-driven methods, among these statistical methods usually employed for time series prediction task, such as AutoRegressive Moving Average (ARIMA) [7], with its variations, Exponential Moving Average (ETS) [8], as well as Machine Learning (ML) based models, most of them employing some type of neural network [9]. Finally, hybrid models combine different physical-based-models with data-driven one to build a comprehensive model that aims to encapsulate the strengths of both types of models [10]. In recent years many papers employing some form of Machine Learning have been published, mainly neural networks of different architecture have successfully been applied to forecast PM. In [11], the authors employed a recursive network known as Long Short-Term Memory (LSTM) using PM concentration values plus meteorological data to provide PM forecast with a horizon of 1–24 h. In design this LSTM model, authors extended classical LSTM architecture exploiting also spatiotemporal data using data from 12 air quality stations located in Beijing; the resulting architecture has been named by authors as LSTME. This peculiar LSTME was able to provide a good accuracy, expressed in term of mean absolute percentage values (MAPE), of 11.93% for forecasts at 1 h ahead and 31.47% for forecasts at 13–24 h ahead. In this paper, a Long Short-Term Neural (LSTM) network has been designed and evaluated. The dataset taken into consideration is publicly available at the Irvine Center for Machine Learning and Intelligent Systems at the University of California (UCI) repository and consists of $PM_{2.5}$ levels as well as atmospheric variables for the city of Beijing. The developed LSTM is able to find useful patterns in previous values of $PM_{2.5}$ that in conjunction to exogenous atmospheric variable provide reliable forecasting of $PM_{2.5}$ in the short

**Table 1** Dataset description

| Variable | Significance & unit measure |
|---|---|
| $PM_{2.5}$ | $PM_{2.5}$ concentration ($\mu g/m^3$) |
| DEWP | Dew point (°C) |
| TEMP | Temperature (°C) |
| PRES | Pressure (hPa) |
| HUMI | Humidity (%) |
| cbwd | Combined wind direction |
| Iws | Cumulated wind speed (m/s) |

period 1–6 h while showing a progressive reduction in accuracy as the forecasting horizon increase. The proposed LSTM has been evaluated with a rolling window validation procedure.

# 2 Data and Methods

## *2.1 Data Description*

The dataset used for training the proposed LSTM model is taken from UCI Repository. It consists of hourly $PM_{2.5}$ concentration values [$\mu g/m^3$] for five Chinese Cites concerning a period that spans from Jan 1st, 2010 to Dec 31st, 2015 for a total of 52,854 records. In this paper, the data concerning Beijing only have been evaluated since this was the most complete dataset and even the more challenging due to the great variability of $PM_{2.5}$ values registered in this city. Table 1 reports a list of variables, with their respective unit of measure, used by the proposed LSTM.

In Fig. 1 is reported a Kernel Density plot for $PM_{2.5}$, with an added rug plot, for the entire dataset. Appear clear that $PM_{2.5}$ values for Beijing show great variability, from a minimum of 1 $\mu g/m^3$ to a maximum of 994 $\mu g/m^3$, a median value of 69 $\mu g/m^3$ and a mean value of 95.90 $\mu g/m^3$. Extremely high values for $PM_{2.5}$, i.e. even greater than 400 $\mu g/m^3$, in Beijing are quite common.

## *2.2 LSTM Model*

Long-Short-Term-Memory (LSTM) neural network is a special type of Recursive Neural Network (RNN). As the name implies, its main feature is its capability to take into account both long and short-term dependency into input sequence, while RNN from which it derives, is not able to take into account long term dependency due to the well-known "vanishing gradient" problem. LSTM was selected to predict $PM_{2.5}$ values as this type of RNN show some peculiar and interesting features: (i)

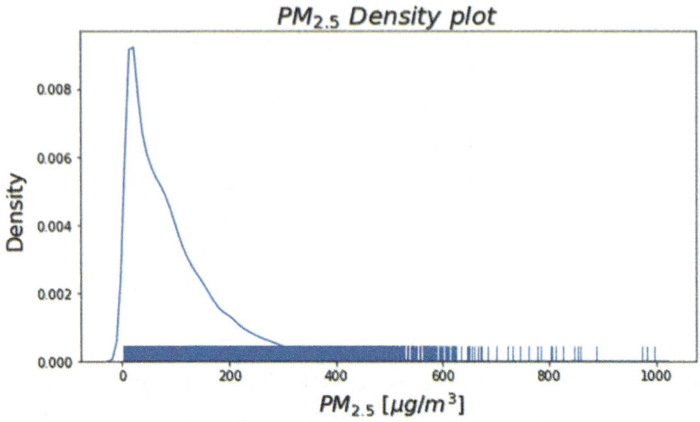


**Fig. 1** Density plot $PM_{2.5}$

are pretty good at extracting patterns from long sequences, (ii) can easily model multivariate forecasting problems, (iii) show great flexibility in modeling multi-step ahead forecasting problem with the ability to set the look-back window used to predict the current step (or steps).

## 2.3 *Accuracy Metrics*

For performance evaluation has been utilized RMSE as described in Eq. (1), this is a common choice resulting in a dimensional accuracy metric that allows to easily compare this work's results with similar works.

$$RMSE = \sqrt{\frac{\sum_{n=1}^{N}(target_n - prediction_n)^2}{N}} \tag{1}$$

## 2.4 *Network Architecture*

The proposed LSTM has been designed and evaluated using Python and KERAS framework, it is made up by an LSTM layer with 80 neurons and activation function RELU, followed by a Dense layer with 24 neurons. For the LSTM layer has been specified a dropout of 0.1 with a recurrent dropout of 0.01. The LSTM has been trained using a batch-size equal to 15 for 100 Epochs (using early stopping). Loss function was set to mean squared error (MSE), while optimizer was set to Adam. The look-back window was set to 3 days. The above hyperparameters have been chosen by a grid search procedure. All numerical data was standardized (resulting in data with mean 0 and standard deviation 1) prior to being feed into the LSTM for training. The unique categorical variable, combined wind direction was converted to numeric using one-hot-encoding. Training made up by 33,094 sample (66.75% entire data set), while 16,483 samples for testing (33.25% entire data set).

### *2.5 Performance Evaluation Methodology (Rolling-Window Validation)*

In machine learning to evaluate model performance the entire data set is split into a train and a test set: the training set used to prepare the model, tuning the hyperparameters, while the test set is used to evaluate it. A common technique to provide a robust estimate of model performance is to use k-fold cross-validation (CV) that repeats this process by systematically splitting the data into k groups, each given a chance to be a held-out model. CV technique cannot be directly used with time series data because it assumes that there is no relationship between the observations and that obviously does not hold for time series where each observation is not independent. For time series K-fold splits have to respect the temporal order in which values were observed to avoid data leakage. In this paper, a methodology known as *rolling-window* validation has been implemented. This procedure providing a more robust estimation of model accuracy performance [12].

## 3 Results and Discussion

Figure 2, summarizes the experimental results obtained by the proposed LSTM in terms of RMSE values.

From results shown in Fig. 2, appear that the proposed LSTM is able to provide satisfactory results, in comparison to what can be found in literature, for a short forecasting horizon, 1–3 h, while exceeding 3 h ahead the RMSE rapidly increase.

## 4 Conclusion

In this paper, an LSTM deep neural network is proposed to forecast $PM_{2.5}$ concentration. The dataset used to evaluate the performance of the proposed model is publically available at the UCI repository. The LSTM provide forecasts into the short term period, 1–24 h, using $PM_{2.5}$ values of previous 3 days as well as some predicted atmospherical variables: humidity, combined wind direction, wind speed. The LSTM network has been evaluated using a rolling window validation methodology (keeping

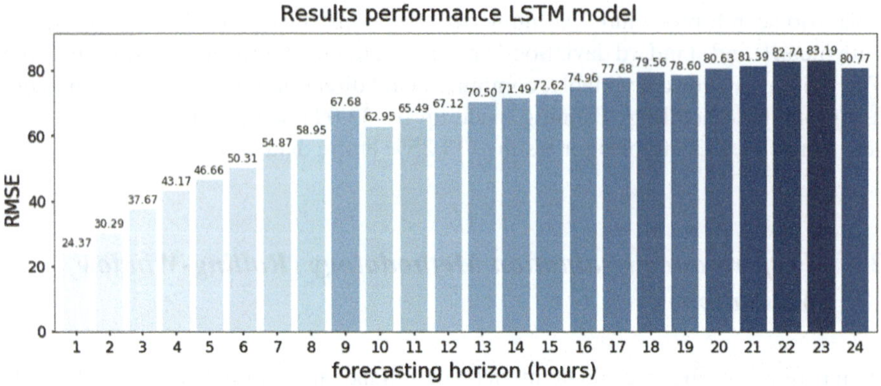


**Fig. 2** Experimental results in term of RMSE

training window fixed in size and without retraining), a type of cross-validation suitable for time series, to provide a more accurate evaluation of model's performance, characterizing the error, expressed in term of RMSE, for a forecasting horizon variable from 1 to 24 h. Performance results, within the forecasting horizon of three hours, are interesting (RMSE = 24.37 $\mu g/m^3$ at 1 h ahead, RMSE = 30.29 $\mu g/m^3$ at 2 h ahead, RMSE = 37.27 $\mu g/m^3$ at 3 h ahead) and qualify the proposed model as a "soft" indicative measurement method in the framework of current EU AQ Monitoring.

## References


1. World Health Organization, Regional Office for Europe (2013) HealtH effects of particulate matter
2. Miller L, Xu X (2018) Ambient $PM_{2.5}$ human health effects-findings in China and research directions. Atmosphere (Basel). 9:1–16. https://doi.org/10.3390/atmos9110424
3. Yang N, Zhang Z, Xue B, Ma J, Chen X, Lu C (2018) Economic growth and pollution emission in China: structural path analysis. Sustain. 10:1–15. https://doi.org/10.3390/su10072569
4. Lin Y, Zou J, Yang W, Li CQ (2018) A review of recent advances in research on $PM_{2.5}$ in China. Int J Environ Res Public Health 15. https://doi.org/10.3390/ijerph15030438
5. Relvas H, Miranda AI (2018) An urban air quality modeling system to support decision-making: design and implementation. Air Qual Atmos Heal 11:815–824. https://doi.org/10.1007/s11869-018-0587-z
6. Buonocore JJ, Dong X, Spengler JD, Fu JS, Levy JI (2014) Using the community multiscale air quality (CMAQ) model to estimate public health impacts of $PM_{2.5}$ from individual power plants. Environ Int 68, 200–208. https://doi.org/10.1016/J.ENVINT.2014.03.031
7. Mehdipour V, Stevenson DS, Memarianfard M, Sihag P (2018) Comparing different methods for statistical modeling of particulate matter in Tehran. Iran Air Qual Atmos Heal 11:1155–1165. https://doi.org/10.1007/s11869-018-0615-z
8. Mahajan S, Chen L-J, Tsai T-C, Mahajan S, Chen L-J, Tsai T-C (2018) Short-term $PM_{2.5}$ forecasting using exponential smoothing method: a comparative analysis. Sensors 18:3223. https://doi.org/10.3390/s18103223

9. Kleine Deters J, Zalakeviciute R, Gonzalez M, Rybarczyk Y (2017) Modeling $PM_{2.5}$ urban pollution using machine learning and selected meteorological parameters. J Electr Comput Eng (2017):1–14. https://doi.org/10.1155/2017/5106045
10. Chen LJ, Ho YH, Lee HC, Wu HC, Liu HM, Hsieh HH, Huang YT, Lung SCC (2017) An open framework for participatory $PM_{2.5}$ monitoring in smart cities. IEEE Access 5:14441–14454. https://doi.org/10.1109/ACCESS.2017.2723919
11. Li X, Peng L, Yao X, Cui S, Hu Y, You C, Chi T (2017) Long short-term memory neural network for air pollutant concentration predictions: method development and evaluation. Environ Pollut 231:997–1004. https://doi.org/10.1016/j.envpol.2017.08.114
12. Bergmeir C, Hyndman RJ, Koo B (2015) A note on the validity of cross-validation for evaluating time series prediction

# A Software System for Predicting Trihalomethanes Species in Water Distribution Networks Using Online Networked Water Sensors

G. Fattoruso, A. Agresta, G. Guarnieri, M. Toscanesi, S. De Vito, M. Fabbricino, M. Trifuoggi and G. Di Francia

**Abstract** Drinking water chlorination reduces the risk of pathogenic infection, but it may be harmful to human health because of trihalomethanes formation. At present, trihalomethanes concentrations are periodically monitored in fixed points along the water distribution network by in situ sampling and laboratory tests. Simulation models combined with online data sources can be useful for reproducing trihalomethanes concentrations in all part of the network and over time. The current challenge is to make reliable the model predictions. At this scope, a novel method and software system has been developed able to reproducing reliable trihalomethanes concentrations, including their species, in all parts of the water network and over time. The system has been tested on the real aqueduct Santa Sofia, disinfected by sodium hypochlorite and monitored by seven networked online multi-parameter sensors.

**Keywords** Trihalomethanes predictions · Online water sensors · Online simulation models · Open source software · QGIS console · Model uncertainty

## 1 Introduction

Drinking water chlorination reduces the risk of pathogenic infection, but it may be harmful to human health because of disinfection by-products (DBPs) formation due to the chlorine's reaction with the natural organic matter (NOM). The by-products include species such as trihalomethanes (THMs) and haloacetic acids (HAAs) that are characterized by recognized toxicity and carcinogenicity so that their concentrations along the entire water distribution network (WDS) have to be maintain below threshold values according to national and European regulation. At present, THMs concentrations are periodically monitored in fixed points along the WDSs by in situ *sampling* and laboratory tests. On the contrary, the dynamic behavior of the WDSs would require to monitor contaminant concentrations in real-time and in all part of

G. Fattoruso (✉) · G. Guarnieri · S. De Vito · G. Di Francia
ENEA, Research Center Portici, Portici (Naples), Italy
e-mail: grazia.fattoruso@enea.it

A. Agresta · M. Toscanesi · M. Fabbricino · M. Trifuoggi
University of Naples, Federico II, Naples, Italy



G. Di Francia et al. (eds.), *Sensors and Microsystems*, Lecture Notes in Electrical Engineering 629, https://doi.org/10.1007/978-3-030-37558-4_62

the WDS in order to timely identify and manage critical changes when they occur, minimizing the population exposure to an unsafe water supply.

At the same time, the maturity of smart water technologies for monitoring WDSs is increasing and the deployment process is becoming easier and faster, moving more and more water utility companies towards an approach "*Smart WAter Network*" (SWAN) for an optimized management of their aqueducts.

Smart, data-driven water technologies have resulted in a rising amount of information being available for operators. In the aftermath of these real-time data sources, which includes online measuring sensors (OMSs) [1], meters and telemetry, the challenge is to make sense of these data effectively and transform it into intelligence to better inform operational and maintenance decisions. A way is to couple the constant stream of data by online monitoring systems with predictive modelling capabilities, obtaining a proactive management of the WDSs.

In recent literature, most of the investigated models capable of reproducing THMs concentrations in WDSs depend on a first water parameter such as residual chlorine as well as organic matter, among others. They are developed using laboratory or field data and investigated on their applicability to real WDSs. The challenging aspect, to be still investigated, is to make reliable their performances when used in real situations, minimizing the uncertainty of their predictions.

A this scope, a novel method and software system has been developed and patented by ENEA [2], able to reproducing reliable residual chlorine and disinfection by products (i.e. THMs and HAAs including their species) concentrations in all parts of the WDS and over time by coupling online networked sensors with simulation modeling and addressing the uncertainty of predictions. The system has been tested on the real aqueduct Santa Sofia, disinfected by sodium hypochlorite and monitored by six networked online multi-parameter sensors.

## 2 The System Architecture

The developed system for the control of water quality in the WDSs provides reliable (near-) real time and predictive estimates of the residual chlorine and THMs, HAAs concentrations, including their species, in all part of the WDS.

Online networked water sensors [3], fixed and/or mobile (i.e. human) that make available continuous and distributed quality/quantitative measurements (including at least chlorine and pressure data) along the aqueduct are a prerequisite because the proposed system provides reliable THMs predictions.

The system includes a mix of software components, developed ad hoc or as extensions of solver engine, integrated within an unique open source GIS-based platform and connected to online networked water sensors data sources. It provides the following capabilities:

- modeling hydraulic behavior and reproducing residual chlorine, THMs and their species concentrations in all parts of the WDSs and over time, assimilating the *live-data* gathered by the online water sensors
- performing an automated and periodic model calibration
- defining optimal sampling schemes by fixed or mobile (human) sensors.

This system intends to provide the control rooms of the water utilities with a comprehensive and effective analytics set of capabilities for identifying, monitoring and tracking residual chlorine and THMs concentrations along the network and consequently activating timely and effective protection actions in order to mitigate the risks on human health of water contamination events as they occur.

## 2.1 The Case Study

The developed system has been tested on the pilot real aqueduct of Santa Sofia (located in Southern Italy), one of trunks of the Acquedotto della Campania Occidentale (ACO), managed da Acqua Campania SpA. The main pipeline runs from the San Prisco reservoir for about 22 km, suppling water to 28 municipalities, with a total discharge equal to 2000 l/s. In San Prisco reservoir, the network is disinfected by 0.2 mg/l of sodium hypochlorite. The aqueduct consists in a main pipeline with NPS 1400 and three pipelines with NPS 800 and one with NPS 700. Then, it is equipped by seven networked online multi-parameter water sensors—ENDETEC-KAPTATM 3000, for chlorine, temperature, pressure, conductivity.

## 2.2 Modeling Capabilities

For ensuring a safe water supply, it is useful to have models able to reproduce contaminant concentrations in all part of a WDS and over time, especially online models so that the operator can use them for timely identifying critical events when they occur. The most important factors influencing THMs formation are operational parameters, such as free chlorine and contact time, pH, environmental conditions, organic content (e.g. Total Organic Carbon—TOC, Dissolved Organic Carbon—DOC) among others. So that the models in literature, grouped in empirical and kinetic, are expressed through a function of these parameters. Furthermore, due to the recent developments of the water technologies (e.g. OMSs), they can be monitored in continuous by online water sensors, installed along the network.

The proposed system provides online modeling capabilities by an extension of the EPANET and EPANET-MSX solver engines enabling to assimilate online pressure and flow data as well as online quality data (e.g. chlorine), and performing extended-period chlorine decay and DBPs formation simulations within the WDS.

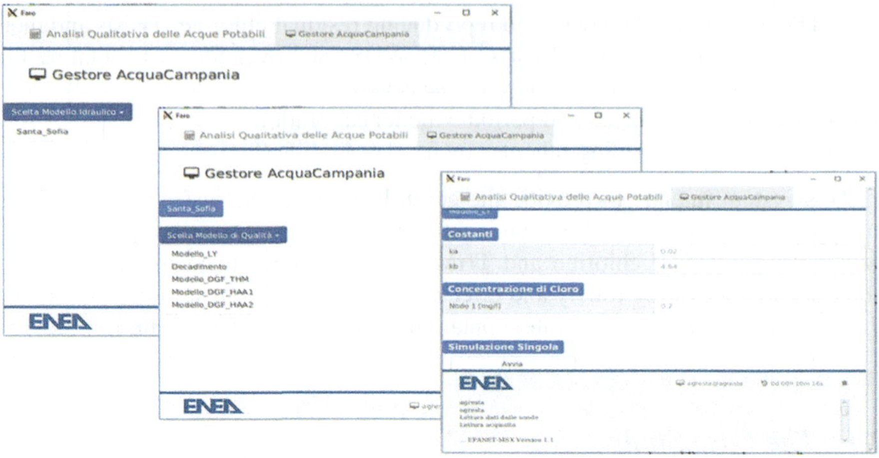


**Fig. 1** The system interface

An important system feature is its interface (Fig. 1), developed ad hoc, that provides a high level of interaction and control from operators. Through this interface it is possible to select the WDS and hence one or more simulation models to be performed for reproducing residual chlorine and THMs concentrations, including the several species, in all part of the network and over time.

The set of available models have to be previously investigated in their ability of reproducing reliable concentrations in real situations. Then, through a setup editor, the operator can create and modify the models setup defining the model parameters. At the same way, the model calibration setup can be edited, specifying, for example, the rate at which the calibration has to be repeated.

For the pilot aqueduct, the system's multi-model component includes the hydraulic model, and the chlorine decay model [4], and two kinetic DBPs formation models [5–7]. The first of them predicts twelve types of DBPs precursors, four THMs species and height HAAs species [5, 7], depending on TOC concentration. The latter simulates the total THMs concentrations as a function of residual chlorine [6]. The hydraulic model is given as input at the water quality models. Similarity, the online data, gathered by the networked water sensors, are assimilated by these models, taking into account the real condition of the system. This realizes a dynamic system modeling component.

Running the model tools, the several output spatio-temporal simulation scenarios are visualized within a QGIS console (Fig. 2.) where can easily be compared, queried and analyzed.

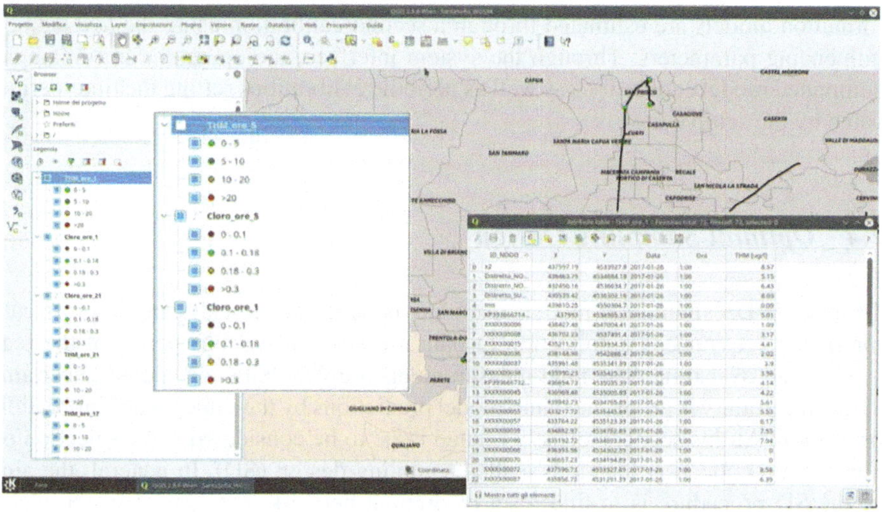


**Fig. 2** QGIS console for visualizing the residual chlorine and THMs predictions in all part of the WDS and over time

## 2.3 *Model Calibration Capabilities*

Before any intended use of the WDS models, it has to be ensured that the models predict the behaviour of the real network with reasonable accuracy. The prediction accuracy can be influenced by several factors. Among others, there are input parameters of the model. An efficient calibration process of the models can significantly reduce the uncertainty of their predictions. This process modifies the models uncertain parameters (e.g. pipe roughness, kinetic constants) until model predictions match closely with measured values on the real system, under a range of operating conditions within acceptable error bounds.

It is widely acknowledge that hydraulic and water quality behavior changes during the seasons due to variable environmental and operational conditions. Thus, periodic model calibrations represent a best practice for ensuring reliable predictions.

For facing these requirements, the system includes a tool, developed ad hoc, for the automated and periodic models calibration that uses the online data gathered by networked water (multi-parameter) sensors. By combining genetic algorithms with model solver engines, the automated calibration procedure generates and evaluates a large number of uncertain parameter sets, solving a optimization problem with an objective function of the weighted least squares of the involved parameters. In particular, by means of a two-steps approach, the developed procedure first calibrates the hydraulics parameters using online pressure and flow measurements. Then, fixing them, the parameters involved within the selected chlorine decay and THMs formation models are estimated through a second calibration using the *live* data of influencing parameters. Through the system interface, the operator can run single automated model calibrations as well as periodic calibrations setting the time-period value by the setup editor.

## 2.4 Optimal Sampling Design

To make WDS simulation models useful, it is necessary to calibrate them. A calibration process needs a collection of data measured in various points of the real system and over time. The field measurements are generally considered uncertain. Their uncertainty is propagated into model predictions by the same model calibration procedure. At this scope, a closely related issue to be considered is the selection of appropriate collection locations, called sampling design (SD). In general, the aim of the SD procedure is to find a set of optimal network locations at which place measurements devices/sensors. Optimal locations are determined with the aim of collecting data that when used for hydraulic and other models calibration will yield the best parameter estimates and predictions.

The developed system provides a tool that performs optimal location schemes of (multi-) water sensors (fixed or mobile sensors). The developed SD procedure use a stochastic approach where the SD problem is formulated and solved as a two-objective optimization problem under parameter uncertainty, by identifying sampling designs that have the optimal trade-off between calibrated model accuracy and the sampling design cost.

Applying this tool at the Santa Sofia aqueduct, the existing water sensor network has been evaluated referred to the hydraulic and chlorine decay models. In Fig. 3, it is evident that the pressure and chlorine predictions accuracy increases, raising the sensor number.

The actual locations of 7 water multi-parameter sensors reduces the residual chlorine predictions uncertainty of about a 40%.

**Fig. 3** Table of uncertainty values associated to hydraulic and residual chlorine decay models performed by the system tool for the Santa Sofia aqueduct

| $\varphi_{idra}$ | $\varphi_{qual}$ | N | K1 | K2 | K3 | K4 | K5 | K6 | K7 |
|---|---|---|---|---|---|---|---|---|---|
| 11.859 | 2.298 | 1 | 0 | 0 | 0 | 0 | 0 | 0 | 1 |
| 5.044 | 1.665 | 2 | 0 | 1 | 0 | 0 | 0 | 0 | 1 |
| 4.035 | 1.320 | 3 | 0 | 1 | 0 | 0 | 1 | 0 | 1 |
| 3.316 | 1.067 | 4 | 1 | 1 | 0 | 0 | 1 | 0 | 1 |
| 2.810 | 0.851 | 5 | 1 | 1 | 1 | 0 | 1 | 1 | 0 |
| 2.435 | 0.599 | 6 | 0 | 1 | 1 | 1 | 1 | 1 | 1 |
| 1.961 | 0.375 | 7 | 1 | 1 | 1 | 1 | 1 | 1 | 1 |

## 3 Conclusion

The proposed System returns THMs predictions in all part of the pilot aqueduct with a mean error of 16%, remaining almost constant over the two annual semesters. The predictions accuracy returned by our system is reasonably acceptable and close to the recent literature values.

Through the set of tools available within the system, the operator of the control room can identify in (near-) real time critical THM concentrations and quantify the single species; locate the single pipelines affected; analyze the spatio-temporal evolution of the phenomenon; defining optimal SD during the emergency; evaluate the behavior of the WDS changing the chlorine dose used in the disinfection; minimize the negative effects on the population when contamination events occur and finally ensure a safe and continuous water service.

## References

1. Di Nardo A et al (2018) On-line measuring sensors for smart water network monitoring. In: Proceedings of international conference HIC 2018, Palermo
2. Patent "Metodo per controllare una concentrazione di trialometani in una rete idrica acquedottistica", N. 102018000006379
3. Fattoruso G et al (2015) A SWE architecture for real time water quality monitoring capabilities within smart drinking water and wastewater network solutions. In: ICCSA: international conference on computational science and its applications, Springer, Cham, pp 686–697
4. Fattoruso Grazia et al (2014) Simulation of chlorine decay in drinking water distribution systems: case study of santa sofia network (Southern Italy). Sens Microsyst Lect Notes Electr Eng 268:467–470
5. Fattoruso Grazia et al (2014) Use of kinetic models for predicting DBP formation in water supply systems. Sens Microsyst Lect Notes Electr Eng 268:471–474
6. Lin Y, Yeh H (2005) J Comput Civil Eng
7. Greca GD, Fabbricino M (2008) Water Sci Technol: Water Supply

# Could the Experience of a Photovoltaic Testing Laboratory be of any Help to the World of Sensors?

**Michele Pellegrino, G. Di Francia, Giovanni Flaminio, Vincenzo Giglio, Giuseppe Nardelli and Arturo Matano**

**Abstract** Sensors should exhibit many different characteristics which have to be provided all along their life time and that requires extensive experimental activity. But the world of sensors is very wide, covering different kind of materials and mechanisms of transduction, with a large variety of types and applications and it is hard to find either Standards or technical Specifications or Protocols these devices shall be complying to. As a consequence it is no one way for classifying sensors, to categorize them and to find either Standards or Technical Specifications suitable for all the sensors. So there are traditional sensors that must comply with applicable Standards (for instance the ASTM International, formerly known as American Society for Testing and Materials, is an international organization that develops and publishes voluntary consensus technical standards for a wide range of materials, products, such sensors) but other more innovative devices where Standards and Technical specifications can be hardly found. Laboratory PVSMART of CR ENEA in Portici, near Naples (I), has been active for so many years for quality and reliability of photovoltaic devices such as solar cells and modules and owns a very deep experience on laboratory artificial ageing processes and compliance to Standards. Although the field could seem quite different, nevertheless this competence might be useful for the world of sensors; in that way in the past the laboratory has already provided technical support to some customer by testing specific detectors. In this paper, the experience of this investigation is summarized and the potentials for a more general use is then proposed.

**Keywords** Sensors · Testing · Photovoltaics

## 1 Introduction

The world of sensors is very wide, covering different kind of materials and mechanisms of transduction, with a large variety of types and applications and it is hard to find either Standards or technical Specifications or Protocols these devices shall be complying to [1]. But they are swiftly finding their way into real-world applications

---

M. Pellegrino (✉) · G. Di Francia · G. Flaminio · V. Giglio · G. Nardelli · A. Matano
ENEA, CR-Portici, P.le E. Fermi 1, 80055 Naples, Italy
e-mail: michele.pellegrino@enea.it

 
G. Di Francia et al. (eds.), *Sensors and Microsystems*, Lecture Notes
in Electrical Engineering 629, https://doi.org/10.1007/978-3-030-37558-4_63

and intensively invading our daily activities so experimental activities to demonstrate their reliability and their compliance as well to the existing Standards, if any, should be necessary. Laboratory PVSMART of CR ENEA in Portici, near Naples(I), has been active for so many years for quality and reliability of photovoltaic devices such as solar cells and modules and owns a very deep experience on laboratory artificial ageing processes and compliance to Standards. Although the field could seem quite different, nevertheless this competence might be useful for the world of sensors; in that way in the past the laboratory has already provided technical support to some customer by testing specific detectors. In this paper, the experience of this investigation is summarized and the potentials for a more general use is then proposed.

The ageing process could be very complicated depending on the real conditions of exposure of the devices and their endurance. Besides it could last too much to provide useful information on experimental scale. For that reason accelerated ageing processes need to be provided at experimental stage. Table 1 describes the main severity mechanisms that are used during the artificial ageing process for photovoltaics.

## 2 Sensors

### *2.1 Classification of Sensors*

The world of sensors is very wide, covering different kind of materials and mechanisms of transduction, with a large variety of types and applications. Different types of sensors are already in our homes, offices, cars etc. working to make our lives easier, for instance by turning on the lights by detecting our presence, adjusting the room temperature, detect smoke fire, make us delicious coffee, open garage doors as soon as our car is near the door and many other tasks. Besides standard sensors used for process measuring such as temperature, pressure, flow, location, to name a few there are a plenty of innovative and exotic now on the market [2]. In Figs. 1 and 2 different examples are shown.

As a consequence is that there is no one way for classifying sensors.

They can be divided into active and passive. Active sensors are those which require an external excitation signal or a power signal. Passive Sensors, on the other hand, do not require any external power signal and directly generates output response. The other type of classification is based on the means of detection used in the sensor. Some of the means of detection are Electric, Biological, Chemical, Radioactive etc. Another classification is based on conversion phenomenon i.e. the input and the output. Some of the common conversion phenomena are Photoelectric, Thermoelectric, Electrochemical, Electromagnetic, Thermooptic, etc. The last classification of the sensors is Analog and Digital Sensors. Analog Sensors produce an analog output

**Table 1** List of tests being in use for photovoltaics

| Severity | Range | Time/cycling/dose | Facilities provided at CR ENEA in Portici |
|---|---|---|---|
| Temperature/relative humidity | 85 °C/85% | 10 cycles/1000 h | Damp heat, humidity freeze and thermal cycling tests controlling the temperature and the relative humidity, from −40 °C ± 1°up to 110 °C ± 1° and from 0–90% respectively. Chamber sizes are 2.5 m as height, 4.5 m as depth and 3 m as width |
| Thermal cycling | 200 cycling from −40 to 85 °C | 50 or 200 cycles | |
| Sun irradiation | 600–1000 $Wm^{-2}$ | Stabilization after different intervals of at least 43 $kWh \cdot m^{-2}$ each over periods when temperature is between 40 and 60 °C | Light exposure of 600–1000 $Wm^{-2}$ under resistive load until *P*max is stable within 2%. Class BBB solar simulator, in accordance with the IEC 60904 |
| UV irradiation | | Parameters adjustable between 20 and 80 °C in temperature and between 20–90% as relative humidity with irradiance on the testing plane as much as 250 $W/m^2$ | UV test to determine the ability of the module to long exposure to ultra violet radiations, especially A and B part of the spectrum. Sizes of chamber are about 2 m × 1.2 m |
| Biasing | | From −1000 V up to 1000 V | |
| Salt spray | pH <7, salinity | | Chamber sizes are 3.8 m. long 1.8 m. wide and 1.9 m. high. The temperature range is between room temperature and 55 °C and the relative humidity range between 50% up to the saturation condition (rain). Nozzles on the floor level can spray a pre-prepared mixture of water upright so that the ensuing fall of the raindrops can impinge the samples |
| Combination of some of the above stress factors | 85 °C, 85 R. H Light and DC bias | | |

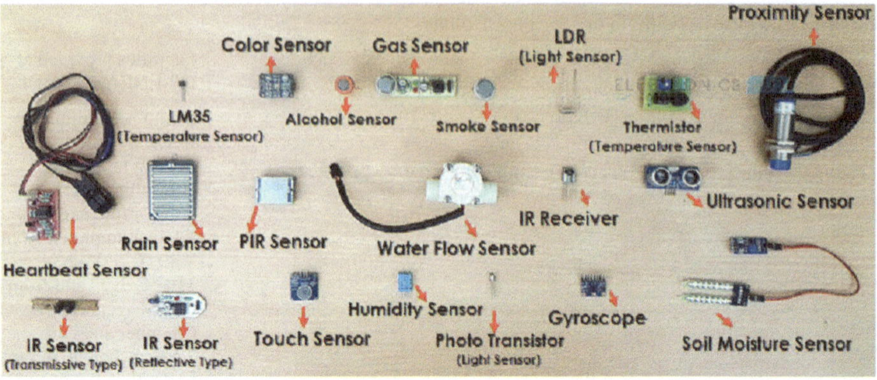


**Fig. 1** A sort of standard sensors for process control

**Fig. 2** Sensors used for physical activity

i.e. a continuous output signal with respect to the quantity being measured. Digital Sensors, in contrast to Analog Sensors, work with discrete or digital data. The data in digital sensors, which is used for conversion and transmission, is digital in nature. Another effect is that it is hard to find either Standards or Technical Specifications suitable for all the sensors.

## 2.2 *Different Types of Sensors*

The following is a list of different types of sensors that are commonly used in various applications. All these sensors are used for measuring one of the physical properties like Temperature, Resistance, Capacitance, Conduction, Heat Transfer etc.

- Temperature Sensor
- Proximity Sensor
- Accelerometer
- IR Sensor (Infrared Sensor)
- Pressure Sensor
- Light Sensor
- Ultrasonic Sensor

- Smoke, Gas and Alcohol Sensor
- Touch Sensor
- Color Sensor
- Humidity Sensor
- Tilt Sensor
- Flow and Level Sensor.

### *2.3 Characteristics of Sensors*

Sensors should exhibit many different characteristics which have to be provided all along their life time. They are as given below:

1. Accuracy
2. Environmental condition—usually has limits for temperature/humidity
3. Range—Measurement limit of sensor
4. Calibration—Essential for most of the measuring devices as the readings changes with time
5. Resolution—Smallest increment detected by the sensor
6. Cost
7. Repeatability—The reading that varies is repeatedly measured under the same environment.

## 3 Photovoltaic World Support

The world of photovoltaics has been quite different. From one side there isn't such a large variety of devices as sensors, from the other PV modules and systems should comply with numberless Standards usually issued by Technical Committee 82 (Solar Photovoltaic Energy System) of IEC International Electrotechnical Commission. Another challenging effort that PV devices has always required to demonstrate has been their reliability; to do that before their lunching to the market the prototypes has to pass many trials with different levels of severity such the ones indicated in Table 1.

## 4 New Frontier Experience

Although the above mechanisms have been ruled and optimized in the field of photovoltaics, nevertheless they can be adapted, even with some modifications, to objects coming from a different world, such as the sensors. Just to provide an evidence, some

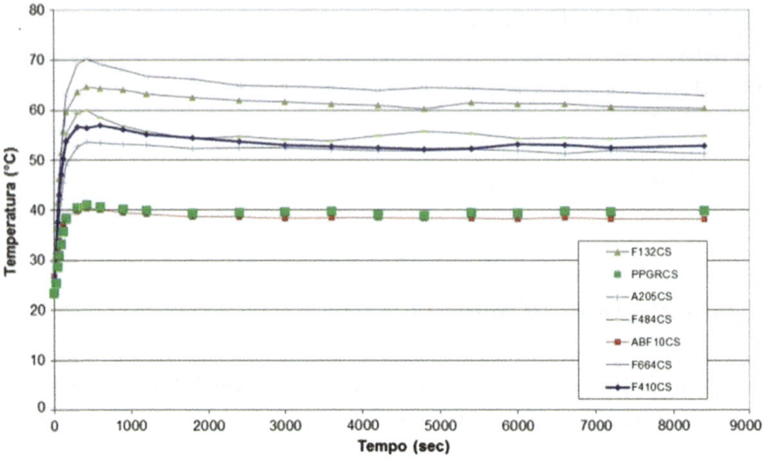


**Fig. 3** Stabilization temperature curves of different samples

samples from aerospace industry were asked to prove their attitude to the excessive heating when exposed to solar radiation during the flight. The goal of the experiment was to investigate the optical behavior of thermal sensors installed on aircrafts, according to different coating materials with different values of optical absorbance material, when deployed on aircraft after absorption of sun radiation. In this case the laboratory exposed the samples to artificial sun radiation at more than 1000 W/m$^2$ for almost two hours while the measurements of the temperatures have been recorded. Figure 3 shows the charts for the samples.

## 5 Conclusion

Conclusion is that, according to the nature of the particular sensor, a combination of severities can be found or agreed with customer to provide useful information on the performance and the endurance of the device.

## References

1. Gardner JW, Bartlett PN (1996) Elsevier Science, Sens Actuators B 60–67
2. ASTM F2362-03 (2013) Standard specification for temperature monitoring equipment ASTM F2174-02 (2015) standard practice for verifying acoustic emission sensor response

# Inkjet Printed Graphene-Based Nanosensors for the Detection of Nitrogen Dioxide

**F. Villani, F. Loffredo, J. P. Santos, I. Sayago, M. Aleixandre, M. L. Miglietta, B. Alfano and T. Polichetti**

**Abstract** Nanotechnology has driven the research toward new advanced materials and innovative fabrication methods in different fields including sensing applications. In this framework, chemical graphene-based nanodevices have been developed and characterized for $NO_2$ detection at low concentrations. The chemiresistors have been manufactured by the digital printing technique inkjet, one of the most attractive deposition methods for miniaturized systems. Graphene, in form of hydro-alcoholic solution (water-isopropanol mixture), has been formulated through Liquid Phase Exfoliation method and used as sensing material-based ink. The devices' response to sub-ppm $NO_2$ concentrations has been measured at different temperatures, from RT up to 200 °C. The electrical characterization of the sensors has highlighted that there is an increase of the response as the $NO_2$ concentration and the heating temperature increase.

**Keywords** Graphene · Inkjet printing · Aqueous dispersion · Chemi-device · Sensor · Nitrogen dioxide

## 1 Introduction

Nitrogen dioxide is one of the main pollutants with a great impact on human health therefore many efforts are spent to develop proper sensor devices with improved sensitivity, selectivity, stability, power consumption and cost. To this aim, the nanotechnology has driven the research toward new advanced materials like graphene and graphene related materials which, thanks to their physical properties, are able to detect gas analytes even at room temperature (RT) [1–3]. In particular, pristine

F. Villani (✉) · F. Loffredo · M. L. Miglietta · B. Alfano · T. Polichetti
Italian National Agency for New Technologies, Energy and Sustainable Economic Development (ENEA), C.R. Portici—P.le E. Fermi 1, 80055 Portici (Naples), Italy
e-mail: fulvia.villani@enea.it

J. P. Santos · I. Sayago · M. Aleixandre
Nanosensors and Intelligent System Group (NoySI), Instituto de Tecnologías Físicas Y de La Informaci6n ITEFI-CSIC, 28006 Madrid, Spain



G. Di Francia et al. (eds.), *Sensors and Microsystems*, Lecture Notes in Electrical Engineering 629, https://doi.org/10.1007/978-3-030-37558-4_64

graphene, although, as other room temperature operating sensing layers, has performances affected by slow response, recovery rate and poor reversibility, which represent some of its major drawbacks, displays a marked sensitivity towards different gases [4].

As further advantage, graphene can be deposited also by printing techniques well addressing miniaturization applications. In this context the inkjet printing (IJP) technology, a deposition method from liquid phase, permits an efficient use of different functional inks thanks to its simultaneous capability of printing and patterning very small areas, so reducing the amount of waste products. These peculiarities of the IJP technique are the main strengths exploited in several electronic applications [5–9], including the sensor devices' field [10–13].

Here we present a feasibility study of graphene-based sensors inkjet printed onto micromechanized substrates, constituted of four microhotplates in a TO-8 package [14]. Graphene, in form of hydro-alcoholic solution, has been made by Liquid Phase Exfoliation (LPE) method [15] and employed as ink. The sensing responses of the devices have been analyzed at different concentrations of nitrogen dioxide and at different heating temperatures.

## 2 Experimental

### 2.1 Ink Formulation

Graphene has been prepared by LPE method starting from commercial graphite powder (Sigma-Aldrich, product 332461). The formulation of LPE graphene-based ink has consisted in dispersing the powder in a mixture of water/isopropanol (IPA/$H_2O$, (1:7)) as solvent and sonicating the obtained dispersion for about 48 h. After that, the subsequent and last process step has been to centrifuge the suspension, in order to remove the larger graphitic structures and the unexfoliated flakes, and then to take the surnatant [15]. The overall process is schematized in Fig. 1.

The final suspension of graphene in hydro-alcoholic solution at the concentration of 0.1 mg/mL has been employed as ink.

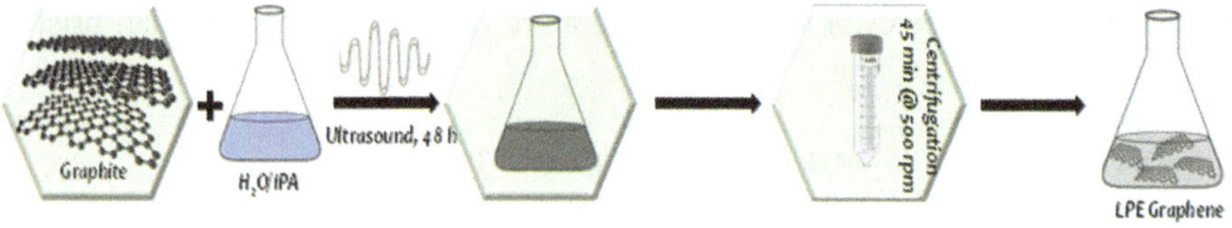


**Fig. 1** Scheme of graphene preparation process

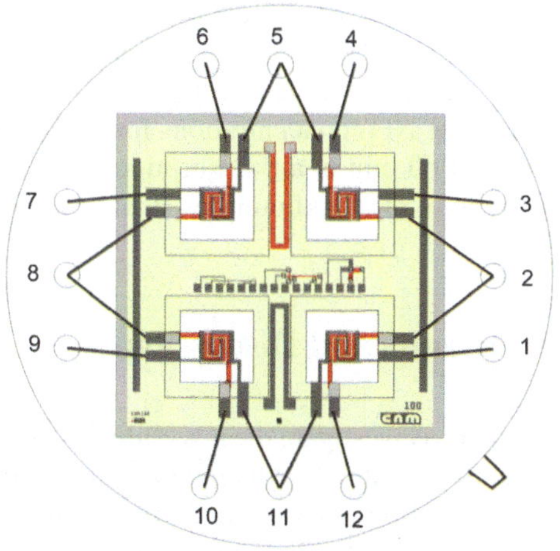


**Fig. 2** Micromechanized substrates with architecture to host four microhotplate devices in a TO-8 package

## 2.2 *Device Fabrication*

The sensor devices have been manufactured onto micromechanized substrates, constituted of four microhotplates in a TO-8 package [14]. In Fig. 2 is displayed the scheme of the micromechanized substrates.

The printing process has been performed by the inkjet equipment Dimatix Materials Printer 2831 (DMP2831) of FUJIFILM—USA. This system uses the piezoelectric drop-on-demand technology to eject droplets through a multi-nozzles printhead.

The printing parameters have been optimized in order to obtain uniform, conductive and reproducible sensing films. The sensing films have been realized by printing multiple overlapped layers with squared pattern. After printing, a thermal treatment has been applied by keeping the devices on hot-plate at 100 °C for 15 min so removing the residual solvent.

## 2.3 *Sensing Characterization*

Tests for sensing analysis upon the analyte $NO_2$ have been carried out by placing the printed devices in a stainless steel cell inside a proper measuring apparatus (WiNOSE 5.0). This equipment consists in a 4-channel electrometer with a scanner card (6517 and 6522, Keithley, Cleveland, OH, USA) operating at a constant bias voltage of 1 V. Voltage for the device heaters has been driven from a 4-channel voltage source through a programmable power supply (V4CH, Ray IE, Mirabel, Spain).

A gas mixing unit (GMU-06, Ray Ingeniería Electrónica, Mirabel, Spain) has been used to generate the desired $NO_2$ concentrations from calibrated cylinders (Praxair, Madrid, Spain). All the measurements have been performed keeping the analyte flow constant (200 mL/min).

The sensing properties of the devices have been investigated measuring their response to different $NO_2$ concentrations varying from 0.1 to 1 ppm for different operating temperatures, from RT up to 200 °C. The sensors have been exposed to increasing $NO_2$ concentrations for 30 min (adsorption time) followed by a recovery phase of 60 min in air (desorption time). Device responses have been calculated as the ratio of the maximum resistance value in $NO_2$, R, to the resistance in air, $R_0$.

## 3 Results and Discussion

In order to investigate the possibility to fabricate graphene-based sensor nanodevices by inkjet technology, the ink has been formulated by dispersing graphite flakes in a hydro-alcoholic mixture and printed onto micromechanized substrates.

The morphological and electrical properties of the printed layers were studied by atomic force microscopy (AFM), scanning electronic microscopy (SEM) and volt-amperometric measurements in an our previous study [11]. Both AFM and SEM analyses highlighted the presence of restacked aggregates formed by planar-structured flakes, randomly distributed, also overlapped in some regions, with cracks and gaps between neighbouring sheets. Despite the rather inhomogeneous material distribution, I-V measurements pointed out evident ohmic behavior of the material.

Starting from this state-of-the-art on the surface and electrical characteristics of IJ-printed aqueous LPE graphene-based films the sensing property of the nanochemiresistors has been tested by exposing them to gas vapours of $NO_2$ at different concentrations from 100 to 1000 ppb by heating the devices from RT up to 200 °C. All the electrical characterizations have been carried out by biasing the single device at 1 V.

The typical sensing responses of printed graphene-based nanodevice as function of $NO_2$ concentration at different heating temperatures and as function of the heating temperatures at fixed $NO_2$ concentration are displayed in Figs. 3 and 4, respectively.

In both graphs the sensing responses have been reported in terms of the inverse of $R/R_0$ due to the p-type nature of the material.

In detail, by observing Fig. 3, for all investigated temperatures, the graphene-based device shows a slow but continuous increase of the response as the $NO_2$ concentration increases. Additionally, from Fig. 4, at fixed $NO_2$ concentration, the response is higher as heating temperature increases: this behavior is more evident at high analyte concentrations.

## 4 Conclusion

In this study we have demonstrated that the inkjet printing technology can be successfully adopted for the fabrication of sensing devices in miniaturized architectures by employing micromechanized substrates.

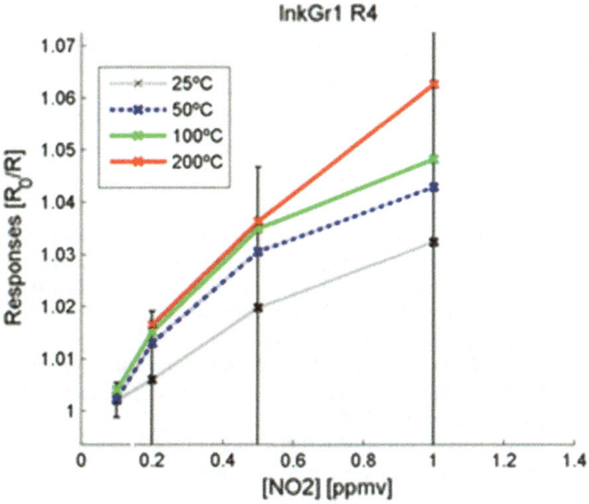


**Fig. 3** Sensing responses of graphene-based device as function of $NO_2$ concentration at different heating temperatures

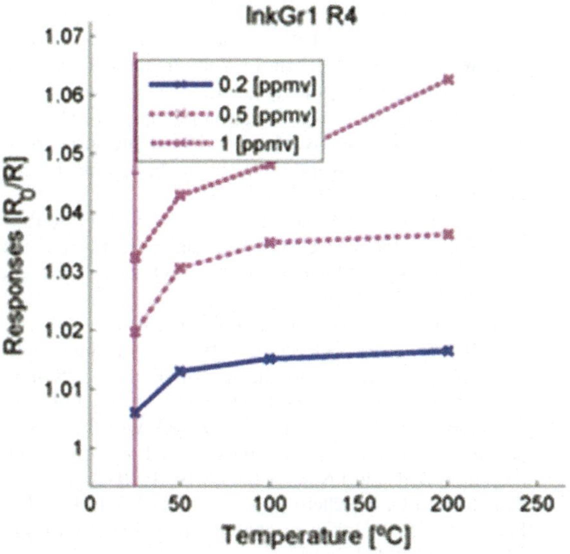


**Fig. 4** Sensing responses of graphene-based device as function of the heating temperatures at fixed $NO_2$ concentration

Graphene, in form of hydro-alcoholic solution, has been made by Liquid Phase Exfoliation method and inkjet printed. The sensing properties of the nanodevices have been investigated as function of the analyte ($NO_2$) concentration (100 ÷ 1000 ppb) by heating the substrate at different temperatures (25 ÷ 200 °C). Sensing measurements towards the target gas have been performed at a constant flow of 200 mL/min, with an exposure window towards analyte of 30 min, followed by a recovery phase of 60 min. The electrical characterization of the sensors, carried out at a constant bias voltage of 1 V, has highlighted that there is an increase of the response as the $NO_2$ concentration and the heating temperature increase.

## References

1. Schedin F, Geim AK, Morozov SV, Hill EW, Blake P, Katsnelson MI, Novoselov KS (2007) Detection of individual gas molecules adsorbed on graphene. Nat Mater (2007):652–5
2. Ricciardella F, Massera E, Polichetti T, Miglietta ML, Di Francia G (2014) A calibrated graphene-based chemi-sensor for sub parts-per-million $NO_2$ detection operating at room temperature. App Phys Lett 104(18):183502(1)–(5)
3. Villani F, Loffredo F, Alfano B, Miglietta ML, Verdoliva L, Alfè M, Gargiulo V, Polichetti T (2019) Graphene-like based-chemiresistors inkjet-printed onto paper substrate. Lect Notes Electr Eng 539:337–343
4. Lu G, Park S, Yu K, Ruoff RS, Ocola LE, Rosenmann D, Chen J (2011) Toward practical gas sensing with highly reduced graphene oxide: a new signal processing method to circumvent run-to-run and device-to-device variations. ACS Nano 5:1154–1164
5. Sowade E, Ramon E, Mitra KY, Martínez-Domingo C, Pedró M, Pallarès J, Loffredo F, Villani F, Gomes HL, Terés L, Baumann RR (2016) All-inkjet-printed thin-film transistors: manufacturing process reliability by root cause analysis. Sci Rep 6:33490
6. Gomes HL, Medeiros MCR, Villani F, Canudo J, Loffredo F, Miscioscia R, Martinez-Domingo C, Ramon E, Sowade E, Mitra KY, Baumann RR, McCulloch I, Carrabina J (2015) All-inkjet printed organic transistors: dielectric surface passivation techniques for improved operational stability and lifetime. Microelectron Reliab 55:1192–1195
7. Borriello C, Miscioscia R, Mansour SA, Di Luccio T, Bruno A, Loffredo F, Villani F, Minarini C (2015) Influence of ligand exchange on the electrical transport properties of PbS nanocrystals. Phys Status Solidi A 212:2677–2685
8. Bruno A, Villani F, Grimaldi IA, Loffredo F, Morvillo P, Diana R, Haque S, Minarini C (2014) Morphological and spectroscopic characterizations of inkjet-printed poly(3-hexylthiophene-2,5-diyl): phenyl-C61-butyric acid methyl ester blends for organic solar cell applications. Thin Solid Films 560:14–19
9. Grimaldi IA, Barra M, Carella A, Di Girolamo F, Loffredo F, Minarini C, Villani F, Cassinese A (2013) Bias stress effects investigated in charge depletion and accumulation regimes for inkjet-printed perylene diimide organic transistors. Synth Met 176:121–127
10. Villani F, Schiattarella C, Polichetti T, Di Capua R, Loffredo F, Alfano B, Miglietta ML, Massera E, Verdoliva L, Di Francia G (2017) Study of the correlation between sensing performance and surface morphology of inkjet-printed aqueous graphene-based chemiresistors for $NO_2$ detection. Beilstein J Nanotechnol 8:1023–1031
11. Schiattarella C, Polichetti T, Villani F, Loffredo F, Alfano B, Massera E, Miglietta ML, Di Francia G (2018) Inkjet printed graphene-based chemiresistive sensors to $NO_2$. Lect Notes Electr Eng 431:111–118
12. Ricciardella F, Alfano B, Loffredo F, Villani F, Polichetti T, Miglietta ML, Massera E, Di Francia G (2015) Inkjet printed graphene-based chemi-resistors for gas detection in environmental conditions. In: Proceedings of the 2015 18th AISEM annual conference, AISEM 2015, 7066858
13. De Girolamo Del Mauro A, Grimaldi IA, Loffredo F, Massera E, Polichetti T, Villani F, Di Francia G (2011) Geometry of the inkjet-printed sensing layer for a better volatile organic compound sensor response. J Appl Polymer Sci 122(6):3644–3650
14. Horrillo MC, Martí J, Matatagui D, Santos JP, Sayago I, Gutiérrez J, Martin-Fernandez I, Ivanov P, Grcia I, Cané C (2011) Single-walled carbon nanotube microsensors for nerve agent simulant detection. Sens Actuators, B Chem 157:253–259
15. Fedi F, Ricciardella F, Polichetti T, Miglietta ML, Massera E, Di Francia G (2014) Exfoliation of graphite and dispersion of graphene in solutions of low-boiling-point solvents for use in gas sensors. Lect Notes Electr Eng 268:143–147

The manufacturer's authorised representative in the EU is Springer Nature Customer Service Centre GmbH, Europaplatz 3, 69115 Heidelberg, Germany. If you have any concerns regarding our products, please contact ProductSafety@springernature.com

Printed and bound by CPI Group (UK) Ltd, Croydon, CR0 4YY
30/09/2026
02225662-0015